PART I

CYTOKINE BIOLOGY OVERVIEW

CYTOKINE EFFECTOR FUNCTIONS IN TISSUES

CYTOKINE EFFECTOR FUNCTIONS IN TISSUES

Edited by

MARIA FOTI
University of Milano-Bicocca, Milan, Italy

MASSIMO LOCATI
Department of Medical Biotechnonology, University of Milan, Milan, Italy
Humanitas Clinical and Research Center, Rozzano, Italy

Academic Press is an imprint of Elsevier
125 London Wall, London EC2Y 5AS, United Kingdom
525 B Street, Suite 1800, San Diego, CA 92101-4495, United States
50 Hampshire Street, 5th Floor, Cambridge, MA 02139, United States
The Boulevard, Langford Lane, Kidlington, Oxford OX5 1GB, United Kingdom

British Library Cataloguing-in-Publication Data
A catalogue record for this book is available from the British Library

Library of Congress Cataloging-in-Publication Data
A catalog record for this book is available from the Library of Congress

ISBN: 978-0-12-804214-4

For Information on all Academic Press publications
visit our website at https://www.elsevier.com/books-and-journals

Publisher: Mica Haley
Acquisition Editor: Linda Versteeg-Buschman
Editorial Project Manager: Timothy Bennett
Production Project Manager: Laura Jackson
Cover Designer: Victoria Pearson Esser

Typeset by MPS Limited, Chennai, India

Contents

I

CYTOKINE BIOLOGY OVERVIEW

II

CYTOKINE TISSUE-SPECIFIC REGULATION

List of Contributors

Hasib Ahmadzai UNSW Australia, Sydney, NSW, Australia

Katie Allen St James's University Hospital, Leeds, United Kingdom

Asma Ayari Center for Infection and Immunity of Lille, Lille, France

N. Ellissa Baskind St James's University Hospital, Leeds, United Kingdom

Francesca Bernuzzi University of Milan-Bicocca, Milan, Italy

Michele Cummings St James's University Hospital, Leeds, United Kingdom

Poulami Datta Krembil Research Institute, Toronto, ON, Canada

Boel De Paepe Ghent University Hospital, Ghent, Belgium

Petr Dubový Masaryk University, Brno, Czech Republic

Sarah L. Field St James's University Hospital, Leeds, United Kingdom

Maria Foti University of Milano-Bicocca, Milan, Italy

Ruth Ganss Harry Perkins Institute of Medical Research, Centre for Medical Research, The University of Western Australia, Perth, WA, Australia

Cristan Herbert UNSW Australia, Sydney, NSW, Australia

Geoffrey R. Hill Royal Brisbane and Women's Hospital, Brisbane, QLD, Australia

Pietro Invernizzi University of Milan-Bicocca, Milan, Italy; University of California Davis, Davis, CA, United States

Rakesh K. Kumar UNSW Australia, Sydney, NSW, Australia

Dobroslav Kyurkchiev Medical University of Sofia, Sofia, Bulgaria

Paige Lacy University of Alberta, Edmonton, AB, Canada

Carlos Marín-Delgado University Hospital of the Nuestra Señora de Candelaria, Santa Cruz de Tenerife, Spain

Kate A. Markey QIMR Berghofer Medical Research Institute, Brisbane, QLD, Australia

Edduin Martín-Izquierdo University Hospital of the Nuestra Señora de Candelaria, Santa Cruz de Tenerife, Spain

Kapoor Mohit Krembil Research Institute, Toronto, ON, Canada; Toronto Western Hospital, Toronto, ON, Canada; University of Toronto, Toronto, ON, Canada

Joe W.E. Moss Cardiff School of Biosciences, Cardiff University, Cardiff, United Kingdom

Hani S. Mousa University of Cambridge, Cambridge, United Kingdom

Juan F. Navarro-González University Hospital of the Nuestra Señora de Candelaria, Santa Cruz de Tenerife, Spain

Nicolas M. Orsi St James's University Hospital, Leeds, United Kingdom

Bandna Pal Krembil Research Institute, Toronto, ON, Canada

Dipak P. Ramji Cardiff School of Biosciences, Cardiff University, Cardiff, United Kingdom

Antonio Rivero-González University Hospital of the Nuestra Señora de Candelaria, Santa Cruz de Tenerife, Spain

Anabel Rodríguez-Muñoz University Hospital of the Nuestra Señora de Candelaria, Santa Cruz de Tenerife, Spain

Anirudh Sharma Krembil Research Institute, Toronto, ON, Canada; Toronto Western Hospital, Toronto, ON, Canada

Paul S. Thomas UNSW Australia, Sydney, NSW, Australia; POWH Clinical School, UNSW, Randwick, NSW, Australia; Prince of Wales Hospital, Randwick, NSW, Australia

Isabelle Wolowczuk Center for Infection and Immunity of Lille, Lille, France

Preface

Cytokines are important mediators of diverse biological processes and are therefore highly regulated in any organism. Cytokines assist the initial recognition of danger or nonself-antigens as they are rapidly induced during host responses to infection, inflammation, and damage. They launch innate host protections and form the bridge that allows a nonspecific response to develop into an acquired immune response. Clearly, any event that affects the biology of these important mediators is likely to influence the overall immune competence of the host. In *Cytokine Effector Functions in Tissues*, experts of cytokine biology bring together their knowledge on various aspects of how modulation of cytokines can affect tissue regulation in the context of health and disease. First, the basic biology of cytokines is reviewed, particularly as this relates to their biological function on tissue regulation.

Various chapters define how differential cytokine modulation determines the type of immune response generated during a specific disease settings and how cytokines may regulate tissue homeostatic interactions.

This sets the step for practical preclinical understanding of the effect of cytokines, including how they function in specific pathological conditions such osteoarthritis, atherosclerosis, diabetes, sarcoidosis, nerve degeneration, skeletal muscle disease, allergies, asthma, and COPD. This group of chapters provides the reader with a broad understanding of the range of cytokine activity in human diseases.

Although cytokines show great promise as therapeutic agents, they exhibit a sustained level of toxicity that need to be taken into account when a cytokine trial is designed. Therefore, we summarized the promising advantages in using cytokines in cancer therapies, their role in the field of transplantation, and the possibility to exploit specific cell types such as mesenchymal stem cells (MSCs) for their ability to produce cytokines that may affect the outcome of cellular therapies. The field is far from being completely covered as the issue is heavily complex and novel aspects of the cytokine biology are rapidly evolving. Nevertheless, it is our truthful hope that this collection of expert reviews will help as a starting point for a more complete investigation by the reader of the role these charming biological regulators play in human health and disease.

Maria Foti and Massimo Locati

CHAPTER

1

Introduction to Cytokines as Tissue Regulators in Health and Disease

Maria Foti

University of Milano-Bicocca, Milan, Italy

OUTLINE

INTRODUCTION

Cytokines are a complex of soluble, cell-signaling proteins that affect the biological function of cells and process such as inflammation, a variety of immune responses, and the wound healing activity. Cytokines include interleukins, chemokines, interferons, and growth factors.

Cytokine Effector Functions in Tissues.
DOI: http://dx.doi.org/10.1016/B978-0-12-804214-4.00019-1

Although similar to hormones in that they act at low levels either systemically or locally, cytokines are unique in the large number of cells they are able to target. Cytokines are not restricted to any single organ or tissue but instead they are synthesized by a variety of hematopoietic cells, including myeloid (e.g., macrophages, dendritic cells (DCs), granulocytes, mast cells), lymphoid cells (e.g., B cell, T cells, natural killer cells) and by nonhematopoietic cells such as endothelial and epithelial cells. Different immune cells can secrete the same cytokines building up complex microenvironments that are functionally pro or antiinflammatory depending on the specific cytokine and cellular target present in that organ. In fact, a role of cytokines in certain organs is not yet fully determined, for example, the role of interleukin-1β (IL-1β) in normal physiology of the central nervous system still remains unclear. Therefore, understanding the basic of this network of interactions in a tissue-specific manner will advance our knowledge on cytokine biology to better exploit them in the context of clinical immunotherapeutic settings.

BIOLOGICAL CHARACTERISTICS OF CYTOKINES

There are over 100 cytokines and many exist in families that share receptor components and signal transduction pathways. Many diseases are characterized by complex cytokine networks that may have a positive or a negative impact on the disease.[1] Cytokines are low molecular weight molecule that possess some common properties: They have a short half-life, they are de novo synthesized and secreted during cellular activation; they are produced by different cells of the immune system and have multiple biological activities depending on the cell type to which they bind, for this reason they are defined as pleiotropic molecules; their biological functions can be redundant, as two different cytokines can have overlapping effects on the target cells; many cytokines can induce or inhibit the production of other cytokines by creating complex network of regulators that are able to finely modulate the biological effects; they may influence the activity of other cytokines, inducing antagonistic or synergistic effects.

They induce their biological effects by activating specific receptors on the cell surface and they can act on the same cells that secreted them (autocrine activity) or on neighboring cells (paracrine activity). In general, cytokines act mainly in the neighboring microenvironment in which they are produced; however, some cytokines (IL-1β and tumor necrosis factor α (TNFα)) may be present in the circulating blood and exert systemic and endocrine effects. Generally, the cellular responses to cytokines are expressed in several hours and require de novo production of mRNA and the synthesis of specific proteins. Cytokines play a critical role in cellular interactions and immune responses. These molecules exert many biological activities which ultimately coordinate multiple aspects of innate and acquired immunity as well as the control of the organs physiology. An important characteristic of cytokines is represented by their ability to regulate various aspects of the host immune response. The study of immunological characteristics of cytokines is complicated by the properties of pleiotropism, redundancy, regulation of the production of other cytokines, synergism, and antagonism. For examples, IL-1 is important for T-cell differentiation and at the same time together with TNFα is able to control the body temperature.

Most cytokines are not produced and do not act alone but they work as part of a complex network in which they regulate both the production activities of other cytokines and their effector activities. For this reason, their study is particularly complicated. The synergism is another important characteristic of certain cytokines. IL-2 and IL-12 act together in modulating the activity of cytotoxic lymphocytes (CTL) and the antagonism is another important characteristic of cytokine appearance. For example, the IL-10 produced by Th2 cells inhibits the activity of Th1 cells. Therefore, it is the overall balance between the different cytokines activities to determine the beginning, the maintenance, and the resolution of the immune and inflammatory responses.

It follows that the study of these mediators in vivo is extremely challenging since it is difficult to determine the exact sequence of events in which the cytokines are produced during the tissue interactions in which cells and mediators cooperate in the different phases of the inflammatory processes.

CYTOKINE FAMILIES

The number of known cytokines is wide and constantly expanding and therefore it is difficult to describe them all individually. However, the biological activity of cytokines is important in terms of molecular and cellular interactions between cells of the immune system and the parenchymal cells. Here, I summarize the main functional class of cytokines categorized based on the biological process they participate in. The key cytokines for each biological process will be described.

Proinflammatory Cytokines

Proinflammatory cytokines are produced by several cell types following interaction with microbes or from tissue-damaging events. These cytokines play a very important role as they are responsible for the activation of the innate and acquired immune response by enhancing the activity of immune system cells. These activities are mainly carried out by altering the expression of molecule important for the presentation of antigens such as *Major Histocompatibility Complex* (MHC) molecules and the costimulatory molecules such as CD80, CD86, and CD40. The cells most involved in these processes are the neutrophils, macrophages, dendritic cells (DCs) and mast cells. In general, proinflammatory cytokines possess pleiotropic functions in innate immunity. They orchestrate acute inflammatory processes locally and, in severe cases, systemically. In addition, they participate in shaping adaptive immunity by skewing the differentiation of naive helper T lymphocytes and by directly affecting the effector functions of different subsets of T and B lymphocytes.

The first cytokines to be produced in this class are IL-1β, IL-6, TNFα, and Interferons gamma (IFNγ). These cytokines are produced and act both on cells that produced them and on other cells of the immune system or the parenchyma.

IL-1α and IL-1β are ubiquitously expressed although monocytes/macrophages are the main producers. Both cytokines display an immune-stimulating and inflammatory activity and the mature form of the proteins as well as its extracellular release requires caspase-1 activity. Currently, IL-1β is considered the master regulator of the innate immunity. Different microbial ligands are able to induce the IL-1β gene either directly or indirectly through the secretion of other cytokines such as TNFα, IFNγ, or IL1 itself. IL-1β, IL-6, and TNFα participate to the induction of the acute phase response. Among different functions, IL-1β is able to induce other cytokines such as IL-6, adhesion molecules to promote cell–cell contact and transendothelial migration of inflammatory cells. IL1α and IL-1β exert similar effect by binding the same receptor complex, the IL1 type I and type II receptor. The IL1 receptors are present in different cells types.

IL-6 is a pleiotropic molecule produced by monocytes, endothelial cells, and from fibroblasts in response to IL-1β and TNFα. It is one of the major players in the induction of the acute phase response by stimulating hepatocytes to release plasmatic proteins important to fight bacterial infections. In addition, IL-6 induce B-cell differentiation and survival, T cells and thymocytes proliferation, and increase NK cell activity. IL-6 release is induced during inflammatory conditions upon stimulation of Toll-like receptors (TLRs) or upon stimulation of cells by IL-1β or TNFα. It binds to interleukin-6 receptor (IL-6R) which is not signaling competent on target cells; however, signaling is initiated upon association of the IL-6/IL-6R complex with a second receptor protein, the glycoprotein (gp) 130. gp130 dimerization leads to the activation of the tyrosine kinase Janus kinase 1 (JAK1).[2] An important characteristic of IL-6 is that it binds to its receptor gp130 only when the complex IL-6/IL-6R is formed. This has the consequence that whereas gp130 is expressed by all cells of the body, IL-6R expression is restricted to few cell types such as hepatocytes, some leukocytes, and some epithelial (e.g., biliary epithelial cells) and nonepithelial cells (e.g., hepatic stellate cells). Therefore, only IL-6R positive cells can respond to IL-6.

Nevertheless, it has been shown that IL-6R can be cleaved by different proteases during a process called shedding[3] and still being able to bind IL-6, broadening the spectrum of target cells. This IL-6 activity was defined IL-6 trans-signaling. As already mentioned, the major inducer of the hepatic acute phase proteins is IL-6 that is secreted by neutrophils, monocytes, and macrophages upon TLR stimulation.[4] Activation of myeloid cells induces the release of IL-1β and TNFα that lead to a massive production of IL-6 from other cells. It is generally believed that while IL-6 classical signaling is crucial for the induction of the acute phase response, IL-6 trans-signaling mediates strong mitogenic signals for T, B cells and hepatocytes, function that it is particular important during liver regeneration. Nevertheless, it needs to be mentioned that there is a growing body of evidence that IL-6 maybe important for the control of metabolic functions.[5] It has been observed that a correlation exists between serum levels of IL-6 and degree of obesity[6] and the development of type 2 diabetes.[7] These data show that in the liver, IL-6 not only regulates glucose metabolism but is also necessary to maintain tissue homeostasis for proper control of metabolic functions. The metabolic function of IL-6 should be further investigated also in other organs and indeed and future studies should consider that IL-6 classic and IL-6 trans-signaling might differentially regulate metabolism in a tissue-specific manner. In summary, IL-6 is a cytokine with pleiotropic functions. Under physiological conditions, it is essential for proper hepatic tissue homeostasis, liver regeneration, infection defense, and fine tuning of metabolic functions. However, persistent activation of the IL-6 pathway seems to be detrimental and can even lead to the development of liver cancer. Although much advancement has been made, there are still many open questions concerning the implication of IL-6 in physiology and pathology. In order to efficiently target only the detrimental effects of IL-6, we need to better understand the effects of IL-6 of different cell types of the liver and other organs.

TNF alpha (TNFα) is a key cytokine with a broad spectrum of biological activities with a key role in a variety of pathological processes. The biologically active native form of TNFα is a homotrimer. TNFα is produced by macrophages, DCs, B cells, and T cells as well as by other types of somatic cells such as endothelial cells, mast cells, neuronal tissue cells, and tumor cells.[8,9]

TNFα operates in various parts of the body as a key cytokine in inflammatory and immune processes.[10] At low concentrations, TNFα has multiple biological properties; it increases the expression of some adhesion molecules (e.g., Intercellular Adhesion Molecule (ICAM-I)), important for the transendothelial migration of lymphocytes toward the site of tissue damage. It stimulates the release of IL-8, IL-1, and IL-6 by macrophages. In the course of bacterial infections, high TNFα concentrations may be present in the blood and they are a powerful mediator of the inflammatory response to bacteria. From this point of view, TNFα shares many of the biological properties of IL-1. TNFα is important in the control of the body temperature and induce the acute phase response proteins. TNFα has a crucial role in host immunity, preventing infection, and the growth of malignant tumors.[11,12] TNFα acts by binding to two receptors, TNFR1 (known as p55TNFR) and TNFR2 (known as p75TNFR). TNFR1 is ubiquitously expressed by a broad array of cell types and contains a death domain in its intracellular portion. The two receptors are differentially expressed in various cells and also differ in terms of their functionally. TNFR1 contains an intracellular 'death domain' that activates Nuclear Factor kappa B (NF-kB) signaling pathways leading to apoptosis.[13] TNFR2 does not contain a death domain and it is expressed in lymphocytes and in different anatomic regions of the brain cells. Its effects include T-cell activation and proliferation via signaling pathways that involve NF-kB, *Activator Protein 1* (AP-1), and the Mitogen-Activated Protein Kinase (MAPKs).[14] Knockdown studies have shown that TNFR2 has a trophic or protective role in neuronal survival.[15]

Similar to most cytokines, TNFα is expressed at very low levels in a healthy adult brain but highly expressed in neurodegenerative brains, and neuroinflammation can be detected years before neuronal death which is partially mediated through TNFR1.[16] High concentrations of TNFα can cause severe disease and therefore TNFα synthesis should be kept at homeostatic levels to combat infections avoiding overproduction of the cytokine that may cause adverse reactions. Indeed, extremely high blood concentrations of TNFα cause disseminated intravascular coagulation, shock, and severe tissue damage. In conclusion, the biology of TNFα whose effects depend on whether it acts in soluble or membrane-bound form and through which of two distinct receptors it signals, underlies the complex nature of this cytokine. More understanding on the regulation of expression of the two TNFα receptors and signaling molecule in primary cells will allow to understand how to control the duality of function of TNFα in organs such the *Central Nervous System (CNS)* in physiological conditions and during disease state.

Type I IFNs were described as proteins that are secreted by virus-infected cells and interfere with virus replication in autocrine and paracrine ways.[17] It is now more appreciated that type I IFNs are cytokines produced in response to a wide variety of microbes such as viral, bacterial, and fungal pathogens as well as parasites.[18] It was later discovered that these molecules can also activate inflammatory cells. There are two forms of interferons, the type I IFNs and type II IFNs. The type I IFNs are produced primarily during viral and microbial infections and include IFNα produced by leukocytes, IFNβ produced by fibroblasts, and IFNk produced by keratinocytes. Type II IFN, better known as

IFN-γ, is produced by activated T and NK cells and share many of the antiviral activity of type I IFNs. However, IFNγ is not structurally related to IFNα and IFNβ as it interacts with specific receptor and possesses immune-regulatory properties that are distinct from type I IFNs.

IFNγ is a key mediator of inflammatory immune responses and it activates mononuclear phagocytes to phagocytose and kill microorganisms. This cytokine belongs to a class of molecules that induce the differentiation of monocytes into macrophages. IFNγ increases the expression of MHC class I and II molecules on the antigen-presenting cells (APC) surface, enhancing the capacity of recognition by the T cells during antigen presentation. IFNγ possesses numerous other biological properties: It directly promotes T and B-cell differentiation and can activate neutrophils and NK cells. Also, IFNγ stimulates the expression of adhesion molecules on different immune cells and increase transendothelial migration to the inflammatory sites. IFNγ is widely distributed playing a pivotal role in host defense.[19–21] The binding of IFNγ to its receptor activates the JAK–signal transducer and activator of transcription (STAT) pathway which modulates the transcriptional activation of several genes and mediates diverse biological responses. The IFNγ receptor consists of two subunits, IFNγ receptor (IFNΓR)1 and IFNΓR2 and each molecule interacts with a member of the JAK family. JAKs phosphorylate receptors and transcriptional coactivators named STATs.[22] IFNγ acts as a master regulator of many inflammatory pathways as it is involved in regulating the production of several molecules involved in the inflammatory response. For example, IFNγ upregulates the production of IL-12, TNFα, and several chemokines in macrophages induced by lipopolysaccharide (LPS).[23] The pleiotropic effects of IFNγ are mediated by modulating a large collection of genes and highlight the importance of identifying and studying IFNγ-regulated genes. Thus, it would be interesting to identify novel genes that are the direct targets of the IFNγ response. Further investigations into the functions of IFNγ-modulated genes will help in understanding their functional effects. Such studies may identify therapeutic targets and interventions during injury or disease.

IL-18 is closely related to IL-1β in that both are first synthesized as inactive precursors, both require caspase-1 for cleavage, and both have decoy receptors. IL-18 is an important component of polarized Th1 cell and NK cell responses. IL-18 primarily induces Th1 responses due to the induction of IFNγ in combination with IL-12. However, in the absence of IL-12, there is a role for Th2[24] and NK cell maturation, cytokine production, and cytotoxicity.[25]

IL-33 is a member of the IL-1 family of cytokines with a growing number of target cells and a plethora of biological functions. The cytokine acts on many different target cells in many organs. Its main effects on innate and adaptive cells, Th2 and innate lymphoid cell-2 (ILC2) cells, and alternatively activated M2 polarized macrophages are consistent with its general function. It can be expressed in a constitutive or inducible manner by several stromal, parenchymal, and hematopoietic cell types. Mature IL-33 signals through the suppression of tumorigenicity 2 (ST2) receptor, which associates with IL-1 accessory protein (IL-1AcP) to induce myeloid differentiation primary response gene MyD88-dependent signaling. IL-33 amplifies innate immunity and inflammatory responses.[26]

IL-36 family members IL-36α (IL-1F6), IL-36β (IL-1F8), and IL-36γ (IL-1F9) bind to IL-1Rrp2 and use IL-1RAcP as a coreceptor. IL-36 is produced by innate immune cells and lymphocytes inducing the production of proinflammatory cytokines, chemokines, and costimulatory molecules, thus promoting Th1 and Th17 cell polarization.[27,28] In general, IL-36 mirrors IL-1 and may serve as an amplifier of innate immunity and a mediator of inflammation in selected tissues such as skin and lung.

Regulatory Cytokines

Interleukin-10 (IL-10) and transforming growth factor-beta (TGF-β) are the two best studied regulatory cytokines. In humans, IL-10 suppresses the expression of MHC class II, costimulatory, and adhesion molecules on monocytes. IL-10 also inhibits the production of proinflammatory cytokines and the activation of T cells via APCs. It possesses important anti-inflammatory factor that controls the response of lymphoid and myeloid cells and thus inhibits immunopathology. This role of IL-10 is extremely important in protecting the host from infection-associated immunopathology, autoimmunity, allergy, sepsis, arthritis, insulitis, and inflammatory bowel disease (IBD). In addition to these activities, IL-10 regulates growth or differentiation of B cells, NK cells, cytotoxic and helper T cells, mast cells, granulocytes, DCs, keratinocytes, and endothelial cells.[29]

IL-10 production was originally ascribed to CD4+ Th2 cells.[30] However, it is now known that IL-10 is released by a wide variety of immune cells, including Th1 and regulatory T (Treg) cells, CD8+ T cells, B cells, macrophages, DCs, neutrophils and eosinophils. The cytokine presents a complex transcriptional regulation and indeed mechanisms such as epigenetic regulation, activation of specific transcription factors and triggering of signaling pathways have been all described.[31] It is clear that deregulation of these tightly regulated processes is associated with detrimental effects of IL-10. Therefore, understanding the regulation of IL-10 expression by different cells is instrumental for the targeted design of immune intervention strategies. Binding of IL-10 to its receptor triggers a series of signaling cascades mediated by the JAK signal transducer and STAT pathway, particularly by STAT3.[32]

Signaling through the IL-10 receptor regulates several steps of the immune response, from decreasing cytokine gene expression to downregulating the expression of MHC-II and thus antigen presentation to T cells.[29,32] Moreover, IL-10 has been shown to prevent apoptosis by activating the phosphatidylinositol-4,5-bisphosphate-3-kinase (PI3K)/Akt cascade and enhancing the expression of antiapoptotic factors as B cell lymphoma-xl (Bcl-xl) and Bcl-2, while attenuating that of caspase-3. It is important to note that the cellular source of IL-10 production is critical to its immunological activities in a cell-specific manner. For example, a critical role of B cell-derived IL-10 has been demonstrated in the mouse model of experimental autoimmune encephalomyelitis (EAE). Mice with a disruption in the Igμ heavy chain (μMT), which results in a lack of B cells, develop a non-remitting form of EAE. Transfer of Wild Type (WT) B cells restored remission, whereas B lymphocytes from IL-10-deficient mice were unable to suppress the disease progression.[33] Therefore, these studies highlight the importance of understanding the biology of IL-10 derived from different cellular origins as it determines its unique range of activities.

TGF-β plays a diverse set of roles in development, cell differentiation, wound healing, and immune regulation. It is now recognized that TGF-β is produced by many cell types, both immune and nonimmune, and is particularly abundant in the gut, immune tissues, and central nervous system. Three TGF-β isoforms have been described: TGF-β1, TGF-β2, and TGF-β3. The expression of each isoform is spatially and temporally distinct. TGF-β1 is expressed abundantly in the immune system, whereas expression of TGF-β2 and TGF-β3 are minimal. TGF-β has been studied extensively in the immune system as a major contributor to immune regulation. TGF-β signals through a heterodimeric receptor composed of TGF receptor I and TGF receptor II that are serine/threonine kinase receptors that induce the phosphorylation of SMAD proteins. SMAD proteins act as transcription factors, resulting in the positive or negative regulation of specific genes. It is well-known that TGF-β can influence the differentiation of naïve CD4 + T cells into effector T cells with different phenotypes. Much less is known about how TGF-β affects effector T cells, particularly at sites of inflammation. TGF-β signaling likely plays a vital role in other cells and tissues, particularly the gut where TGF-β expression is abundant and our immune system is tuned based on the interaction between the gut with the microbiota. Finally, TGF-β is highly expressed in the CNS and is known to play a vital role in CNS development and function. TGF-β has both stimulatory and inhibitory activity; however, it is generally considered as a negative modulator of other cytokines especially those proinflammatory. The TGF-β inhibits the proliferation of T cells, the activation of CTL cells, and macrophages. Despite a detailed understanding of the signaling principles and the effectors of its regulation, the TGF-β pathway remains rather enigmatic as it is still unclear how ligands can transmit context and concentration-dependent signals through a misleadingly simple signaling pathway. The number of different regulatory proteins that modulate the TGF-β signaling pathway at different steps underlines the complex mode by which a specific biological outcome may be generated. A major challenge at the current stage is to understand how to integrate the diverse regulatory circuits to explain the multifunctional nature of the TGF-β pathway during cellular differentiation and in the maintenance of tissues homeostasis.

IL-37 is a newly described cytokine that broadly inhibits inflammation by reducing several signaling kinases and augmenting antiinflammation genes. IL-37 can be induced by several TLR agonists and proinflammatory cytokines such as IL-1β, TNFα, and IFNγ in peripheral blood mononuclear cells (PBMCs) and DCs.[34] The mature IL-37b can translocate into the nucleus via a caspase-1-dependent process. Consequently, it is possible that IL-37b may act as a cytokine with both intracellular and extracellular functionality.

A neutralizing monoclonal anti-IL-37-treated human PBMCs from healthy donors can increase the inflammatory cytokines production, such as IL-1β, IL-6, and TNF-α.[35] Monocytes-induced DCs (moDCs) from normal donors stimulated with LPS can upregulate the expression of IL-1β, IL-6, TNF-α, and reactive oxygen species (ROS). Collectively, these data suggest that IL-37 plays a negative role in human PBMCs, M1 macrophages, and moDCs-induced Th1/Th17 cell response.[36]

Cytokines Important in T Lymphocyte Differentiation (Th1, Th2, Th22)

T lymphocytes can differentiate into different effector cells based on their cytokine profile that it is dependent on APC, the type of inflammatory response, the costimulatory molecules presented with the antigen, and on the other cytokines that are present into the same environment.[37] CD4+ naïve T cells can differentiate into Th1, Th2, Th9, Th17, Th22, and follicular effector T cells as well as different subsets of Treg cells. Based on their cytokine profile, interaction with other cell types, growth factors, and chemokines, these subsets can drive different type of inflammatory responses.

Th1 cells are best characterized for their production of IL-2, IL-12, IL-23, IL-15, IL-16, IL-17, IL-18, IFNγ, and TNFα. The cytokines produced by Th1 cells, such as IFNγ, TNFα, and IL-2, are able to induce macrophage activation, enhance the ability of antigen presentation, increase the expression of Fc-epsilon receptors, and cause the production of toxic oxygen radicals. These activities will result in an efficient killing of intracellular and extracellular pathogens and promote the long-term immunity against these microorganisms. Th1 cells predominantly produce IFNγ and are important for protective immune responses to intracellular viral and bacterial infections. Th1 cells are also responsible for the induction of some autoimmune diseases. In turn, IFNγ is important for the differentiation of IFNγ-producing Th1 cells although IL-12 is also critical for this process.

Th2 cells produce IL-4, IL-5, IL-9, and IL-13. In addition, thymic stromal lymphopoietin (TSLP), IL-25, IL-31, and IL-33 contribute to the development and intensity of Th2 responses and inflammation. Th2 cells mediate host defense against extracellular parasites including helminthes. Similarly, when stimulated by a cognate antigen presented by accessory cells in the presence of IL-4, naive CD4 + T cells differentiate into IL-4-producing Th2 cells. CD25 (IL-2Rα) expression is higher in Th2 cells than in Th1 cells such higher expression of CD25 may confer hyperresponsiveness to IL-2. The key effector cytokines include IL-4, IL-5, IL-9, IL-13, IL10, IL-25, and amphiregulin. IL-4 is a major cytokine involved in allergic inflammation. It is involved in IgE switching and secretion by B cells. IL-4 also upregulates low-affinity IgE receptor (FcεRI) on B lymphocytes and mononuclear phagocytes, and also high-affinity IgE receptor (FcεRII) on mast cells and basophils with subsequent degranulation of the cells and release of several active metabolites, including histamine and serotonin. IL-4 also induces the increase of several other proinflammatory mediators, including IL-6, granulocyte-macrophage colony-stimulating factor (GM-CSF), vascular cell adhesion molecule-1 (VCAM-I) adhesion molecule. IL-5 mainly targets eosinophils and its precursors since these cells have relatively higher amounts of IL-5R expressed on their surface and subsequently leads to their activation with upregulation of CD11b and inhibition of apoptosis.[38] IL-9 participates actively in the immunopathogenesis of asthma. It activates the function of several cells, including mast cells, B cells, eosinophils, neutrophils as well as airway epithelial cells. IL-13, through the activation of cell-mediated immunity, helps in the elimination of intracellular pathogens, such as *Leishmania*. It also plays a major role in the induction of allergic asthma through activation of eosinophils, enhanced mucus secretion, and airway hyperresponsivity. IL-25 is a member of the IL-17 family of cytokines. It is structurally similar to IL-17 but functionally different. It promotes Th2 responses. It induces increased mucus production, eosinophilia, IgE switching, and enhanced Ig secretion, as a result of upregulation of IL-4, IL-5, and IL-13, thereby amplifying aTh2 response.

Initially characterized as a subset of Th2 cells, cells producing IL-9 are now classified as Th9 cells as a distinct subset of CD4 + T cells. TGF-β was found to divert the differentiation of Th2 towards the development of Th9 cells. Moreover, TGF-β in combination with IL-4 directly induces the differentiation of Th9 cells.[39] However, more research needs to be directed to get more understandings about the Th9 cells, before being classified as a distinct lineage of CD4 + cells.

Th17 cells are recently discovered T cellular subsets that enabled a better definition of the inflammatory process that it is occurring during autoimmunity and the immune response to the extracellular infections. Th17 cells are characterized by the production of IL-17A, IL-17F, IL-6, IL-8, TNF-a, IL-22, and IL-26. There is still an ongoing discussion whether or not Th17 and Th22 cells subsets exist as the main Th22 cell cytokine, the IL-22, can be also produced by Th17 cells.[40] IL-6, IL-21, IL-23, and TGF-β are the major signaling cytokines involved in Th17 cells differentiation and retinoic acid receptor-related orphan receptor gamma-T (RORγt) is the master regulator. The differentiation process can be split into different stages, including the differentiation stage mediated by TGF-β and IL-6, the self-amplification stage by IL-21, and the stabilization stage by IL-23.

Follicular helper (Tfh) T cells are C-X-C motif receptor-5 (CXCR5+) expressing cells and are located in follicular areas of lymphoid tissue, where they contribute in the development of antigen-specific B-cell immunity.[41,42] IL-6 and IL-21 are the main cytokines involved in the differentiation process.[43]

Treg cell subsets have distinct phenotypes. Treg exists as natural thymus-derived subset with expressed forkhead box P3 (FOXP3), and as peripheral-induced Treg cells, which arise from naïve CD4 + CD25 cells after antigen priming in a relevant cytokine environment. Their main effector cytokines include IL-10, TGF-β, and IL-35. IL-10 is a potent inhibitory cytokine with the ability to suppress proinflammatory response and thus limits tissue damage by the inflammatory process. IL10 and TGF-β potently suppress IgE production, thereby showing their important role in attenuating allergic inflammation. IL-10 is a well-recognized antiinflammatory cytokine that shows immune-stimulatory capacity acting as a growth and differentiation factor for NK cells and B lymphocytes. The dual role of IL10 depends on its tissue level, the nature of the target cells, the activating signals, the timing, and the sequence of cytokine networks in which it is produced.

Cytokines Important in B Cells Differentiation and Activation

B cells are the only cell type capable of differentiating into antibody-producing cells known as plasma cells.[44] In addition, they can present antigens to T cells and produce cytokines. Naive B cells can develop into distinct cytokine-producing subsets that influence CD4 + T cell responses. Various cytokine-producing B cell subsets have now been identified that modulate the polarization of CD4+ T cell responses in vivo. By secreting cytokines, B cells can guide the development of lymphoid tissues, shape effector T cell responses, and negatively regulate immunity. Apart from lymphoid tissue-inducer (LTi) cells, lymphoid organogenesis can be controlled by B cell-derived lymphotoxin α1β2. In the spleen during development B cell-derived LTα1β2 and TNFα are necessary for the development of the follicular DCs (FDCs).[45–48] IL-2 producing B cells promote the development of memory T cells in response to certain pathogens such as *Heligmosomoides polygyru*.[49] B cells can also stimulate Th1 responses by releasing IFNγ as it has been shown in the case of *Salmonella enterica* and *Listeria monocytogenes* infection.[50] In mouse model of atherosclerosis, B cells exert pathogenic responses through the production of GM-CSF which stimulates the accumulation of IL-12-expressing DCs in the spleen and finally results in an increased accumulation of T cells and macrophages in atherosclerotic lesions.[51] Interestingly, B cells can produce large amounts of IL-6 following activation. In EAE, B cell-derived IL-6 increases disease pathogenesis by promoting the activation of Th1 cells and Th17 cells. B cells can positively contribute to the polarization of effector T-cell responses and the formation of memory T cells through cytokine production. However, B cells can also inhibit inflammatory immune responses through the production of IL-10 and IL-35. These B cells activities has been studied in various diseases, including Th1 or Th17 cell-mediated autoimmune diseases such as EAE, Th2 cell-mediated pathologies (such as ulcerative colitis and allergic airway inflammation), systemic autoimmune diseases (such as systemic lupus erythematosus (SLE)), and bacterial infections (such as with *Salmonella enterica* or *Listeria monocytogenes*).[52–54] In conclusion, the understanding of the biology of B cell-derived cytokines is becoming to be appreciated only recently and progress has also been made in identifying the molecular mechanisms that control cytokine production by human B cells and in defining how these cytokines are deregulated in autoimmune diseases.

CYTOKINES AND TISSUE HOMEOSTASIS

Cytokines play a key role in the maintenance of tissue homeostasis, although in most of the tissues cytokines became strongly induced following injuries.

In the context of mucosal immunity, cytokines regulate mucosal barrier function, influence immune cell migration and activation, and impact cellular metabolism to maintain a homeostatic environment. Cytokine dysregulation underlies the pathogenesis of multiple inflammatory disorders including Crohn's disease (CD), ulcerative colitis, and celiac disease. The main sources of cytokines in the intestine include transient and resident lymphocytes, mononuclear cells (macrophages and DCs), ILCs, polymorphonuclear cells as well as intestinal epithelial cells and stromal cells. IL-22 is one of the most intestinally active cytokines; it plays a key role in regulating mucosal barrier function and maintaining intestinal homeostasis. This cytokine has both proinflammatory and antiinflammatory functions but studies suggest a protective role for IL-22 in the gut. IL-22 is a member of the IL-10 cytokine family and is produced by ILCs, DCs, neutrophils, and various subtypes of CD4 + T cells, most notably Th17, Th1, and Th22 cells.[55,56] IL-22 binds a heterodimeric receptor complex composed of two subunits, IL-10R2 which is expressed on a broad array of tissues and IL-22R1 which is expressed in a reduced number of tissues including the skin, lung, and intestine.[57] Therefore, the effects of IL-22 are important to barrier surfaces including the mucosal epithelia of the digestive system.[58] IL-22 signaling restores barrier function during inflammation as shown in mouse models of colitis. In the intestine, IL-22 promotes wound healing, tissue regeneration, and epithelial cell proliferation.[59]

In the thymus, all types of cells are able to produce cytokines either spontaneously or after stimulation. The main producers of cytokines in the thymus are thymic epithelial cells (TEC) and thymocytes. Some cytokines have paracrine effect and other shows an autocrine effect. For example, IL-7 is produced by TEC or stromal fibroblasts and induces thymocyte growth and differentiation whereas IFNγ is produced by thymocytes and induces TEC activation. An example of cytokine of this type is IL-2, for which the producers and targets are thymocytes. Interestingly, differences in the function of the cytokine system in thymus and peripheral compartments of the immune system can be appreciated. In the periphery, cytokines and their receptors are usually induced in thymus the synthesis of cytokines and expression of their receptors depends on cell–cell interactions. The principal role of thymic cytokines is to support constant processes such as the migration and the development of thymocytes, but not usually those processes that are inducible such as the inflammation or the immune response as it is observed in the periphery. As a consequence, the functions of some cytokines in the thymus can be significantly different from those in the periphery. For example, proinflammatory cytokines act in the thymus as molecules that mediate thymocyte or TEC activation, differentiation, or proliferation. Cytokines such as IFNγ and IL-4 mediate the cellular interactions between thymocyte and TEC and play a role in maintaining the thymocyte homeostatic population instead of activating and regulating the immune response. Therefore, to understand the tissue physiology it will be important to perform detailed analysis within a tissue to reveal cytokine impact in the organ physiology as a whole.

T and B-cell numbers have a constant size despite daily migration of cells from the thymus to the periphery. Although hormones, antigen receptors, and regulatory cells may participate to maintain this process, cytokines are the principal signals for maintaining lymphocyte homeostasis. The best characterized cytokines for lymphocytes are IL-7 and IL-15 that are produced by nonlymphoid cells[60,61] and are essential for survival of several lymphoid subsets.[62,63]

Generally, other tissues are regulated by cytokines during homeostatic conditions, for example, IL-1 and IL-6 are known to be expressed in the spleen, liver, and peripheral blood leukocytes of normal individuals.[64] Nevertheless, their role in normal physiology is not yet well understood.

In the brain, circulating leukocytes were shown to play an indispensable role in maintenance and repair of the CNS. Blood-derived monocytes were shown to promote tissue recovery following acute CNS injury[65] and help amyloid β (Aβ) clearance in Alzheimer's disease.[66] Specific T-cell populations and T cell-derived cytokines were suggested to mediate neuroprotective mechanisms in physiology.[67–69] While the blood–brain barrier (BBB) is impermeable under physiological conditions, recent studies suggest that immune cells infiltrate the CNS trough a specialized interface, the so-called blood–cerebrospinal fluid barrier (BCSFB) formed by the choroid plexus (CP) of the brain.[70] It is clear now, that the BBB and the BCSFB display different functions. The BBB and the meningeal layers, which are based on endothelial tight junctions, do not normally allow immune cell passage and that breaching these barriers by leukocytes is considered a sign of pathology. On the other hand, epithelium-based barriers, such as the CP, are immunologically active surfaces, well positioned to appreciate and respond to signals from compartments that they isolate (blood and CSF, respectively), and well-equipped to mediate controlled immune cell trafficking.[71,72] Therefore, it is now proposed that cytokines through these structures are able to regulate CNS physiological functions.

In the context of brain physiology, it is known that type I and type II IFNs are important players in the cytokine network regulating the CNS by controlling immunity at the borders of the brain. Nevertheless, opposing roles ascribed to these cytokines and their outcomes under various pathological conditions have been reported. Cytokine signaling controls the expression of brain-derived neurotrophic factor (BDNF) and insulin-like growth factor 1 (IGF-1) in the CP and they are important for maintenance of cognitive function and neurogenesis. IFNγ is produced by immune cells located within the CP stroma, predominantly by Th1 cells and possibly by NK cells, potent producers of IFNγ, reported to reside at the CP. IFNγ stimulates the CP epithelium to express leukocyte trafficking determinants including integrins and chemokines.[73] In the absence of damage, IFNγ-dependent CP activity licenses entry of a little number of leukocytes to the CSF; absence of IFNγ-signaling in mice deficient in IFNγ receptor (IFNγR) shows reduced leukocyte count in the CSF and interestingly, is correlated with premature cognitive decline in mice during adulthood.[74] In pathological settings, an altered local cytokine environment may modulate IFNγ availability and signaling at the CP, and thereby affect leukocyte trafficking to the CNS.[71] Leukocyte trafficking to the CNS promotes repair and physiological processes that are not observed in mice lacking IFNγR. In contrast, chronic degenerative conditions of the CNS, such as *Amyotrophic lateral sclerosis (ALS)* and alzheimer disease (AD) which could benefit from immune cell recruitment to the CNS, have been recently linked with poor availability of IFNγ at the CP, resulting in diminished CP gateway activity, and thus with loss of immune support needed for repair. On the other hands, in autoimmune diseases such as relapsing remitting multiple sclerosis (RRMS), the presence of outnumbers of IFN-γ producing CNS-specific CD41 T cells is associated with unrestrained infiltration of leukocytes to the CNS through both the CP and the permeable BBB, a phenomenon considered to be central to the disease pathology.[75]

Concerning type I IFNs, it is known that they can both have a positive and a negative role on a broad range of immune activities.[76] In the brain, local production of type I IFNs has been observed on the CP epithelium in neurodegeneration conditions such as AD or during human aging.[74,77,78] In preclinical models, brain homeostasis such decreased aging-related inflammation could be restored by IFN-I blockage within the brain tissue.[74] Nevertheless, IFN-I is used to ameliorate neuroinflammation in RRMS when injected in the periphery pointing out that the cytokines has different effects whether the administration is local or at the systemic level. In this context, it is believed that IFN-I produced within the brain and delivered systemically appears to activate several mechanisms that indirectly decrease leukocyte infiltration responsible for reducing disease severity. Interestingly, IFN-I seems to affect only the RRMS and not the progressive forms. The progressive forms of the disease are believed to be sustained by proinflammatory innate immune cells, microglia, and macrophages that are already present within the CNS.[75] In conclusion, the role of IFN-I and II is very complex and we are far from understanding their role during the physiology and pathology of the brain.

Cytokines are central regulator of another complex organ, the liver—important for its metabolic and detoxifying functions. The central function of the liver covers homeostasis and inflammatory responses; the organ allows continuous blood supply from hepatic arteries and from the gastrointestinal tract via the portal vein. Circulating blood cells from the innate or adaptive immune system are present through a network of sinusoids allowing contact to a variety of intrahepatic cell populations, such as hepatocytes, endothelial cells, Kupffer cell (KC; liver resident macrophages) or lymphocyte populations, hepatic stellate cells (mainly NK cells), and others. Communication between these cells types and the regulation of hepatic functions are primarily achieved by cytokines. Among the various cytokines relevant for liver homeostasis and injury, TNF-α and IL-6 show a prominent role on liver physiology. Studies in patients and preclinical models have powerfully implicated that death receptor ligands (TNFα and FAS-L) are currently involved in the induction of apoptosis and in triggering damage of the liver. In many circumstances, hepatic failure might derive from an imbalance between damaging and protective signals tightly regulated under physiologic conditions. As previously reported, TNFα and related cytokines are key players in liver homeostasis as they can activate both proapoptotic (caspases) and antiapoptotic pathways (NF-kappaB) in hepatocytes. Liver parenchymal and nonparenchymal survival pathways like NF-kappaB or IL-6 suppress a protective function in many experimental liver disease models and thus are an attractive target for a pharmacological intervention. However, many experimental models display dual functions as positive or opposing effects.

The liver is especially enriched in KC, NK, and Natural killer T (*NKT*) cells. The healthy adult liver has an active and complex cytokine environment, which includes basal expression of proinflammatory (IL-2, IL-7, IL-12, IL-15 and IFNγ) and antiinflammatory (IL-10, IL-13 and TGFβ) cytokines.[79,80] This cytokine network exists in the absence of infection or pathological inflammation and presumably arises through normal physiological processes within the healthy liver. Inflammation and inflammatory mediators, generated by liver-resident immune cell populations and nonhematopoietic cells, play essential roles in maintaining local liver and systemic homeostasis. These homeostatic roles must be acknowledged when considering the role of inflammation and liver-resident immune cell populations in liver disease and pathology. Liver inflammation should be considered as a highly dynamic and complex network of responses to maintain organ and also systemic homeostasis in healthy subjects. Resident myeloid cells contribute to the maintenance of hepatic tolerance. KCs respond to bacterial endotoxin by producing antiinflammatory cytokines such as IL-10 and prostaglandins.[81,82] These molecules downregulate expression of costimulatory molecules on APCs, preventing activation of CD4 + T cells and limiting the development of the adaptive immune response. Presentation of particulate antigens by KCs preferentially expands IL-10 producing Treg cell populations inducing antigen-specific tolerance.[83] The tolerogenic liver environment is further maintained by the presence of regulatory myeloid populations such as myeloid-derived suppressor cells (MDSCs). MDSCs mediate their suppressive activity through the production of the IL-10 and TGFβ.[84] The liver also has a central role in sensing and responding to inflammatory signals from other sites in the body. During the early stages of extrahepatic inflammation, cytokines produced by immune cells enter the blood stream and are detected by liver hepatocytes which initiate the systemic acute phase response. Hepatocytes increase acute phase protein production and synthesize IL-6 which acts to amplify the acute phase response.[85] The acute phase proteins drive a range of mechanisms throughout the body, including leukocytosis in the bone marrow, changes in the brain that mediate pyrexia and alter behavior, and massive immune cell infiltration to the site of initial inflammation. At the same time, the acute phase proteins include a range of processes to limit excessive inflammation. It is clear that inflammatory mechanisms in the healthy liver maintain local organ and systemic homeostasis. Complex cytokines networks that need to be fully clarified regulate these inflammatory mechanisms.

CYTOKINES AS GROWTH FACTORS

Cytokines that are able to stimulate the growth and differentiation of hematopoietic progenitor cells in the bone marrow are referred to as colony-stimulating factors (CSF). The family has essentially three members, named macrophage (M)-CSF (also CSF-1), granulocyte (G)-CSF (also CSF-3), and GM-CSF (also CSF-2). Many phagocytes can potentially be controlled by these three CSFs. However, a number of other cytokines, such as IL-3, IL-5, and IL-34, are also involved in hematopoietic cellular differentiation and have specialized roles for the induction of specific tissue macrophages (e.g., IL-34 is important for Langerhans cells in the epidermis).[86,87]

GM-CSF is produced by different cells such as epithelial cells, endothelial cells, fibroblasts, stromal cells, and hematopoietic cells. GM-CSF production has mainly been attributed to these cells following their stimulation in vitro through pathogen-associated molecular patterns (PAMPs) or specific cytokines. However, under physiological conditions, the source of GM-CSF is less clearly defined. This cytokine stimulates antibody-mediated cytolysis of tumor cells induced by neutrophils and acts in an autocrine and paracrine manner; since it is not present in the circulation it is believed not to possess endocrine activity. M-CSF exists in three different biologically active dimeric isoforms (soluble, extracellular-matrix-anchored, and cell-associated isoforms).[88–90] GM-CSF and G-CSF are secreted as monomeric proteins.[91] The cognate receptor for M-CSF consists of an extracellular region composed by five immunoglobulin domains, a transmembrane part, and an intracellular tyrosine kinase domain whose phosphorylation leads to the activation of several signaling pathways. The intracellular signaling pathways of survival (mTOR), chemotaxis (PI3K, Src, Syk2), and cellular differentiation (PLC-γ2 and Erk1/2) are all activated by the cytokine GM-CSF.[92] The GM-CSF receptor (GM-CSFR) is a heterodimeric complex consisting of a GM-CSF-specific α chain and a signal-transducing β subunit which is shared with the IL-3 and IL-5 receptors.[93] The G-CSF receptor (G-CSFR) is a single transmembrane protein that homodimerizes upon G-CSF binding. GM-CSF signaling triggers JAK2 and STAT-5, which appear to control the differentiation and the inflammatory signature,[94] whereas PI3K signaling promotes proliferation and survival.[95] Structural differences but also the divergent spatiotemporal expression patterns of CSFs and their cognate receptors explain their specific biological functions. Whereas M-CSF is broadly expressed and can be detected in the circulation, G-CSF is detected at lower concentrations and GM-CSF is virtually absent.[96,97] In conclusion, M-CSF and G-CSF are important for steady-state myelopoiesis and are also involved in inflammatory processes. GM-CSF plays only a minor role in the physiological support of myeloid cells but it is a major mediator of tissue inflammation.

IMMUNOMODULATION BY CYTOKINES IN HUMAN DISEASES

A fundamental insight gained over more than three decades is that diverse cytokines are crucial drivers of different diseases. Cytokine-targeting therapies developed on the basis of these discoveries have dramatically changed the treatment options for patients with different inflammatory-based disease such as rheumatoid arthritis (RA), psoriasis, IBD, and others.[98] The classic, well-known proinflammatory cytokines include TNFα, IL-1β, IL-6, IFNγ, and IL-17 among others. These proinflammatory cytokines are involved in the differentiation and activation of pathogenic T cells (i.e., Th1 or Th17), the migration into the target organ, the process of neovascularization (angiogenesis), the development and activation of osteoclasts, and the process of bone damage during the course of autoimmune arthritis for example. The classic anti-inflammatory cytokines include IL-4 and IL-10 which display immunosuppressive activities.

The main target cytokines of interest for the treatment of RA are: TNFα, IL-1β, IL-6, IL-17, and IL-23. Various biological molecules that target cytokines or their receptors are composed by specifically engineered soluble recombinant cytokine receptors, monoclonal antibodies, and fusion-proteins containing the cytokine binding domain linked to the Fc portion of human IgG1. Although extremely effective, the above mentioned molecules have some restrictions. These include the unresponsiveness of a proportion of patients; the short-lived effect necessitating frequent injections; gradual loss of sensitivity to the therapeutic agent with time; the likelihood of disease exacerbation if therapy stopped for a long time; the high cost; and the increased risk of infections because of systemic immunosuppression. Given the limitations of anti-TNFα and anti-IL-1β therapy, continued efforts are directed towards targeting other cytokines (e.g., IL-6 and IL-17). It is now believed that combination therapy with different biologics might offer an advantage over a single agent. Paradoxically, TNFα, which represents a prototypic proinflammatory mediator, can exhibit antiinflammatory effects under certain conditions. A comprehensive understanding of the role of cytokines in the pathogenesis of the diseases such as RA has led to the development of new therapeutic approaches aimed at specifically neutralizing the cytokine or inhibiting cytokine signaling pathways. Unlike generalized immunosuppression, selective targeting of a particular pathogenic cytokine offers a distinct advantage for the treatment of autoimmune diseases. Given the importance of TNFα, IL-6, IL-23, and IL-17 in the development and pathogenesis of autoimmune diseases, lots of efforts have been dedicated on targeting these cytokines to treat these diseases. Accumulating preclinical and clinical studies show that blocking TNFα, IL-6, IL-23, IL-17, or their corresponding receptors by use of neutralizing antibodies is highly effective in the treatment of multiple autoimmune diseases such as psoriasis, RA, and IBD. In particular, targeting the IL-23–IL-17 axis has been the most successful strategy for the treatment of psoriasis in the past decade.

Another successful targeted cytokine is TNFα and indeed it is one of most widely studied and clinically tested cytokines in the treatment of autoimmune diseases. To date, anti-TNFα molecules are still the best-selling cytokine-targeting drug type on the market. Although anti-TNFα therapies failed in the treatment of cancer and sepsis, the first clinical trial of a TNFα inhibitor succeeded in the treatment of RA.[99] This achievement subsequently led to five approved TNFα-blocking biologicals: Infliximab, etanercept, adalimumab, certolizumab, and golimumab.[100] The data suggested the use of all these five TNFα-blocking biologicals in the treatment of autoimmune diseases associated with overproduction of TNFα, such as psoriasis, RA, multiple sclerosis (MS), and IBD.[101,102] However, not all patients have responded equally to each medication. Despite the striking positive effects of blockade of TNFα in psoriasis, psoriatic arthritis, and RA, anti-TNFα agents worsened disease in patients with MS[103] and etanercept failed to treat patients with IBD.[104] The differential effects of TNFα blockade might be due to the heterogeneity of RA, MS, and IBD. Indeed, TNFα-blocking biologicals have been reported to be effective in the treatment of RA but about one-third of patients with RA treated with anti-TNFα agents did not respond. Recent studies indicate that both the timing and the duration of TNFα expression are important in determining the pathogenic roles of TNFα. Consequently, determining the appropriate phase of disease at which to interfere is crucial. In addition, systemic blockade of soluble TNFα and membrane-bound TNFα by using neutralizing antibodies has been indicated to enhance cancer risk and susceptibility to infections.[105]

IL-6 is another key cytokine that has role in the development of autoimmune models such as EAE, collagen-induced arthritis (CIA), RA, IBD, and psoriasis. Therefore, targeting IL-6 holds therapeutic potential in the treatment of these autoimmune diseases. And indeed, multiple neutralizing antibodies targeting IL-6 and its receptor have been developed. To date, tocilizumab (humanized IL-6R-specific monoclonal antibody (mAb)) has been approved in combination with methotrexate. Siltuximab (chimeric IL-6-specific mAb) has been approved for multicenter trials of Castleman's disease treatment.[106] IL-6 plays a critical role in host defense against microbial infections and plays a protective role during neural and liver injury. Long-term treatment of IL-6 neutralization may induce important side-effects as it may increase susceptibility to bacterial and viral infections or may cause mortality in patients with ischemic neural damage or alcoholic cirrhosis. The discovery of the IL-23–IL-17 immune axis has helped to defining the Th17 cells lineage. This new T-cell subset plays a key role in the pathogenesis of multiple autoimmune diseases as Th17 cells are one of major producers of IL-17. Given the role of IL-23 in Th17 differentiation, neutralizing antibodies have been developed that target IL-23. The anti-p40 subunit therapy targeting IL-12 was first evaluated in patients with CD and provided benefit in this disease. It was shown that in patients with psoriasis, levels of mRNA for the p40 and for the p19 of IL-23 were higher in lesions of psoriasis patients than in nonlesional and normal skin, whereas mRNA for p35 of IL-12 was present but decreased in lesional skin. The evidence suggests that targeting the p40 of IL-12 indeed inhibits IL-23 and that inhibiting p40 might be effective to treat psoriasis too. Therefore, two therapeutic mAbs targeting the p40 subunit, ustekinumab and briakinumab, were used in treatment of psoriasis. Ustekinumab also showed benefits in the treatment of moderate to severe CD when it was administered as maintenance therapy but failed to prevent inflammation in patients with MS. To date, ustekinumab has been approved to treat psoriasis and psoriatic arthritis. Although IL-23 is redundant in host defense to many pathogens, it is important for the host responses against *Candida* infections. Therefore, treatment of autoimmune diseases with IL-23 antagonists carries the risk of impaired host defense responses to these types of pathogens. IL-23 has important properties in tumor surveillance as it increases tumor cell proliferation, survival, and invasion by activating STAT3. Therefore, the risk of antagonizing the

IL-12/IL-23 p40 subunit is also of particular concern because it may interfere with tumor surveillance activity.

Finally, IL-17 plays a crucial role in the pathogenesis of multiple autoimmune diseases such as psoriasis, RA, MS, IBD, and myocarditis and it has been one of the best targets proposed for the treatment of autoimmune diseases. Several monoclonal antibodies targeting IL-17A and IL-17RA have been developed. Anti-IL-17 antibodies are effective for psoriasis and clinical trial targeting psoriatic arthritis and RA are still ongoing. Similar to targeting IL-23, the rates of adverse events including infections need to be considered due to a crucial role of IL-17A in the control of extracellular but not intracellular bacterial infections and fungi.

RA, psoriasis, and colitis have been successfully treated with the successful drugs such as anticytokines Abs. Although the potential of inflammatory cytokines and their receptors as drug targets is clearly established, an important question is how to select key actors in specific autoimmune diseases or the cytokine paths that should be best targeted. Therefore, it will be crucial to understand the molecular mechanisms of each cytokine in different disease settings and which cellular system it acts in order to select a specific inflammatory target with better efficacy and safety profiles in the treatment of each autoimmune disease. Since not all the reagents functioned as expected, the heterogeneity and complexity of human autoimmune diseases need to be reconsidered. Also, it should be considered that cytokines are critical mediators to maintain effective host defense or prevent tumorigenesis. Therefore, blocking cytokines might make patients more susceptible to infection or risk at developing cancer.

CONCLUSION

Cytokines are key mediators during tissue homeostasis and immune responses regulation. They mediate defense mechanisms to antigen contact between cells and tissues, thus determining the quality and nature (humoral, cellular, or cytotoxic) of immune responses. A role of cytokines on tissue homeostasis regulation is also beginning to be appreciated. Cytokines are expressed in a context of cell and tissue specificity and therefore the understanding of the underlying molecular mechanisms and regulations in a tissue-specific manner will be important. Unregulated levels of cytokines are central mediators of many inflammatory diseases. So, targeting these molecules using recombinant proteins, recombinant soluble receptors, or antibodies against cytokines has demonstrated success in treating patients with autoimmune diseases which are refractory to other therapies such as glucocorticoids treatments. However, systemic cytokine blocking carries a number of serious limitations such as hematopoiesis alterations and increased risks of infections. In addition, the pleiotropic nature of most cytokines makes it almost impossible to inhibit their signaling cascade in a long-term therapy without predictive severe complications. Therefore, new approaches based on tissue or cells-restricted biologics which maintain the cytokine activity in other organs, are highly requested.

References

1. Dinarello CA. Historical insights into cytokines. *Eur J Immunol*. 2007;37(Suppl 1): S34–S45.
2. Kishimoto T. Interleukin-6: From basic science to medicine—40 years in immunology. *Annu Rev Immunol*. 2005;23:1–21.
3. Mullberg J, Schooltink H, Stoyan T, Gunther M, Graeve L, Buse G, et al. The soluble interleukin-6 receptor is generated by shedding. *Eur J Immunol*. 1993;23:473–480.
4. Gabay C, Kushner I. Acute-phase proteins and other systemic responses to inflammation. *N Engl J Med*. 1999;340:448–454.
5. Matsubara T, Mita A, Minami K, Hosooka T, Kitazawa S, Takahashi K, et al. Pgrn is a key adipokine mediating high fat diet-induced insulin resistance and obesity through il-6 in adipose tissue. *Cell Metab*. 2012;15:38–50.
6. Weiss R, Dziura J, Burgert TS, Tamborlane WV, Taksali SE, Yeckel CW, et al. Obesity and the metabolic syndrome in children and adolescents. *N Engl J Med*. 2004;350:2362–2374.
7. Pradhan AD, Manson JE, Rifai N, Buring JE, Ridker PM. C-reactive protein, interleukin 6, and risk of developing type 2 diabetes mellitus. *JAMA*. 2001;286:327–334.
8. Vassalli P. The pathophysiology of tumor necrosis factors. *Annu Rev Immunol*. 1992;10:411–452.
9. Sethi G, Sung B, Aggarwal BB. Tnf: A master switch for inflammation to cancer. *Front Biosci*. 2008;13:5094–5107.
10. Carswell EA, Old LJ, Kassel RL, Green S, Fiore N, Williamson B. An endotoxin-induced serum factor that causes necrosis of tumors. *Proc Natl Acad Sci USA*. 1975;72:3666–3670.
11. Bruns H, Meinken C, Schauenberg P, Harter G, Kern P, Modlin RL, et al. Anti-tnf immunotherapy reduces cd8 + t cell-mediated antimicrobial activity against mycobacterium tuberculosis in humans. *J Clin Invest*. 2009;119:1167–1177.
12. Li H, Luo K, Pauza CD. Tnf-alpha is a positive regulatory factor for human vgamma2 vdelta2 t cells. *J Immunol*. 2008;181:7131–7137.
13. Yang L, Lindholm K, Konishi Y, Li R, Shen Y. Target depletion of distinct tumor necrosis factor receptor subtypes reveals hippocampal neuron death and survival through different signal transduction pathways. *J Neurosci*. 2002;22:3025–3032.
14. Faustman D, Davis M. Tnf receptor 2 pathway: Drug target for autoimmune diseases. *Nat Rev Drug Discov*. 2010;9:482–493.
15. Shen Y, Li R, Shiosaki K. Inhibition of p75 tumor necrosis factor receptor by antisense oligonucleotides increases hypoxic injury and beta-amyloid toxicity in human neuronal cell line. *J Biol Chem*. 1997;272:3550–3553.
16. Cheng X, Yang L, He P, Li R, Shen Y. Differential activation of tumor necrosis factor receptors distinguishes between brains from alzheimer's disease and non-demented patients. *J Alzheimers Dis*. 2010;19:621–630.
17. Isaacs A, Lindenmann J. Virus interference. I. The interferon. *Proc R Soc Lond B Biol Sci*. 1957;147:258–267.

18. Pontiroli F, Dussurget O, Zanoni I, Urbano M, Beretta O, Granucci F, et al. The timing of ifnbeta production affects early innate responses to listeria monocytogenes and determines the overall outcome of lethal infection. *PLoS One*. 2012;7:e43455.
19. Boehm U, Klamp T, Groot M, Howard JC. Cellular responses to interferon-gamma. *Annu Rev Immunol*. 1997;15:749–795.
20. Schroder K, Hertzog PJ, Ravasi T, Hume DA. Interferon-gamma: An overview of signals, mechanisms and functions. *J Leukoc Biol*. 2004;75:163–189.
21. Young HA, Bream JH. Ifn-gamma: Recent advances in understanding regulation of expression, biological functions, and clinical applications. *Curr Top Microbiol Immunol*. 2007;316:97–117.
22. Platanias LC. Mechanisms of type-i- and type-ii-interferon-mediated signalling. *Nat Rev Immunol*. 2005;5:375–386.
23. Brewington R, Chatterji M, Zoubine M, Miranda RN, Norimatsu M, Shnyra A. Ifn-gamma-independent autocrine cytokine regulatory mechanism in reprogramming of macrophage responses to bacterial lipopolysaccharide. *J Immunol*. 2001;167:392–398.
24. Nakanishi K, Yoshimoto T, Tsutsui H, Okamura H. Interleukin-18 regulates both th1 and th2 responses. *Annu Rev Immunol*. 2001;19:423–474.
25. Micallef MJ, Ohtsuki T, Kohno K, Tanabe F, Ushio S, Namba M, et al. Interferon-gamma-inducing factor enhances t helper 1 cytokine production by stimulated human t cells: Synergism with interleukin-12 for interferon-gamma production. *Eur J Immunol*. 1996;26:1647–1651.
26. Oboki K, Ohno T, Kajiwara N, Arae K, Morita H, Ishii A, et al. Il-33 is a crucial amplifier of innate rather than acquired immunity. *Proc Natl Acad Sci USA*. 2010;107:18581–18586.
27. Vigne S, Palmer G, Lamacchia C, Martin P, Talabot-Ayer D, Rodriguez E, et al. Il-36r ligands are potent regulators of dendritic and t cells. *Blood*. 2011;118:5813–5823.
28. Vigne S, Palmer G, Martin P, Lamacchia C, Strebel D, Rodriguez E, et al. Il-36 signaling amplifies th1 responses by enhancing proliferation and th1 polarization of naive cd4 + t cells. *Blood*. 2012;120:3478–3487.
29. Moore KW, de Waal Malefyt R, Coffman RL, O'Garra A. Interleukin-10 and the interleukin-10 receptor. *Annu Rev Immunol*. 2001;19:683–765.
30. Fiorentino DF, Bond MW, Mosmann TR. Two types of mouse t helper cell. Iv. Th2 clones secrete a factor that inhibits cytokine production by th1 clones. *J Exp Med*. 1989;170:2081–2095.
31. Gabrysova L, Howes A, Saraiva M, O'Garra A. The regulation of il-10 expression. *Curr Top Microbiol Immunol*. 2014;380:157–190.
32. Murray PJ. Understanding and exploiting the endogenous interleukin-10/stat3-mediated anti-inflammatory response. *Curr Opin Pharmacol*. 2006;6:379–386.
33. Fillatreau S, Sweenie CH, McGeachy MJ, Gray D, Anderton SM. B cells regulate autoimmunity by provision of il-10. *Nat Immunol*. 2002;3:944–950.
34. Massague J, Wotton D. Transcriptional control by the tgf-beta/smad signaling system. *EMBO J*. 2000;19:1745–1754.
35. Li S, Neff CP, Barber K, Hong J, Luo Y, Azam T, et al. Extracellular forms of il-37 inhibit innate inflammation in vitro and in vivo but require the il-1 family decoy receptor il-1r8. *Proc Natl Acad Sci USA*. 2015;112:2497–2502.
36. Ye Z, Wang C, Kijlstra A, Zhou X, Yang P. A possible role for interleukin 37 in the pathogenesis of behcet's disease. *Curr Mol Med*. 2014;14:535–542.
37. Lloyd CM, Saglani S. T cells in asthma: Influences of genetics, environment, and t-cell plasticity. *J Allergy Clin Immunol*. 2013;131:1267–1274.
38. Martinez-Moczygemba M, Huston DP. Biology of common beta receptor-signaling cytokines: Il-3, il-5, and gm-csf. *J Allergy Clin Immunol*. 2003;112:653–665.

39. Veldhoen M, Uyttenhove C, Van Snick J, Helmby H, Westendorf A, Buer J, et al. Transforming growth factor-beta 'reprograms' the differentiation of t helper 2 cells and promotes an interleukin 9-producing subset. *Nat Immunol.* 2008;9:1341–1346.
40. Wawrzyniak M, Ochsner U, Wirz O, Wawrzyniak P, van de Veen W, Akdis CA, et al. A novel, dual cytokine-secretion assay for the purification of human th22 cells that do not co-produce il-17a. *Allergy.* 2016;71:47–57.
41. Vinuesa CG, Tangye SG, Moser B, Mackay CR. Follicular b helper t cells in antibody responses and autoimmunity. *Nat Rev Immunol.* 2005;5:853–865.
42. Breitfeld D, Ohl L, Kremmer E, Ellwart J, Sallusto F, Lipp M, et al. Follicular b helper t cells express cxc chemokine receptor 5, localize to b cell follicles, and support immunoglobulin production. *J Exp Med.* 2000;192:1545–1552.
43. Vogelzang A, McGuire HM, Yu D, Sprent J, Mackay CR, King C. A fundamental role for interleukin-21 in the generation of t follicular helper cells. *Immunity.* 2008;29:127–137.
44. Manz RA, Hauser AE, Hiepe F, Radbruch A. Maintenance of serum antibody levels. *Annu Rev Immunol.* 2005;23:367–386.
45. Endres R, Alimzhanov MB, Plitz T, Futterer A, Kosco-Vilbois MH, Nedospasov SA, et al. Mature follicular dendritic cell networks depend on expression of lymphotoxin beta receptor by radioresistant stromal cells and of lymphotoxin beta and tumor necrosis factor by b cells. *J Exp Med.* 1999;189:159–168.
46. Fu YX, Huang G, Wang Y, Chaplin DD. B lymphocytes induce the formation of follicular dendritic cell clusters in a lymphotoxin alpha-dependent fashion. *J Exp Med.* 1998;187:1009–1018.
47. Gonzalez M, Mackay F, Browning JL, Kosco-Vilbois MH, Noelle RJ. The sequential role of lymphotoxin and b cells in the development of splenic follicles. *J Exp Med.* 1998;187:997–1007.
48. Tumanov A, Kuprash D, Lagarkova M, Grivennikov S, Abe K, Shakhov A, et al. Distinct role of surface lymphotoxin expressed by b cells in the organization of secondary lymphoid tissues. *Immunity.* 2002;17:239–250.
49. Wojciechowski W, Harris DP, Sprague F, Mousseau B, Makris M, Kusser K, et al. Cytokine-producing effector b cells regulate type 2 immunity to h. Polygyrus. *Immunity.* 2009;30:421–433.
50. Barr TA, Brown S, Mastroeni P, Gray D. Tlr and b cell receptor signals to b cells differentially program primary and memory th1 responses to salmonella enterica. *J Immunol.* 2010;185:2783–2789.
51. Hilgendorf I, Theurl I, Gerhardt LM, Robbins CS, Weber GF, Gonen A, et al. Innate response activator b cells aggravate atherosclerosis by stimulating t helper-1 adaptive immunity. *Circulation.* 2014;129:1677–1687.
52. Bach JF. The effect of infections on susceptibility to autoimmune and allergic diseases. *N Engl J Med.* 2002;347:911–920.
53. Fillatreau S, Gray D, Anderton SM. Not always the bad guys: B cells as regulators of autoimmune pathology. *Nat Rev Immunol.* 2008;8:391–397.
54. Shen P, Roch T, Lampropoulou V, O'Connor RA, Stervbo U, Hilgenberg E, et al. Il-35-producing b cells are critical regulators of immunity during autoimmune and infectious diseases. *Nature.* 2014;507:366–370.
55. Li LJ, Gong C, Zhao MH, Feng BS. Role of interleukin-22 in inflammatory bowel disease. *World J Gastroenterol.* 2014;20:18177–18188.
56. Fumagalli S, Torri A, Papagna A, Citterio S, Mainoldi F, Foti M. Il-22 is rapidly induced by pathogen recognition receptors stimulation in bone-marrow-derived dendritic cells in the absence of il-23. *Sci Rep.* 2016;6:33900.
57. Wolk K, Kunz S, Witte E, Friedrich M, Asadullah K, Sabat R. Il-22 increases the innate immunity of tissues. *Immunity.* 2004;21:241–254.

58. Sabat R, Ouyang W, Wolk K. Therapeutic opportunities of the il-22-il-22r1 system. *Nat Rev Drug Discov*. 2014;13:21–38.
59. Pickert G, Neufert C, Leppkes M, Zheng Y, Wittkopf N, Warntjen M, et al. Stat3 links il-22 signaling in intestinal epithelial cells to mucosal wound healing. *J Exp Med*. 2009;206:1465–1472.
60. Kitazawa H, Muegge K, Badolato R, Wang JM, Fogler WE, Ferris DK, et al. Il-7 activates alpha4beta1 integrin in murine thymocytes. *J Immunol*. 1997;159:2259–2264.
61. Ge Q, Palliser D, Eisen HN, Chen J. Homeostatic t cell proliferation in a t cell-dendritic cell coculture system. *Proc Natl Acad Sci U S A*. 2002;99:2983–2988.
62. Lodolce JP, Boone DL, Chai S, Swain RE, Dassopoulos T, Trettin S, et al. Il-15 receptor maintains lymphoid homeostasis by supporting lymphocyte homing and proliferation. *Immunity*. 1998;9:669–676.
63. Tan JT, Dudl E, LeRoy E, Murray R, Sprent J, Weinberg KI, et al. Il-7 is critical for homeostatic proliferation and survival of naive t cells. *Proc Natl Acad Sci USA*. 2001;98:8732–8737.
64. Tovey MG. The expression of cytokines in the organs of normal individuals: Role in homeostasis. A review. *J Biol Regul Homeost Agents*. 1988;2:87–92.
65. London A, Itskovich E, Benhar I, Kalchenko V, Mack M, Jung S, et al. Neuroprotection and progenitor cell renewal in the injured adult murine retina requires healing monocyte-derived macrophages. *J Exp Med*. 2011;208:23–39.
66. Naert G, Rivest S. A deficiency in ccr2 + monocytes: The hidden side of alzheimer's disease. *J Mol Cell Biol*. 2013;5:284–293.
67. Kipnis J, Gadani S, Derecki NC. Pro-cognitive properties of t cells. *Nat Rev Immunol*. 2012;12:663–669.
68. Gadani SP, Cronk JC, Norris GT, Kipnis J. Il-4 in the brain: A cytokine to remember. *J Immunol*. 2012;189:4213–4219.
69. Ottum PA, Arellano G, Reyes LI, Iruretagoyena M, Naves R. Opposing roles of interferon-gamma on cells of the central nervous system in autoimmune neuroinflammation. *Front Immunol*. 2015;6:539.
70. Ransohoff RM, Engelhardt B. The anatomical and cellular basis of immune surveillance in the central nervous system. *Nat Rev Immunol*. 2012;12:623–635.
71. Schwartz M, Baruch K. The resolution of neuroinflammation in neurodegeneration: Leukocyte recruitment via the choroid plexus. *EMBO J*. 2014;33:7–22.
72. Shechter R, London A, Schwartz M. Orchestrated leukocyte recruitment to immune-privileged sites: Absolute barriers versus educational gates. *Nat Rev Immunol*. 2013;13:206–218.
73. Kunis G, Baruch K, Rosenzweig N, Kertser A, Miller O, Berkutzki T, et al. Ifn-gamma-dependent activation of the brain's choroid plexus for cns immune surveillance and repair. *Brain*. 2013;136:3427–3440.
74. Baruch K, Deczkowska A, David E, Castellano JM, Miller O, Kertser A, et al. Aging. Aging-induced type i interferon response at the choroid plexus negatively affects brain function. *Science*. 2014;346:89–93.
75. Dendrou CA, Fugger L, Friese MA. Immunopathology of multiple sclerosis. *Nat Rev Immunol*. 2015;15:545–558.
76. Gonzalez-Navajas JM, Lee J, David M, Raz E. Immunomodulatory functions of type i interferons. *Nat Rev Immunol*. 2012;12:125–135.
77. Baruch K, Rosenzweig N, Kertser A, Deczkowska A, Sharif AM, Spinrad A, et al. Breaking immune tolerance by targeting foxp3(+) regulatory t cells mitigates alzheimer's disease pathology. *Nat Commun*. 2015;6:7967.
78. Mesquita SD, Ferreira AC, Gao F, Coppola G, Geschwind DH, Sousa JC, et al. The choroid plexus transcriptome reveals changes in type i and ii interferon responses in a mouse model of alzheimer's disease. *Brain Behav Immun*. 2015;49:280–292.

79. Kelly AM, Golden-Mason L, Traynor O, Geoghegan J, McEntee G, Hegarty JE, et al. Changes in hepatic immunoregulatory cytokines in patients with metastatic colorectal carcinoma: Implications for hepatic anti-tumour immunity. *Cytokine*. 2006;35:171–179.
80. Golden-Mason L, Kelly AM, Doherty DG, Traynor O, McEntee G, Kelly J, et al. Hepatic interleuklin 15 (il-15) expression: Implications for local nk/nkt cell homeostasis and development. *Clin Exp Immunol*. 2004;138:94–101.
81. Knolle P, Schlaak J, Uhrig A, Kempf P, Meyer zum Buschenfelde KH, Gerken G. Human kupffer cells secrete il-10 in response to lipopolysaccharide (lps) challenge. *J Hepatol*. 1995;22:226–229.
82. Callery MP, Mangino MJ, Flye MW. Kupffer cell prostaglandin-e2 production is amplified during hepatic regeneration. *Hepatology*. 1991;14:368–372.
83. Heymann F, Peusquens J, Ludwig-Portugall I, Kohlhepp M, Ergen C, Niemietz P, et al. Liver inflammation abrogates immunological tolerance induced by kupffer cells. *Hepatology*. 2015;62:279–291.
84. Gabrilovich DI, Nagaraj S. Myeloid-derived suppressor cells as regulators of the immune system. *Nat Rev Immunol*. 2009;9:162–174.
85. Moshage H. Cytokines and the hepatic acute phase response. *J Pathol*. 1997;181:257–266.
86. Greter M, Lelios I, Pelczar P, Hoeffel G, Price J, Leboeuf M, et al. Stroma-derived interleukin-34 controls the development and maintenance of langerhans cells and the maintenance of microglia. *Immunity*. 2012;37:1050–1060.
87. Wang Y, Bugatti M, Ulland TK, Vermi W, Gilfillan S, Colonna M. Nonredundant roles of keratinocyte-derived il-34 and neutrophil-derived csf1 in langerhans cell renewal in the steady state and during inflammation. *Eur J Immunol*. 2016;46:552–559.
88. Hamilton JA. Gm-csf as a target in inflammatory/autoimmune disease: Current evidence and future therapeutic potential. *Expert Rev Clin Immunol*. 2015;11:457–465.
89. Hamilton JA, Anderson GP. Gm-csf biology. *Growth Factors*. 2004;22:225–231.
90. Wicks IP, Roberts AW. Targeting gm-csf in inflammatory diseases. *Nat Rev Rheumatol*. 2016;12:37–48.
91. Souza LM, Boone TC, Gabrilove J, Lai PH, Zsebo KM, Murdock DC, et al. Recombinant human granulocyte colony-stimulating factor: Effects on normal and leukemic myeloid cells. *Science*. 1986;232:61–65.
92. Yeung YG, Stanley ER. Proteomic approaches to the analysis of early events in colony-stimulating factor-1 signal transduction. *Mol Cell Proteomics*. 2003;2:1143–1155.
93. Kitamura T, Hayashida K, Sakamaki K, Yokota T, Arai K, Miyajima A. Reconstitution of functional receptors for human granulocyte/macrophage colony-stimulating factor (gm-csf): Evidence that the protein encoded by the aic2b cdna is a subunit of the murine gm-csf receptor. *Proc Natl Acad Sci USA*. 1991;88:5082–5086.
94. Croxford AL, Lanzinger M, Hartmann FJ, Schreiner B, Mair F, Pelczar P, et al. The cytokine gm-csf drives the inflammatory signature of ccr2 + monocytes and licenses autoimmunity. *Immunity*. 2015;43:502–514.
95. van de Laar L, Coffer PJ, Woltman AM. Regulation of dendritic cell development by gm-csf: Molecular control and implications for immune homeostasis and therapy. *Blood*. 2012;119:3383–3393.
96. Panopoulos AD, Watowich SS. Granulocyte colony-stimulating factor: Molecular mechanisms of action during steady state and 'emergency' hematopoiesis. *Cytokine*. 2008;42:277–288.
97. Ushach I, Zlotnik A. Biological role of granulocyte macrophage colony-stimulating factor (gm-csf) and macrophage colony-stimulating factor (m-csf) on cells of the myeloid lineage. *J Leukoc Biol*. 2016;100:481–489.

98. O'Shea JJ, Gadina M, Schreiber RD. Cytokine signaling in 2002: New surprises in the jak/stat pathway. *Cell*. 2002;109(Suppl:):S121–S131.
99. Feldmann M, Maini RN. Lasker clinical medical research award. Tnf defined as a therapeutic target for rheumatoid arthritis and other autoimmune diseases. *Nat Med*. 2003;9:1245–1250.
100. Croft M, Benedict CA, Ware CF. Clinical targeting of the tnf and tnfr superfamilies. *Nat Rev Drug Discov*. 2013;12:147–168.
101. Leonardi CL, Powers JL, Matheson RT, Goffe BS, Zitnik R, Wang A, et al. Etanercept as monotherapy in patients with psoriasis. *N Engl J Med*. 2003;349:2014–2022.
102. Lipsky PE, van der Heijde St DM, Clair EW, Furst DE, Breedveld FC, Kalden JR, et al. Infliximab and methotrexate in the treatment of rheumatoid arthritis. Anti-tumor necrosis factor trial in rheumatoid arthritis with concomitant therapy study group. *N Engl J Med*. 2000;343:1594–1602.
103. Tnf neutralization in MS: Results of a randomized, placebo-controlled multicenter study. The lenercept multiple sclerosis study group and the university of British Columbia MS/MRI analysis group. *Neurology* 1999;**53**:457-65.
104. Sandborn WJ, Hanauer SB. Antitumor necrosis factor therapy for inflammatory bowel disease: A review of agents, pharmacology, clinical results, and safety. *Inflamm Bowel Dis*. 1999;5:119–133.
105. van Horssen R, Ten Hagen TL, Eggermont AM. Tnf-alpha in cancer treatment: Molecular insights, antitumor effects, and clinical utility. *Oncologist*. 2006;11:397–408.
106. van Rhee F, Wong RS, Munshi N, Rossi JF, Ke XY, Fossa A, et al. Siltuximab for multicentric castleman's disease: A randomised, double-blind, placebo-controlled trial. *Lancet Oncol*. 2014;15:966–974.

PART II

CYTOKINE TISSUE-SPECIFIC REGULATION

CHAPTER

2

Physiology and Pathophysiology of Adipose Tissue-Derived Cytokine Networks

Asma Ayari and Isabelle Wolowczuk

Center for Infection and Immunity of Lille, Lille, France

OUTLINE

Although cytokines are most frequently studied as discrete variables, emerging data highlights the importance of elucidating the regulation

Cytokine Effector Functions in Tissues.
DOI: http://dx.doi.org/10.1016/B978-0-12-804214-4.00001-4

and functional consequences of complex cytokine networks at the organ and/or tissue level, as well as in interorgan connections.

Beside its primary role in long-term storage of energy excess, the adipose tissue (AT) has also important endocrine functions through a complex network of secreted factors (called adipokines), which are key mediators in interorgan crosstalks; thereby enabling the organism to adapt to diverse physiological or pathophysiological situations such as aging, starvation, excessive/insufficient calorie intake, stress, cancer, or infection.

The purpose of the present chapter is to review the current knowledge on the functional pleiotropism exhibited by the AT, which notably relies on its ability to synthesize and release a variety of local and systemic signals acting in networks or cascades.

We will first define and describe some basic facts about the AT. Then we will propose a general scheme of cytokine networking in the AT and its role both in physiological and in pathological (i.e., obesity) states, focusing on two major adipokines: leptin and adiponectin. Finally we will show the most salient data supporting the notion that the AT—through its secretions—may play a largely neglected and underestimated role in lung development and respiratory diseases such as bacterial and viral pulmonary infections.

BASIC FACTS ABOUT ADIPOSE TISSUE HETEROGENEITY

The biological diversity of AT—both in terms of differences between depots and in the type of cells which are present within the tissue—has lately become a fundamental issue due to AT role in health and disease. Indeed chronic and excessive growth of AT (such as encountered in obesity) plays an important role in the development of metabolic syndrome so as, strikingly, its lack, loss, or abnormal anatomical distribution (such as encountered in lipodystrophy).[1] A mechanism whereby AT may foster the development of metabolic syndrome is through the release of different adipose-derived bioactive molecules having distinct metabolic and inflammatory effects on the AT itself and on distant organs.

Different Flavors of Adipose Tissue: White, Brown, Beige/Brite, and Pink

In mammals, ATs were formerly classified into two main types characterized by different anatomical locations, morphological/histological structures, functions, secretions, and regulations[2]: The white adipose tissue (WAT; which stores energy as triglycerides for release as free fatty acids (FFAs) during fasting periods) and the brown adipose tissue (BAT; which dissipates energy and generates heat to maintain thermal homeostasis).

However, recent emerging evidence supports the existence of two additional categories of AT: The beige/brite (brown-in-white) AT (which corresponds to the development of brown-like adipocytes with thermogenic properties within WAT depots[3]) and the pink AT (which arises during pregnancy and corresponds to adipocyte-derived milk-producing mammary gland alveolar epithelial cells[4]). These reversible transdifferentiation properties of adipocytes (from white to brown for beige/brite adipocytes and from white to milk secretory epithelial cells for pink adipocytes) add to the enormous plasticity of the AT for volume and cell-type/number variations.

White Adipose Tissue: Distribution Matters

The WAT develops at multiple anatomical sites with major depots residing in the subcutaneous and the visceral regions. The subcutaneous adipose tissue (SCAT) is found just beneath the skin—constituting the major cellular component of the hypodermis—where it acts as a barrier against dermal infections, an insulator to prevent heat loss and a cushion to protect against external mechanical stress. The visceral adipose tissue (VAT) is found around all vital organs within the peritoneal cavity and the rib cage (i.e., omental, mesenteric, mediastinal, and epicardial AT). The distribution of VAT and SCAT differs from individuals and is dependent of several factors such as age, nutrition, gender, and genetics.[5–7] Importantly each anatomical depot differs in metabolic and hormonal profiles and, thus, has different physiological roles. In 1956, Vague was the first to notice that visceral obesity (android or "apple shape") is more frequently associated to metabolic disturbances than peripheral obesity (gynoid or "pear shape").[8] Since this initial observation numerous epidemiological and physiological studies have documented a close association between excessive VAT accumulation and a cluster of different metabolic diseases.[9] Recently a large-scale metaanalysis showed that different AT distribution patterns have distinct genetic components; with genes implicated in adipogenesis, angiogenesis, and insulin resistance playing a key role in determining AT distribution.[7]

Among the differences that could account for the association between AT distribution and metabolic diseases, the most obvious is SCAT and VAT respective anatomical localizations. The "portal hypothesis" proposes that the liver, via the portal vein circulation, is directly exposed to VAT-released metabolites such as FFAs and adipokines.[10] However, data obtained in humans[11] and from murine models[12,13] suggest that this might be an over simplification. In fact the divergent metabolic effects of excessive SCAT and VAT also rely on intrinsic, cell-autonomous differences between both tissues. Indeed, comparative analyses of gene expression and secretome of SCAT and VAT revealed major differences between the two AT depots.[14–16]

Visceral and Subcutaneous Adipose Tissues Have Different Adipokine Secretion Patterns: A Focus on Leptin and Adiponectin

Since the report of tumor necrosis factor alpha (TNFα) production by the AT in 1993[17] and the discovery of the leptin and adiponectin hormones the following years,[18,19] AT has been recognized as a major endocrine organ secreting a wide array of hormones, cytokines, growth factors, and vasoactive substances; collectively called adipokines. Adipokines exert pleiotropic effects on different tissues/organs such as AT, brain, skeletal muscle, heart, liver, pancreas, intestine, bone, reproductive organs, blood vessels, and lungs; thus regulating several vital functions like appetite, insulin sensitivity and secretion, fat distribution, lipid and glucose metabolism, endothelial function, blood pressure, hemostasis, neuroendocrine functions and, last but not least, immunity.[20] To date hundreds adipokines constituting the adipose secretome have been identified; including the hormones leptin and adiponectin, and the proinflammatory cytokines TNFα, interleukin (IL)-6, and monocyte chemoattractant protein (MCP)-1. Most adipokines identified so far are proinflammatory and are upregulated in the obese state, while only few have antiinflammatory activities (adiponectin).

There is evidence that adipokines are differently expressed in, and secreted from, SCAT and VAT,[14–16] reinforcing the opposite role of these depots in the development of obesity and its comorbidities.[8,9] Generally the expression of proinflammatory adipokines is higher in VAT whilst leptin expression is higher in SCAT. Several comprehensive reviews on AT secretions have been already published,[21,22] we have thus chosen to focus here on the two prototypical adipokines leptin and adiponectin for four reasons. First, leptin and adiponectin are two of the most abundant circulating adipokines. Second, leptin and adiponectin have opposite inflammatory properties (respectively, pro- and antiinflammatory) and both play pivotal roles in the pathogenesis of obesity. Third, leptin and adiponectin secretions differ between SCAT and VAT depots. This is an important issue since, as mentioned previously; visceral adiposity is associated with low grade, chronic inflammation that contributes to the development of obesity-associated diseases, whereas subcutaneous obesity is not. Fourth, as will be evoked in the last part of the chapter, leptin and adiponectin have been reported to participate in lung homeostasis as well as in the pathophysiology of several pulmonary diseases; paving the way for a secret talk between AT and lungs within a "fat-to-lung axis."

In vivo removal or transplantation of AT in mice allowed to assess SCAT and VAT respective metabolic features and highly suggested that AT functionality per se, rather than anatomical site of accumulation, is the major determinant for the association with metabolic disorders.[13,23,24] Beside adipogenic potential and ability to store and release selective lipids, the other intrinsic factor that differentiates SCAT and VAT is adipokine secretion, notably of leptin and adiponectin. However, it should be borne in mind that the secretome of whole AT (SCAT vs VAT) is not simply the sum of each of its components but rather the result of paracrine interactions between the multiple cell-types within the AT.

The discovery of leptin, the product of the obesity gene (*ob*),[18] led to the description of a previously unknown bidirectional communication network between AT and brain. Indeed leptin, secreted from AT into the bloodstream, can act in the central nervous system (CNS) to inhibit food intake by signaling the extent of AT mass.[25] Even if the AT is the major site of leptin production (more particularly the mature adipocytes within), the hormone is also produced in lower amounts by other tissues[26] such as BAT, stomach, skeletal muscle, placenta, possibly the brain, and lungs. If SCAT and VAT depots both express the *ob* gene, there are differences between sites in the relative *ob* mRNA expression levels. Indeed, SCAT seems to have an exclusive role in leptin secretion, since its mass correlates with plasma leptin levels, both in rodents[27] and humans.[28] Importantly, blood leptin levels are markedly increased in obesity, as noted in human studies and in different animal models of obesity.[25]

Initially considered as a satiety factor through its direct action on anorexigenic hypothalamic neurons, leptin has also substantial effect on food intake, activity-independent thermogenesis, glucose and lipid metabolism, insulin secretion, hematopoiesis, fetal growth, behavior, reproduction, and immunity.[26] Leptin exerts its effects via the Ob-R receptor[29] encoded by the *db* gene, which has several different splicing variants in mouse (from a to f); the full-length isoform (Ob-Rb) mediating most of the actions of leptin, notably in the CNS.[25] Beside specific regions of the brain (such as choroid plexus, leptomeninges, and hypothalamic arcuate nucleus)[30] Ob-R is also expressed in several tissues, including e.g., AT,[31] lungs[32] as well as on innate and adaptive immune cells.[33–35]

Despite its pivotal role in body weight regulation, through food intake inhibition and stimulation of energy expenditure by increased thermogenesis, leptin is much more than a simple adipostat. Indeed leptin participates to immune homeostasis;[33–35] as revealed by thymic atrophy, T-lymphocyte diversity constriction, and immunodeficiency occurring in leptin-deficient (*ob/ob*) and leptin receptor-deficient (*db/db*) mice. Leptin can also modulate the onset and progression of inflammatory immune responses. Structurally wise leptin resembles proinflammatory cytokines such as IL-6 and IL-12, thus its overall action on the immune system is a proinflammatory effect; activating proinflammatory cells, promoting T-helper 1 (Th1) responses and mediating the production of other proinflammatory cytokines. Moreover, the expression of leptin and its receptor are upregulated by proinflammatory signals such as TNFα and IL-1. This is in accordance with the observation that blood leptin levels are increased during infection and acute inflammation. In contrast to acute stimulation of the inflammatory system, chronic inflammation causes a reduction in leptin levels. As such, leptin links metabolism to inflammation and immunity; inspiring the recent emergence of a new fascinating field of investigation: Immunometabolism.[36]

Identified in 1995,[19] adiponectin (also referred to as AdipoQ, Acrp30, apM1, or GBP28) is predominantly synthesized and secreted by adipocytes. In fact, AdipoQ is one of the most abundant adipokines considering its concentration in plasma relative to many other hormones. Posttranslational modifications result in three secreted isoforms that may vary in efficacy regarding their effects on target tissues: Low-, medium-, and high-molecular weight (LMW, MMW, and HMW, respectively) complexes.[37] The latter is considered the most bioactive and proinflammatory isoform in regulating insulin resistance. From a structural point of view, AdipoQ is related to the complement 1q/TNFα superfamily and contains a collagen-repeat domain. Mainly expressed by adipocytes, AdipoQ is also expressed at the mRNA and/or protein level by BAT, liver, skeletal muscle, bone marrow, and lungs.[38]

Contrary to leptin, blood adiponectin levels are high in lean individuals and markedly decreased with increasing obesity; especially AdipoQ HMW multimer levels.

Adiponectin modulates a wide range of metabolic processes such as food intake and energy expenditure, glucose and lipid metabolism, and insulin sensitivity.[39] In addition to its metabolic actions, AdipoQ is also reported to possess antiatherogenic and antiinflammatory properties.[40] Such effects are mediated by three different receptor types: AdipoR1 (almost ubiquitously expressed and abundantly so in skeletal muscle and brain), AdipoR2 (highly present in liver and AT), and the more recently found nonclassical receptor T-Cadherin (CDH13, expressed in endothelial, epithelial, and smooth muscle cells).[41] While AdipoQ is mostly produced by the AT, its receptors are expressed in AT as well as in brain hippocampus, hypothalamus, and brainstem, as well as in, e.g., liver, skeletal muscle, colon, and lungs.[42]

Adiponectin is a predominantly antiinflammatory adipokine that inhibits proinflammatory cytokines (TNFα, IL-6) while also stimulating the production of antiinflammatory cytokines (IL-10, IL-1 receptor antagonist (IL-1RA)) by different innate and adaptive immune cells.[40] In return, proinflammatory cytokines down regulate AdipoQ expression in adipocytes; resulting in decreased blood adiponectin levels. However, under certain conditions (i.e., asthma, arthritis), AdipoQ may have proinflammatory effects as well.[43]

Thus, balanced production of these two countering adipokines has to be maintained to ensure proper metabolic and immune homeostasis. Dysregulation of this equilibrium may signify the early development of inflammatory diseases such as obesity and any attempt to temper this axis may represent an opportunity to correct disease processes.

ADIPOSE TISSUE AUTOCRINE, PARACRINE, AND ENDOCRINE CROSS-TALK SCENARIOS

Adipokines can act in an autocrine/paracrine or endocrine manner thereby putting the AT at the center of a complex and multidirectional network of crosstalks within the tissue itself as well as between organs such as liver, intestine, brain, bone, pancreas, skeletal muscle, and lungs. In the present section, we will summarize the autocrine/paracrine and endocrine functions of two of the most abundant circulating adipokines: leptin and adiponectin.

Autocrine and Paracrine Functions of Leptin and Adiponectin

The long and short isoforms of the Ob-R are expressed in adipocytes[31]; suggesting that leptin may have direct autocrine/paracrine effects on adipocyte development and function. Few studies have been reported that examine the autocrine/paracrine actions of leptin on adipocytes. A recent review, however, focused on the effects of leptin on adipocyte metabolism.[44] Briefly, in vivo, in vitro, and ex vivo studies showed that leptin has direct inhibitory effects on adipocyte lipogenesis and insulin responsiveness whilst enhancing lipolysis; consequently leading to a reduction in adipocyte size. In vivo leptin treatment markedly reduces AT triglyceride stores. Additionally, leptin administration to leptin-deficient (*ob/ob*) and lean mice increases glycerol release. In in vitro experimental settings, leptin has been reported to inhibit accumulation of lipids in adipocytes by increasing triglyceride turnover, inhibiting de novo lipogenesis and promoting FFAs oxidation. However, in human adipocytes from either lean or obese healthy subjects, leptin treatment failed to modulate lipolysis.[45] If leptin reduces adipocyte size, its direct effect on adipocyte number and/or preadipocyte differentiation into lipid-laden mature adipocytes is still debated. Leptin also interferes with insulin signaling and decreases insulin-mediated glucose uptake in mouse or rat adipocytes. As was noted for lipolysis, these findings are not reflected in human cells.[45]

In addition, leptin can also modulate the secretory function of adipocytes, as mainly shown in in vitro models. It has been proposed that leptin, at least in part, regulates its own release by AT, based on findings in mice deficient in Ob-R; which present high circulating leptin levels. Furthermore, leptin stimulates adiponectin expression and release by adipocytes and murine AT both ex vivo and in vitro.[46] Finally, in view of AT expansion in obesity, leptin is known to promote angiogenesis in several experimental models.[47]

In contrast to leptin, adiponectin levels negatively correlated with obesity and its metabolic complications. Adiponectin receptors are expressed in adipocytes; therefore it is tempting to speculate paracrine effect of adiponectin in AT and these effects are regulated at the level of adiponectin receptor gene expression. Indeed, it has been reported that AdipoQ increases basal glucose uptake in adipocytes and enhances insulin-stimulated glucose uptake.[48] Overexpression of adiponectin in 3T3-L1 adipocytes resulted in increased expression of GLUT-4 and enhanced insulin-mediated glucose uptake. In 3T3-L1 cells, adiponectin overexpression promotes preadipocyte proliferation and accelerated the expression of adipogenic transcription factors and adipogenesis. Adiponectin exerts antiinflammatory effects on adipocytes.[43] As a result, release of a number of proinflammatory cytokines (IL-6, IL-8, and MCP-1) by adipocytes is significantly suppressed. Like leptin, AdipoQ is also able to induce angiogenesis.[49]

Besides mature adipocytes and their precursors, AT also contains several other cell-types including fibroblasts, endothelial cells, and immune cells. During the progression of obesity, in addition to changes in adipocyte number, size, and secretory function (in which leptin and adiponectin play important roles, cf. supra), modifications in the number and activity of AT immune cells also occurred. The dramatic changes in AT immune cell composition and function arising during the development of obesity have been extensively reviewed.[50–52] Specifically, adipocytes—via their secretions—interact with certain immune cells and directly regulate their activation and proliferation state.

Leptin has proinflammatory functions and plays important roles in the immune system.[33–35] Ob-R is ubiquitously expressed on the surface of both innate (neutrophils, macrophages, mast cells, eosinophils, natural killer (NK) cells, innate NKT (iNKT) cells) and adaptive (T cells such as regulatory T cells (Tregs), and Th17 cells, B cells) immune cells present in the AT. Leptin induces IL-6 and TNFα synergizing with LPS (lipopolysaccharide) in monocytes, a relevant cell type involved in inflammatory response. Leptin is also able to modulate the survival of autoreactive $CD4^+$ T cells and the activity of regulatory T cells.

Adiponectin also exerts relevant actions on the innate and adaptive immune systems.[40,43] There is a consensus that AdipoQ has antiinflammatory effects on several immune cell types such as monocytes and macrophages, as well as on the systemic regulation of T cell responses. Adiponectin inhibits IL-6 and TNFα production by macrophages and increases the production of antiinflammatory factors such as IL-10 or IL-1RA by monocytes, macrophages, and dendritic cells. Thus adiponectin-mediated effects on AT immune cell function and phenotype may contribute to its role in reducing inflammation within the tissue.

Endocrine Functions of Leptin and Adiponectin

Leptin primarily identified function is to inform the brain of changes in AT mass and Ob-R was found in hypothalamic nuclei involved in food intake control.[25] However, Ob-R is also expressed in several peripheral tissues including skeletal muscle, liver, pancreas, and lungs, as well as endothelial cells and immune cells. Moreover, besides the AT, the stomach, placenta, brain, and lungs are able to synthesize leptin. The appreciation of leptin as a major regulator of many, if not all, endocrine systems such as food intake, reproduction, growth behavior, mood, bone metabolism has been frequently reviewed.[53,54]

Adiponectin is mainly produced by AT and acts to reduce insulin resistance but, alike leptin, it also exerts actions on other tissues that expressed either AdipoR1 or AdipoR2 such as reproductive tissues, skeletal muscle, liver, and lungs. The endocrine effects of AT-derived AdipoQ are conventionally believed to regulate many physiological processes and several studies have reported strong correlations between adiponectin levels and various disease states such as obesity and diabetes, cardiovascular disease, certain types of cancer, hepatic fibrosis, reproduction, bone mass density, and inflammation.[55] In addition, emerging evidence suggests that adiponectin can also be expressed and secreted by skeletal muscle, cardiomyocytes, liver, and lungs.[56]

FROM ADIPOSE TISSUE TO LUNGS: THE ADIPOKINE CONNECTION

Beyond storing energy and participating to inflammatory processes, AT is also a key player in the regulation of several lung homeostatic processes such as development, respiratory system mechanics, breathing regulation, and immunity. The association between AT and lungs is still incompletely understood, however, adipokines—notably leptin and adiponectin—have been reported to have substantial respiratory effects.

As previously mentioned, leptin is also synthesized and secreted by fetal lung tissue. In addition leptin receptors, including the signaling isoform Ob-Rb, are expressed in the fetal lungs of a variety of mammalian species; suggesting a role for leptin in lung growth and/or maturation during prenatal development.[57] Indeed, in vitro studies using fetal rat lung tissue showed that leptin treatment induced the division and maturation of type II pneumocytes; thus favoring surfactant protein synthesis and secretion, which will improve lung compliance.[58] Leptin is also involved in lung postnatal development, as revealed by decreased alveolar number and delayed alveolar enlargement in leptin-deficient *ob/ob* mice.[59]

In human, Ob-Rb is expressed by human bronchial and alveolar epithelial cells, bronchial smooth muscle cells, and bronchial submucosa.[60] Overweight and obese subjects, with high blood leptin levels, are more likely to have respiratory symptoms than lean individuals, even in the absence of demonstrable lung disease. Indeed obesity is often accompanied by pulmonary diseases such as obstructive sleep apnea hypopnea syndrome (OSAHS), obesity hypoventilation syndrome (OHS), chronic obstructive pulmonary disease (COPD), and asthma. While it has been proposed that OSAHS might correspond to a central leptin-resistant state due to impaired leptin transport through the blood brain barrier,[61] accumulating evidence suggest that leptin is involved in the local inflammatory response occurring in the airways of COPD patients. Indeed leptin levels are elevated in COPD patients during acute exacerbation. Smoking, which is the leading cause of COPD, also increases the expression of leptin/leptin receptor in lungs. It has also been reported that leptin may augment allergic airway responses partly through the induction of eosinophil accumulation and the enhancement of inflammatory processes in the lung.

Leptin has been reported to increase cell proliferation, differentiation, survival, migration, and invasion responses in several in vitro and in vivo cancer model systems; mainly through promotion of tumor vascularization. However, the role of leptin in lung cancer initiation and progression remains controversial. Terzidis et al. reported that elevated serum leptin concentration in nonsmall-cell lung cancer (NSCLC) patients may favor the tumorigenesis process.[62] Others, however, showed that mice deficient in leptin or its receptor had increased number of metastatic lung carcinoma cells, which was decreased upon leptin administration.[63]

Leptin is also involved in lung infectious diseases such as pneumonia, tuberculosis, and flu. Following intratracheal challenge with *Klebsiella pneumoniae*, leptin levels were increased in whole lung homogenate, bronchoalveolar fluid, and serum.[64,65] In addition, leptin-deficient *ob/ob* mice were more susceptible to *K. pneumoniae* infection than control mice. Leptin-deficient mice also displayed reduced survival after *Streptococcus pneumoniae* infection and leptin administration improved pulmonary bacterial clearance and survival.[66] However, it has to be noted that other authors failed to detect differences between *ob/ob* and controls regarding lung inflammation and bacterial outgrowth during pneumoniae. Leptin plays a role in the early immune response to pulmonary infection with *Mycobacterium tuberculosis* through favoring the proinflammatory Th1 response and suppressing the antiinflammatory Th2 response. A chronic deficiency in leptin leads to an altered host defense against *M. tuberculosis*.[67] These observations might explain the increased susceptibility of lean individuals to develop active tuberculosis compared to obese patients who are able to better control the infection.[68]

Besides its role during bacterial infections of the lung, leptin has also been reported to play a key role in pneumoniae post-Influenza A virus (IAV) infection through its direct role on neutrophils.[69] In diet-induced obese mice infected with the H1N1pdm09 IAV strain, preexisting high levels of circulating leptin were shown to contribute to the development of severe lung injury.[70] Radigan et al. reported reduced viral clearance and worse outcome of Influenza A virus (A/WSN/33 [H1N1]) infection in leptin receptor-deficient mice.[71] Indeed epidemiological data showed that, during H1N1 2009 outbreak, obese or morbidly obese patients were more susceptible to infection and exhibited higher death rates among all patients.[72]

Adiponectin and its receptors are both expressed on multiple pulmonary cell-types; which make lungs a target tissue for adiponectin signaling. Similar to leptin, AdipoQ is involved in lung development, as deduced from abnormal postnatal alveolar development in normal-weight adiponectin-deficient mice.[73]

Allergen bronchoprovocation in sensitized mice reduced serum adiponectin levels as well as the expression of adiponectin receptor mRNAs in the lung.[74] Likewise, adiponectin attenuates allergic airway inflammation and airway hyperresponsiveness in mice but not in human, through the inhibition of proinflammatory cytokines.[75] Adiponectin levels have been found to be elevated in COPD patients compared to controls.[76] To support its antiinflammatory activity; it was shown that alveolar macrophages from AdipoQ-deficient mice display increased production of TNFα. In a dose- and time-dependent manner, it was demonstrated that AdipoQ reduced cytotoxic effects of TNFα and IL1β and induced the expression of the antiinflammatory IL-10 cytokine in lung epithelial cells; thus improving cell viability and decreasing apoptosis.[77] Overexpression of adiponectin has been shown to protect mice from developing pulmonary arterial hypertension in response to inflammation and hypoxia through its antiinflammatory and antiproliferative properties.[78] Importantly, adiponectin administration reversed the detrimental effects of obesity on the lung endothelium via increasing the expression of vascular barrier-enhancing molecules; thus attenuating lung injury.[79] Whether this protective effect of AdipoQ on acute lung injury could be translated to human is still questioned. In a population of critically ill patients requiring mechanical ventilation 21% of whom had Acute Respiratory Distress Syndrome (ARDS), high serum adiponectin levels at admission were associated with increased mortality.[80] Therefore it remains unclear whether circulating adiponectin is really involved in the pathogenesis of ARDS, in humans.

Adiponectin was shown to facilitate the uptake of apoptotic cells by macrophages and increase the production of IL-8 in the presence of LPS; thus suggesting its role as modulator of the immune response.[81] It was reported that several tumor cell lines express adiponectin receptors, thus suggesting a potential effect of AdipoQ on tumor cells. However, studies evaluating the association between adiponectin and lung cancer are still scarce. Petridou et al. showed in patients with lung cancer that adiponectin levels were not significantly different from controls, but rather lower concentrations were detected in advanced disease stage; thus suggesting that AdipoQ might be a potential marker of lung cancer progression.[82] In advanced NSCLC patients, serum adiponectin levels did not show significant differences when compared to controls, thus not allowing the use of this parameter for overall survival estimation.[83]

Finally regarding the role of adiponectin in pulmonary infectious diseases, it has been reported that high AdipoQ levels might indicate activity and severity of *M. tuberculosis* infection.[84] However, additional studies are required to understand the role of adiponectin during lung viral or bacterial infections.

CONCLUDING REMARKS AND FUTURE PERSPECTIVES

The WAT produces a network of various adipokines—leptin and adiponectin being the best-studied of these—that function to regulate its own microenvironment and to communicate with organs such as, brain, muscle, pancreas, bone, intestine, heart, and liver, as well as with both innate and adaptive immune cells (Fig. 2.1A).

The association between leptin and adiponectin—which have respectively pro- and antiinflammatory activities—and pulmonary diseases, is still largely unexplored. Nevertheless, the existence of an AT–lung axis has been suggested. Here we focused on the current literature reporting

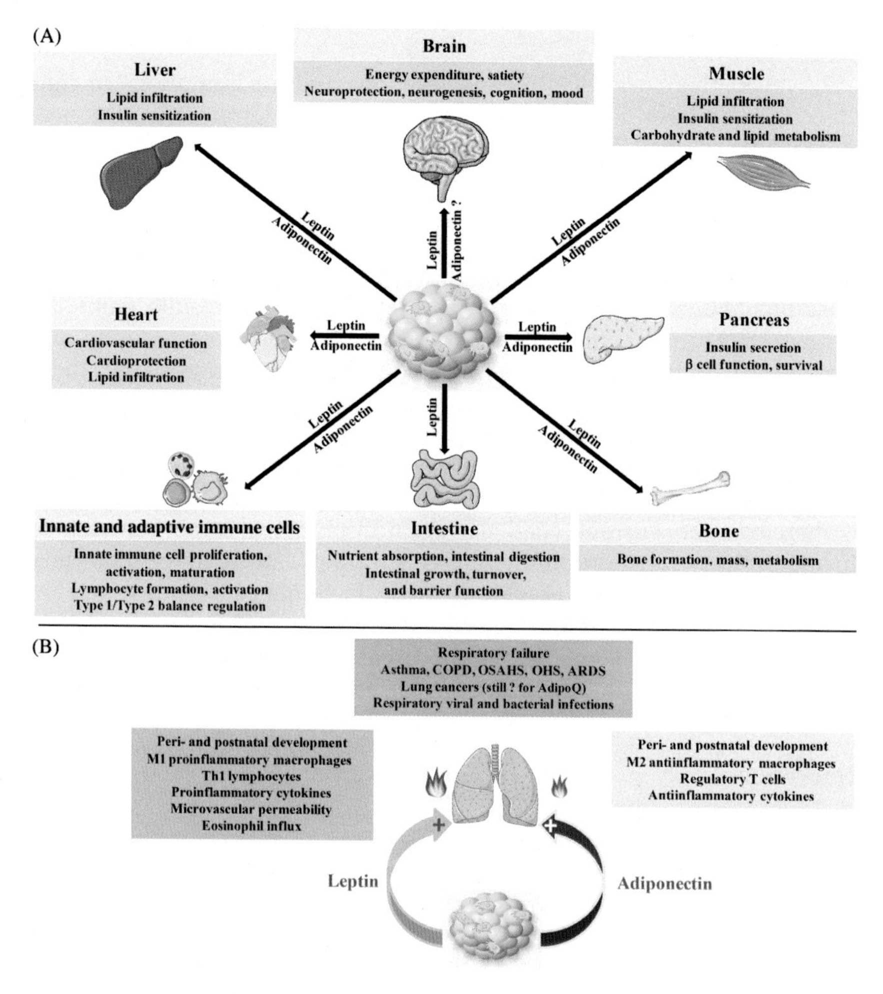

FIGURE 2.1 **Intertwined interactions between the white adipose tissue (WAT) and remote organs and tissues via leptin and adiponectin.** (A) *Leptin and adiponectin in the inter-organ communication network.* Adipose tissue (AT)-derived leptin and adiponectin regulate the function of distant organs and tissues as diverse as brain, skeletal muscle, pancreas, bone, intestine, heart, and liver. Importantly, both adipokines also modulate innate and adaptive immune responses, notably through their proinflammatory and antiinflammatory properties, respectively. Of note: To our knowledge, contrary to leptin,[85] AT-derived adiponectin has never been reported to have a role on the gastrointestinal tract; and its role in brain is still debated since controversial results have been reported regarding its capacity to cross the blood brain barrier.[86,87] (B) *The fat-lung axis.* In this communication network, the WAT influences the lungs by secreting, e.g., leptin (proinflammatory) and adiponectin (antiinflammatory); as summed up in the lateral boxes. In the upper box are listed the lung diseases that have been shown to be associated with leptin or adiponectin (see text for details and abbreviations). Of note: The role of adiponectin in lung carcinogenesis remains controversial.[88] However, adiponectin and leptin seem to have opposing roles in cancer development; similar to their functions in metabolic disease and other diseases.

on the role of leptin and adiponectin in the pathogenesis of lung diseases (Fig. 2.1B).

However, there are several considerations and/or limitations to consider when addressing that specific point. (1) Both leptin and adiponectin are not secreted from AT only; thus defining the contribution of lungs versus AT production on the observed effects is a challenging issue. In addition, leptin and adiponectin can be produced by multiple cell types; (2) Leptin and adiponectin functions in the lung may not necessarily be similar to those well described in the AT or the brain; (3) Substantial study design differences and conflicting results from the existing literature preclude definitive conclusions on the role leptin and adiponectin could play in lung physiology and pathophysiology; these differences include species, gender, age, genetics, and ethnic differences in adipokine levels; (4) Given the vast number of AT-derived adipokines, their cumulative roles as classes of adipokines should also be questioned.

Therefore the current challenge in adipokine research remains to elucidate their function in their tissue targets, at both cellular and molecular levels. This knowledge may facilitate the future use of adipokines—such as leptin and adiponectin—as pharmacotherapies, drug targets, and/or predictors of disease initiation or progression, notably concerning lung inflammatory diseases.

References

1. Grundy SM. Adipose tissue and metabolic syndrome: too much, too little or neither. *Eur J Clin Invest* 2015;**45**:1209–17.
2. Saely CH, Geiger K, Drexel H. Brown versus white adipose tissue: a mini-review. *Gerontology* 2012;**58**:15–23.
3. Wu J, Boström P, Sparks LM, Ye L, Choi JH, Giang AH, et al. Beige adipocytes are a distinct type of thermogenic fat cell in mouse and human. *Cell* 2012;**150**:366–76.
4. Morroni M, Giordano A, Zingaretti MC, Boiani R, De Matteis R, Kahn BB, et al. Reversible transdifferentiation of secretory epithelial cells into adipocytes in the mammary gland. *PNAS* 2004;**101**:16801–6.
5. Fried SK, Lee MJ, Karastergiou K. Shaping fat distribution: new insights into the molecular determinants of depot- and sex-dependent adipose biology. *Obesity* 2015;**23**:1345–52.
6. Fu J, Hofker M, Wijimenga C. Apple or pear: size and shape matter. *Cell metab* 2015; **21**:507–8.
7. Shungin D, Winkler TW, Croteau-Chonka DC, Ferreira T, Locke AE, Mägi R, et al. New genetic loci link adipose and insulin biology to body fat distribution. *Nature* 2015;**518**:187–96.
8. Vague J. The degree of masculine differentiation of obesities: a factor determining predisposition to diabetes, atherosclerosis, gout, and uric calculous disease. *Am J Clin Nutr* 1956;**4**:20–34.
9. Kwok KHM, Lam KSL, Xu A. Heterogeneity of white adipose tissue: molecular basis and clinical implications. *Exp Mol Med* 2016;**48**:e215.
10. Bjorntorp P. "Portal" adipose tissue as a generator of risk factors for cardiovascular disease and diabetes. *Arteriosclerosis* 1990;**10**:493–6.

11. Capeau J, Magre J, Lascols O, Caron M, Bereziat V, Vigouroux C, et al. Diseases of adipose tissue: genetic and acquired lipodystrophies. *Biochem Soc Trans* 2005;**33**: 1073–7.
12. Kim JY, van de Wall E, Laplante M, Azzara A, Trujillo ME, Hofmann SM, et al. Obesity-associated improvements in metabolic profile through expansion of adipose tissue. *J Clin Invest* 2007;**117**:2621–37.
13. Gavrilova O, Marcus-Samuel B, Graham D, Kim JK, Shulman GI, Castle AL, et al. Surgical implantation of adipose tissue reverses diabetes in lipoatrophic mice. *J Clin Invest* 2000;**105**:271–8.
14. Pardo M, Roca-Rivada A, Seoane LM, Casanueva FF. Obesidomics: contribution of adipose tissue secretome analysis to obesity research. *Endocrine* 2012;**41**:374–83.
15. Alvehus M, Buren J, Sjöström M, Goedecke J, Olsson T. The human visceral fat depot has a unique inflammatory profile. *Obesity* 2010;**18**:879–83.
16. Roca-Rivada A, Alonso J, Al-Massadi O, Castelao C, Peinado JR, Seoane LM, et al. Secretome analysis of rat adipose tissues shows location-specific roles for each depot type. *J Proteomics* 2011;**74**:1068–79.
17. Hotamisligil GS, Shargill NS, Spiegelman BM. Adipose tissue expression of tumor necrosis factor-alpha: direct role in obesity-linked insulin resistance. *Science* 1993;**259**: 87–91.
18. Zhang Y, Proenca R, Maffei M, Barone M, Leopold L, Friedman JM. Positional cloning of the mouse obese gene and its human homologue. *Nature* 1994;**372**:425–32.
19. Scherer PE, Williams S, Fogliano M, Baldini G, Lodish HF. A novel serum protein similar to C1q, produced exclusively in adipocytes. *J Biol Chem* 1995;**270**:26746–9.
20. Fasshauer M, Blüher M. Adipokines in health and disease. *Trends Pharmacol Sci* 2015;**36**:461–70.
21. Cao H. Adipocytokines in obesity and metabolic disease. *J Endocrinol* 2014;**220**:47–59.
22. Rodriguez A, Ezquerro S, Mendez-Gimenez L, Becerril S, Frühbeck G. Revisiting the adipocyte: a model for integration of cytokine signaling in the regulation of energy metabolism. *Am J Physiol Endocrinol Metab* 2015;**309**:691–714.
23. Thorne A, Lonnqvist F, Apelman J, Hellers G, Arner P. A pilot study of long-term effects of a novel obesity treatment: omentectomy in connection with adjustable gastric banding. *Int J Obes Relat Metab Disord* 2002;**26**:193–9.
24. Klein S, Fontana L, Young VL, Coggan AR, Kilo C, Patterson BW, et al. Absence of an effect of liposuction on insulin action and risk factors for coronary heart disease. *New Engl J Med* 2004;**350**:2549–57.
25. Friedman JM, Halaas JL. Leptin and the regulation of body weight in mammals. *Nature* 1998;**395**:763–70.
26. Ahima RS, Osei SY. Leptin signaling. *Physiol Behav* 2004;**81**:223–41.
27. Hantschel C, Wagener A, Neuschl C, Schmitt AO. Age- and depot-specific adipokine responses to obesity in mice. *Health* 2012;**4**:1522–9.
28. Montague CT, Prins JB, Sanders L, Digby JE, O'Rahilly S. Depot- and sex-specific differences in human leptin mRNA expression: implications for the control of regional fat distribution. *Diabetes* 1997;**46**:342–7.
29. Tartaglia LA, Dembski M, Weng X, Deng N, Culpepper J, Devos R, et al. Identification and expression cloning of a leptin receptor, OB-R. *Cell* 1995;**83**:1263–71.
30. Lein ES, Hawrylycz MJ, Ao N, Ayres M, Bensinger A, Bernard A, et al. Genome-wide atlas of gene expression in the adult mouse brain. *Nature* 2007;**445**:168–76.
31. Siegrist-Kaiser CA, Pauli V, Juge-Aubry CE, Boss O, Pernin A, Chin WW, et al. Direct effects of leptin on brown and white adipose tissue. *J Clin Invest* 1997;**100**:2858–64.
32. Tsuchiya T, Shimizu H, Horie T, Mori M. Expression of leptin receptor in lung: leptin as a growth factor. *Eur J Pharmacol* 1999;**365**:273–9.
33. Naylor C, Petri Jr. WA. Leptin regulation of immune responses. *Trends Mol Med* 2016; **22**:88–98.

34. Matarese G. Leptin and the immune system: how nutritional status influences the immune response. *Eur Cytokine Netw* 2000;**11**:7–14.
35. La Cava A, Matarese G. The weight of leptin in immunity. *Nat Rev Immunol* 2004;**4**: 371–9.
36. Pearce EL, Pearce EJ. Metabolic pathways in immune cell activation and quiescence. *Immunity* 2013;**38**:633–43.
37. Maeda K, Okubo K, Shimomura I, Funahashi T, Matsuzawa Y, Matsubara K. cDNA cloning and expression of a novel adipose specific collagen-like factor, apM1 (adipose most abundant gene transcript 1). *Biochem Biophys Res Commun* 1996;**221**:286–9.
38. Summer R, Little FF, Ouchi N, Takemura Y, Aprahamian T, Dwyer D, et al. Alveolar macrophage activation and emphysema-like phenotype in adiponectin-deficient mice. *Am J Physiol Lung Cell Mol Physiol* 2008;**294**:1035–42.
39. Weyer C, Funahashi T, Tanaka S, Hotta K, Matsuzawa Y, Pratley RE, et al. Hypoadiponectinemia in obesity and type 2 diabetes: close association with insulin resistance and hyperinsulinemia. *J Clin Endocrinol Metab* 2001;**86**:1930–5.
40. Ohashi K, Ouchi N, Matsuzawa Y. Anti-inflammatory and anti-atherogenic properties of adiponectin. *Biochimie* 2012;**94**:2137–42.
41. Yamauchi T, Kamon J, Ito Y, Tsuchida A, Yokomizo T, Kita S, et al. Cloning of adiponectin receptors that mediate antidiabetic metabolic effects. *Nature* 2003;**423**:762–9.
42. Miller M, Cho JY, Pham A, Ramsdell J, Broide DH. Adiponectin and functional adiponectin receptor 1 are expressed by airway epithelial cells in chronic obstructive pulmonary disease. *J Immunol* 2009;**182**:684–91.
43. Fantuzzi G. Adiponectin and inflammation: consensus and controversy. *J Allergy Clin Immunol* 2008;**121**:326–30.
44. Harris RBS. Direct and indirect effects of leptin on adipocyte metabolism. *Biochim Biophys Acta* 2014;**1842**:414–23.
45. Aprath-Husmann I, Röhrig K, Gottschling-Zeller H, Skurk T, Scriba D, Birgel M, et al. Effects of leptin on the differentiation and metabolism of human adipocytes. *Int J Obes (Lond)* 2001;**25**:1465–70.
46. Zhang W, Della-Fera MA, Hartzell DL, Hausman D, Baile CA. Adipose tissue gene expression profiles in ob/ob mice treated with leptin. *Life Sci* 2008;**83**:35–42.
47. Serra-Honigmann MR, Nath AK, Murakami C, Garcia-Cardena G, Papapetropoulos A, Sessa WC, et al. Biological action of leptin as an angiogenic factor. *Science* 1998;**281**: 1683–6.
48. Yamauchi T, Kamon J, Minokoshi Y, Ito Y, Waki H, Uchida S, et al. Adiponectin stimulates glucose utilization and fatty-acid oxidation by activating AMP-activated protein kinase. *Nat Med* 2002;**8**:1288–95.
49. Ouchi N, Kobayashi H, Kihara S, Kumada M, Sato K, Inoue T, et al. Adiponectin stimulates angiogenesis by promoting cross-talk between AMP-activated protein kinase and Akt signaling in endothelial cells. *J Biol Chem* 2004;**279**:1304–9.
50. Makki K, Froguel P, Wolowczuk I. Adipose tissue in obesity-related inflammation and insulin resistance: cells, cytokines and chemokines. *ISRN Inflamm* 2013;**2013**. ID139239, 12 pages.
51. Apostolopoulos V, de Courten MP, Stojanovska L, Blatch GL, Tangalakis K, de Courten B. The complex immunological and inflammatory network of adipose tissue in obesity. *Mol Nutr Food Res* 2016;**60**:43–57.
52. Brestoff JR, Artis D. Immune regulation of metabolic homeostasis in health and disease. *Cell* 2015;**161**:146–60.
53. Margetic S, Gazzola C, Pegg GG, Hill RA. Leptin: a review of its peripheral actions and interactions. *Int J Obes (Lond)* 2002;**26**:1407–33.
54. Kelesidis T, Kelesidis I, Chou S, Mantzoros CS. The role of leptin in human physiology: emerging clinical applications. *Ann Intern Med* 2010;**152**:93–100.

55. Robinson K, Prins J, Venkatesh B. Adiponectin biology and its role in inflammation and clinical illness. *Crit Care Med* 2011;**15**:221–30.
56. Dalamaga M, Diakopoulos KN, Mantzoros CS. The role of adiponectin in cancer: a review of current evidence. *Endocr Rev* 2012;**33**:547–94.
57. Malli A, Papaioannou K, Gourgoulianis ZD. The role of leptin in the respiratory system: an overview. *Respir Res* 2010;**11**:152–68.
58. Kirwin SM, Bhandari V, Dimatteo D, Barone C, Johnson L, Paul S, et al. Leptin enhances lung maturity in the fetal rat. *Pediatr Res* 2005;**60**:200–4.
59. Huang K, Rabold R, Abston E, Schofield B, Misra V, Galdzika E, et al. Effects of leptin deficiency on postnatal lung development in mice. *J Appl Physiol* 2008;**105**:249–59.
60. Nair P, Radford K, Fanat A, Janssen LJ, Peters-Golden M, Cox PG. The effects of leptin on airway smooth muscle responses. *Am J Respir Cell Mol Biol* 2008;**39**:475–81.
61. Atwood CW. Sleep related hypoventilation. *Chest* 2005;**128**:1079–81.
62. Terzidis A, Sergentanis TN, Antonopoulos G, Syrigos C, Efremidis A, Polyzos A, et al. Elevated serum leptin levels: a risk factor for non-small-cell Lung Cancer? *Oncology* 2009;**76**:19–25.
63. Mori A, Sakurai H, Choo MK, Obi R, Koizumi K, Yoshida C, et al. Severe pulmonary metastasis in obese and diabetic mice. *Int J Cancer* 2006;**119**:2760–7.
64. Mancuso P, Gottschalk A, Phare MS, Peters-Golden M, Luckacs NW, Huffnagle GB. Leptin deficient mice exhibit impaired host defense in Gram-negative pneumoniae. *J Immunol* 2002;**168**:4018–24.
65. Moore SI, Huffnagle GB, Chen GH, White ES, Mancuso P. Leptin modulates neutrophils phagocytosis of *Klebsiella pneumoniae. Infect Immun* 2003;**71**:4182–5.
66. Hsu A, Aronoff DM, Phipps J, Goel D, Mancuso P. Leptin improves pulmonary bacterial clearance and survival in ob/ob mice during pneumococcal pneumonia. *Clin Exp Immunol* 2007;**150**:332–9.
67. Wieland CW, Florquin S, Chan ED, Leemans JC, Weijer S, Verbon A, et al. Animal model of *Mycobacterium tuberculosis* infection in leptin-deficient ob/ob mice. *Int Immunol* 2005;**17**:1399–408.
68. Leung CC, Lam TH, Chan WM, Yew WW, Ho KS, Leung G, et al. Lower risk of tuberculosis in obesity. *Arch Intern Med* 2007;**167**:1297–304.
69. Ubags ND, Vernooy JH, Burg E, Hayes C, Berment J, Dilli E, et al. The role of leptin in the development of pulmonary neutrophilia in infection and acute lung injury. *Crit Care* 2014;**42**:e143–51.
70. Zhang AJX, To KKW, Li C, Lau CCY, Poon VKM, Chan CCS, et al. Leptin mediates the pathogenesis of severe 2009 pandemic Influenza A (H1N1) infection associated with cytokine dysregulation in mice with diet-induced obesity. *J Infect Dis* 2013;**207**:1270–80.
71. Radigan KA, Morales-Nebreda L, Soberanes S, Nicholson T, Nigdelioglu R, Cho T, et al. Impaired clearance of Influenza A virus in obese, Leptin receptor deficient mice is independent of leptin signaling in the lung epithelium and macrophages. *PLoS One* 2014;**9**:e108138.
72. Jain S, Chaves SS. Obesity and influenza. *Clin Infect Dis* 2011;**53**:422–4.
73. Thyagarajan B, Jacobs DR, Smith LJ, Kalhan R, Gross MD, Sood A. Serum adiponectin is positively associated with lung function in young adults, independent of obesity: the CARDIA study. *Respir Res* 2010;**11**:176 (6 pages).
74. Shin JH, Kim JH, Lee WY, Shim JY. The expression of adiponectin receptors and the effects of adiponectin and leptin on airway smooth muscle cells. *Yonsei Med J* 2008;**49**:804–10.
75. Shore SA, Terry RD, Flynt L, Xu A, Hug C. Adiponectin attenuates allergen-induced airway inflammation and hyperresponsiveness in mice. *J Allergy Clin Immunol* 2006;**118**:389–95.

76. Bianco A, Mazzarella G, Turchiarelli V, Nigro E, Corbi G, Scudiero O, et al. Adiponectin an attractive marker for metabolic disorder in Chronic Obstructive Pulmonary Disease (COPD). *Nutrients* 2013;**5**:4115–25.
77. Nigro E, Scudiero O, Sarnataro D, Mazzarella G, Sofia M, Bianco A, et al. Adiponectin affects lung epithelial A549 cell viability counteracting TNFα and IL-1β toxicity through AdipoR1. *Int J Biochem Cell Biol* 2013;**45**:1145–53.
78. Summer R, Little FF, Ouchi N, Takemura Y, Aprahamian T, Dwyer D, et al. Alveolar macrophage activation and an emphysema-like phenotype in adiponectin-deficient mice. *Am J Physiol Lung Cell Mol Physiol* 2008;**294**:1035–42.
79. Shah D, Romero F, Duong M, Wang N, Paudyal B, Suratt BT, et al. Obesity-induced adipokine imbalance impairs mouse pulmonary vascular endothelial function and primes the lung for injury. *Sci Rep* 2015;**5**:11362 (13 pages).
80. Palakshappa JA, Anderson BJ, Reilly JP, Shashaty MGS, Ueno R, Wu Q, et al. Low plasma levels of adiponectin do not explain acute respiratory distress syndrome risk: a prospective cohort study of patients with severe sepsis. *Crit Care* 2016;**20**:71 (9 pages).
81. Saijo S, Nagata K, Nakano Y, Tobe T, Kobayashi Y. Inhibition by adiponectin of IL-8 production by human macrophages upon co-culturing with late apoptotic cells. *Biochem Biophys Res Commun* 2005;**334**:1180–3.
82. Petridou ET, Mitsiades N, Gialamas S, Angelopoulos M, Skalkidou A, Dessypris N, et al. Circulating adiponectin levels and expression of adiponectin receptors in relation to lung cancer: two-case control studies. *Oncology* 2007;**73**:261–9.
83. Karapanagiotou EM, Toschatzis EA, Dilana KD, Tourkantonis I, Gratsias I, Syrigos KN. The significance of leptin, adiponectin, and resistin serum levels in non-small cell lung cancer (NSCLC). *Lung Cancer* 2008;**61**:391–7.
84. Keicho N, Matsushita I, Tanaka T, Shimbo T, Le Hong NT, Sakurada S, et al. Circulating levels of adiponectin, Leptin, Fetuin-A and Retinol binding protein in patients with tuberculosis: markers of metabolism and inflammation. *PLoS One* 2012;**7**: e0038703 (10 pages).
85. Yarandi SS, Hebbar G, Sauer CG, Cole CR, Ziegler TR. Diverse roles of leptin in the gastrointestinal tract: modulation of motility, absorption, growth, and inflammation. *Nutrition* 2011;**27**:269–75.
86. Qi Y, Takahashi N, Hileman SM, Patel HR, Berg AH, Pajvani UB, et al. Adiponectin acts in the brain to decrease body weight. *Nat Med* 2004;**10**:524–9.
87. Spranger J, Verma S, Göhring I, Bobbert T, Seifert J, Sindler AL, et al. Adiponectin does not cross the blood-brain barrier but modifies cytokine expression of brain endothelial cells. *Diabetes* 2006;**55**:141–7.
88. Barb D, Pazaitou-Panayiotou K, Mantzoros CS. Adiponectin: a link between obesity and cancer. *Expert Opin Investig Drugs* 2006;**15**:917–31.

CHAPTER

3

Cytokine Networks in the Ovary

Nicolas M. Orsi, Sarah L. Field, N. Ellissa Baskind, Katie Allen, and Michele Cummings

St James's University Hospital, Leeds, United Kingdom

OUTLINE

Cytokine Effector Functions in Tissues.
DOI: http://dx.doi.org/10.1016/B978-0-12-804214-4.00002-6

INTRODUCTION

Cytokines contribute to all aspects of reproduction, although our understanding of their roles and interactions in ovarian physiology remains patchy.[1,2] Cytokines such as interleukins (ILs) and those of the transforming growth factor (TGF)-β family mediate the dialog between the oocyte and its somatic cell complement and regulate follicle survival and apoptosis.[1,3,4] Intraovarian immune effector cells, such as macrophages and lymphocytes, also secrete cytokines, including interferon (IFN)-γ, tumor necrosis factor (TNF)-α, granulocyte colony-stimulating factor (G-CSF), IL-1, IL-6, IL-8, monocyte chemoattractant protein (MCP)-1, and granulocyte macrophage colony-stimulating factor (GM-CSF), all of which participate in oocyte development, ovulation, and steroidogenesis.[5–8] Intrafollicular cytokines diffuse to create chemotactic gradients and/or act locally, often at low concentrations, and with a short half-life. This chapter explores cytokines' involvement in ovarian function, using pathological examples to underscore their physiological roles.

OVARIOGENESIS AND FUNCTIONAL ACTIVATION

Ovarian Development

Ovarian development commences with ovarian medulla and germinal epithelium emergence from the genital ridge and mesonephros, with primordial germ cells (PGCs) arriving as yolk sac endoderm colonists. PGC migration to the gonadal anlage is controlled by stem cell factor (SCF), with dorsal mesentery SCF-positive autonomic nerve fibers guiding their ascent.[9] The bone morphogenic protein (BMP) system is intimately involved both in murine PGC formation[10,11] and in human postmigratory PGC apoptosis (especially BMP-4).[12] IL-6 family members (leukemia inhibitory factor (LIF), oncostatin M, and ciliary neurotrophic factor) and fibroblast growth factors (FGFs) also influence murine early germ cell development.[13,14] LIF, SCF, and insulin-like growth factor (IGF)-1 inhibit PGC apoptosis in mice, an effect countered by TGF-β, which points to a complex signaling system within the fetal ovary.[15]

Puberty

The onset of puberty is directed by critical interactions between nutrition and the hypothalamic–pituitary–gonadal (HPG) axis. Adipocyte-derived leptin governs this process by signaling to the HPG axis (stimulating gonadotrophin-releasing hormone (GnRH) and luteinizing hormone (LH) production) so that host nutritional status (based on critical body weight/percentage body fat) can support reproductive function[16]: Age at menarche is inversely proportional to serum leptin concentrations.[17] Indeed, leptin gene mutations result in hypogonadotrophic hypogonadism, which occurs as failed pubertal development/primary amenorrhea,[18] and leptin/leptin receptor deficient mice are similarly infertile.[19] Although the mechanisms behind these interactions are unclear, evidence suggests that kisspeptin may be upregulated by leptin, in turn triggering the GnRH pulsatility that drives folliculogenesis.[20]

FOLLICULAR DEVELOPMENT

The generation of viable oocytes is a complex process which takes circa 4 months in women, comprising primordial follicle recruitment, granulosa cell (GC)/theca cell (TC) proliferation, oocyte maturation, and steroidogenesis.[21,22] Each stage involves an oocyte, GC, and TC paracrine dialog mediated by multiple hormones and cytokines.[22–25]

Primordial Follicle Recruitment

This first stage is driven by a cytokine/growth factor-mediated interplay between the oocyte and GC/TC precursors.[21] Perioocytic pre-GCs (Fig. 3.1) express SCF, basic FGF (bFGF/FGF-2), FGF-7, and LIF, which promote the transition to primary follicles in vitro in rodents and goats.[26–32] In women, follicular development is under BMP system control.[1,4,33] BMP-15 is expressed in adult ovary primordial follicles but not in their fetal/peripubertal counterparts, pointing to a role in primordial pool recruitment.[34] Related to BMPs, oocyte-derived growth differentiation factor (GDF)-9 is essential to oocyte, GC, and TC development across all mammals.[34] Indeed, *gdf-9* mutant mice exhibit a folliculogenetic failure to progress beyond the primary follicle stage allied with incomplete TC layer development.[35] IL-16 also promotes the primordial to primary follicle transition in vitro, while TGF-β family members act as repressive agents.[36] By contrast, anti-Müllerian hormone (AMH) suppresses resting primordial follicles[37,38]: Murine AMH knockout models have an increased rate of primordial follicle recruitment, suggesting that AMH may prevent premature primordial follicle pool depletion.[37]

Primary to Antral Follicle Development

Preantral follicle growth involves oocyte-derived GDF-9 and BMP-15, GC activins, TC BMP-4 and -7, and TGF-β produced by GCs and TCs[1,27,39,40] (Fig. 3.2). SCF and BMP-15 operate through a negative feedback loop which regulates GC population expansion, thereby enabling the

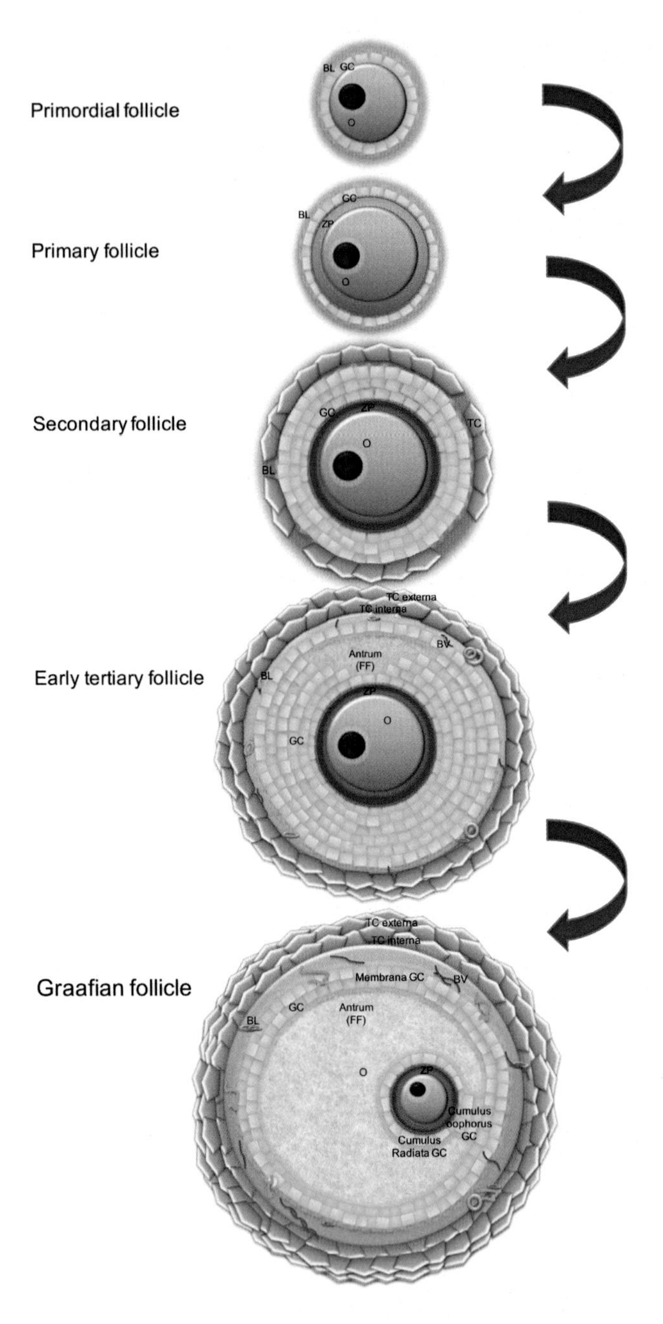

FIGURE 3.1 Schematic representation of development of primordial follicle to preovulatory Graafian follicle. *BL*, basement lamina; *O*, oocyte; *GC*, granulosa cell; *ZP*, zona pellucida; *TC*, theca cell; *FF*, follicular fluid; *BV*, blood vessel/capillary; and *CT*, connective tissue (loose).

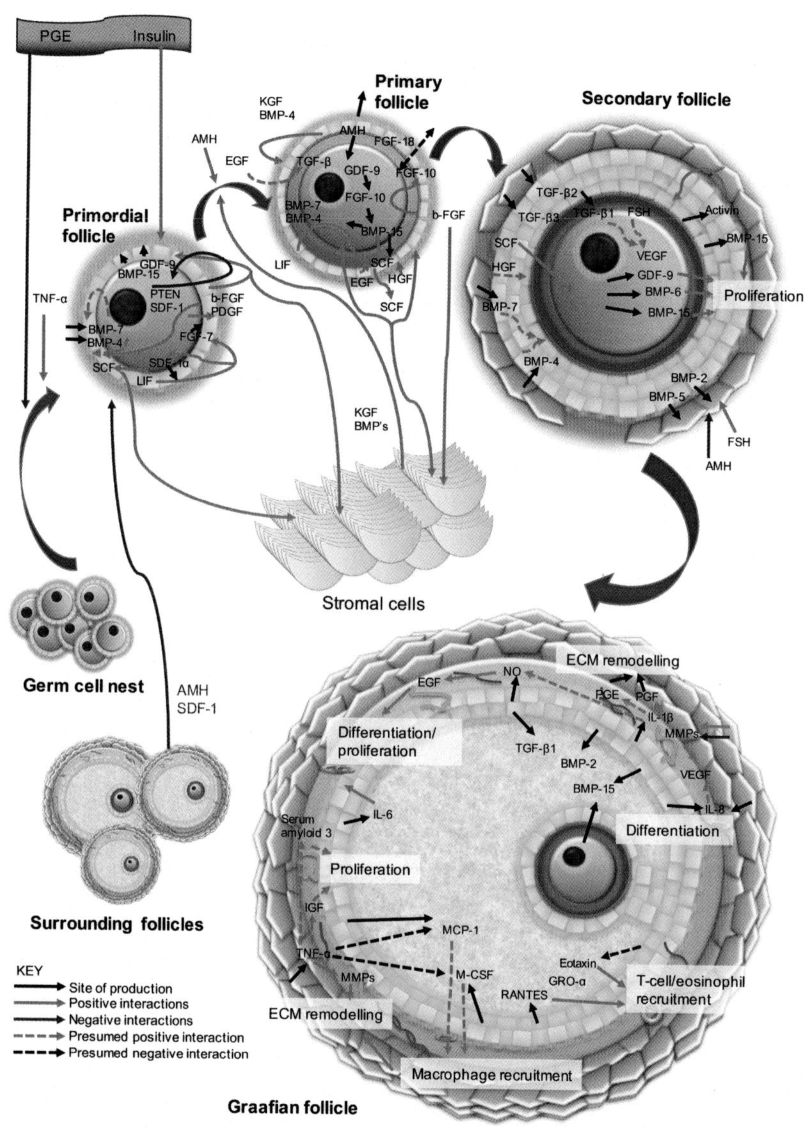

FIGURE 3.2 Schematic representation of the predominant regulatory factors governing primordial follicle transition to the preovulatory follicle in the ovary. The combined actions of primordial follicles themselves, surrounding stromal cells, follicles, and endocrine factors are responsible for primordial follicle activation. Cytokines modulate all stages of folliculogenesis, with the number of known cytokines involved increasing toward ovulation (interactions based on human folliculogenesis with inferences from animal models). Key for previously undefined mediators (see main text for others): *P4*, progesterone; *SDF-1*, stromal derived factor-1; *KGF*, keratinocyte growth factor; *PDGF*, platelet derived growth factor; *PTEN*, phosphatase and tensin homolog; *HGF*, hepatocyte growth factor; *NO*, nitric oxide; *PGE*, prostaglandin E; *MMPs*, matrix metalloproteinases; *GRO-α*, growth regulated oncogene; *RANTES*, regulated upon activation normal T cell expressed and secreted; and *M-CSF*, macrophage-colony stimulating factor.

oocyte to control its own follicle's growth rate.[41,42] Further development requires GDF-9 and vascular endothelial growth factor (VEGF) synergy.[43] GDF-9 also stimulates preantral follicle growth, promotes GC-cumulus cell differentiation and, later, suppresses GC progesterone production in rodents and cattle via induction of steroidogenic factors and prostaglandins.[44–48] Together with epidermal growth factor (EGF), FGFs, and leptin, VEGF orchestrates follicular neoangiogenesis allied to preantral theca/follicular development in many species.[49–53] Most of these cytokines increase GC SCF production, in turn promoting oocyte growth and theca interstitial cell proliferation, as demonstrated in rats.[51,54–56] Angiogenesis is regulated independently across follicles, possibly via the selective local production of pro-/antiangiogenic VEGF isoforms.[57]

Antral Follicle Growth and Selection

Follicle stimulating hormone (FSH) coordinates follicular growth beyond the small antral stage.[58] Rodent models suggest that oocyte-derived GDF-9, BMP-6 and -15, as well as GC-derived activin and BMP-6, modulate FSH-dependent development through GC proliferation and differentiation into cumulus cells.[1,23,59] Unsurprisingly, BMP-15 and GDF-9 inhibition impairs folliculogenesis in mice, although BMP-15 alone is necessary to support development to ovulation.[44,60,61] BMP-15 also regulates the number of recruitable follicles in rats by selectively inhibiting GC FSH-receptor expression, similarly to humans.[42,62] Other BMPs interact with the activin/inhibin system and estrogen signaling to rescue antral follicles from atresia.[1,63,64] Leptin fulfills an analogous role and also inhibits antral follicle steroidogenesis.[65–68]

Again, AMH is antagonistic and reduces preantral/small antral follicle sensitivity to FSH.[69] Various TGF-β isoforms are also involved: GC FSH-receptor expression is promoted by the concerted action of TGF-β1 and 2.[70] TGF-β1 also acts synergistically with FSH to stimulate VEGF production and promote angiogenesis.[71] Moreover, TGF-β1 inhibits oestradiol biosynthesis in FSH-responsive follicles as part of dominant follicle selection.[72] Concurrent oocyte nuclear maturation involves SCF, which is required for antrum formation and subsequent expansion.[73,74] SCF acts with hepatocyte growth factor to promote TC proliferation and GC steroidogenesis.[75] Other cytokines, including TNF-α and IL-6, influence GC proliferation and/or survival by regulating the balance between proliferation and apoptosis/atresia with the input of nerve growth factor, TNF-R1/2 ratio, and IL-6 soluble receptor (IL-6sR) (Fig. 3.3).

Ovulation

Ovulation is a cytokine-coordinated process involving leukocyte recruitment, microcirculatory vasomotion, and extracellular matrix

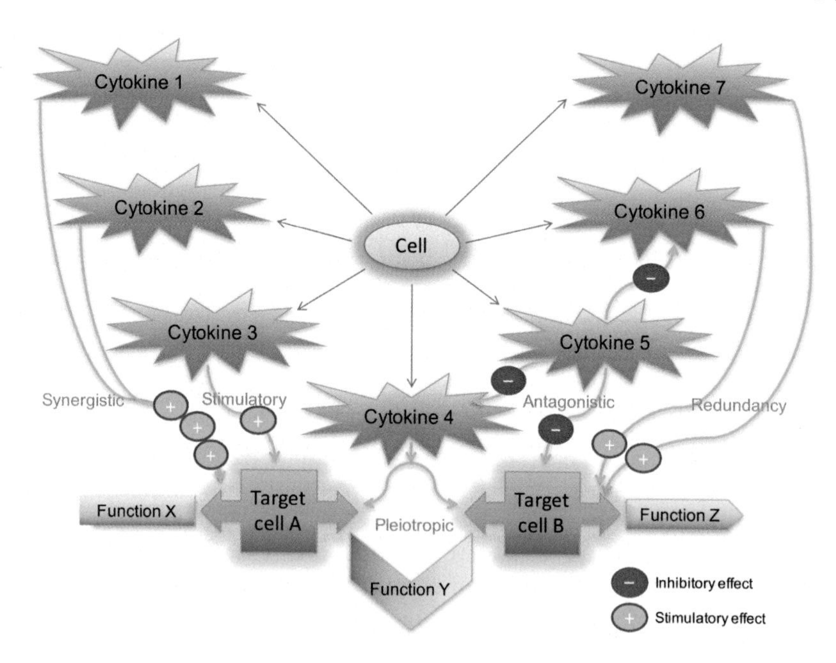

FIGURE 3.3 The multiple functions of cytokines. Different cell types can contribute to these effects enabling a variety of cells to drive any given process. Functions for X, Y and Z may include up- or down-regulation of the expression of membrane proteins (including cytokine receptors), secretion of effector molecules (histamine, antibodies, and cytokines), cellular proliferation, chemotaxis of neutrophils, monocytes and T cells, cellular differentiation, inflammation, phagocytosis, apoptosis, or elimination of pathogens.

(ECM) remodeling geared toward follicle rupture and oocyte-cumulus complex extrusion.[8,76–79] Recruited immune effector cells secrete proteases, angiogenic, and vasoactive agents[7,77]: Macrophages remodel the ECM through matrix metalloproteinase (MMP) production and phagocytose atretic GCs[80]; lymphocytes produce various cytokines,[8] while monocyte lineage, mast, TCs, and GCs produce neutrophil chemotactic IL-8.[81,82] IL-8 levels increase with follicle size, promoting leukocyte recruitment for ovulation.[7,83,84] Other chemotactic agents such as regulated on activation normal T cell expressed and secreted (RANTES), eotaxin, growth regulated oncogene (GRO)-α, MCP-1, and macrophage colony-stimulating factor (M-CSF) also contribute to this process.[85–95]

Response to the preovulatory LH surge is partly under BMP control in mice: BMP-2, -4, -6, and -7 modulate follicular responses to LH and prevent premature luteinization by reducing LH receptor expression.[4,63,96,97] BMP-6 and -7 relieve the block on follicular maturation in human GCs via GDF-3 (a BMP inhibitor) induction.[98] The LH surge also induces the expression of EGF-like amphiregulin (AREG), epiregulin (EREG), and betacellulin.[99,100] The resultant increase in prostaglandin (PG) E_2 initiates a positive feedback loop whereby PGE_2 stimulates cumulus cell AREG and EREG production. These act auto/paracrinally to promote meiotic resumption and, together with IL-6, cumulus expansion.[101] Moreover, TGF-β isoforms and their downstream signaling pathways are responsible for cumulus expansion and follicle rupture.[102,103] Oocyte-derived GDF-9 and BMP-15 stimulate cumulus cell EGFR expression in mice, enabling these cells to respond to LH-induced signaling.[104,105] The site of follicle rupture is determined in part by IL-1β, which modulates GC prostanoid production, and TNF-α, which promotes localized ovarian surface–follicular interface apoptosis and ECM remodeling.[5,106–108]

Luteinization, Luteolysis, and Early Pregnancy

Luteinization completes follicular life and provides progestogenic support for early pregnancy.[109] Corpus luteum formation involves bFGF, leptin, stromal derived factor (SDF)-1α, and VEGF.[110–113] Leptin stimulates luteinized GC estrogen production in women[114] and reduces progesterone synthesis in insulin-stimulated bovine TCs,[66] accounting for the impact of increased serum and follicular fluid leptin concentrations in women with obesity-induced infertility.[115] Leptin may also program abnormal AMH signaling by luteinized GCs, thereby resulting in ovarian dysfunction.[116] Macrophages provide the phagocytic clearance of apoptotic luteal cells and secrete steroidogenic cytokines such as IL-8.[8,117] TGF-β isoforms also modulate the luteinizing effects of the LH surge.[102,103] The onset of luteolysis instead features decreased perifollicular blood flow with resultant hypoxia which induces variable VEGF expression across species.[118,119]

WHEN THINGS GO AWRY: PATHOLOGY SUPPORTS PHYSIOLOGY

Altered cytokine profiles and deregulated inflammatory responses characterize various ovarian pathologies and underscore cytokines' physiological functions. These will therefore be briefly reviewed.

Polycystic Ovary Syndrome

Polycystic ovary syndrome (PCOS) is a complex, heterogeneous endocrine disorder affecting 5%–10% of women of reproductive age associated with Type 2 diabetes and obesity which features oligo-/amenorrhea, anovulation, hirsutism, acne (reflecting high androgen levels), polycystic ovaries, and infertility. In PCOS, an altered IGF system correlates directly with abnormal follicular development, impaired steroidogenesis, inadequate oocyte maturation, and anovulation.[120] More specifically, IGFs stimulate GC mitosis and steroidogenesis and inhibit apoptosis, accounting for the effects of altered IGF expression, hyperinsulinaemia, and GC insulin resistance.[121] The characteristic hyperandrogenaemia and arrested folliculogenesis hinge on changes in IGF-1 bioavailability caused by perturbations in serum and follicular fluid IGFBP-1.[122–124] Macrophage inflammatory protein (MIP)-1α betrays the inflammatory pathophysiology underlying PCOS given its role in inducing IL-1, IL-6, and TNF-α production.[125,126] Women with PCOS have elevated serum leptin levels and may exhibit altered leptin sensitivity.[127] While leptin can stimulate the HPG axis, supraphysiological levels impair folliculogenesis,[128] possibly via deregulation of GnRH pulsatility and LH secretion,[129] altered ovarian steroidogenesis,[67,68] and perturbed perifollicular blood flow.[130] Indeed, hyperinsulinaemia in PCOS contributes to hyperthecotic stromal VEGF hypersecretion[131] which diverts blood flow away from the leading follicle, resulting in uninhibited growth of the entire cohort.[132] Increased stromal vascularity also causes abnormal theca interna growth and a resultant increase in androgen biosynthesis.[133] Similarly, impaired GC estrogen biosynthesis is attributed to altered EGF and TNF-α profiles in PCOS.[5,123,134,135] FGF is also implicated in PCOS through its pro-angiogenic properties and effects on GC/TC proliferation.[136] It may contribute to blocking antral follicle growth through the GC aromatase inhibition, in turn resulting in disordered follicular growth, follicular arrest, and anovulation.[137] The high serum and follicular fluid AMH concentrations reported in PCOS likely reflect the increased development of small antral follicles and follicular arrest as well as increased AMH production by individual follicles.[138–140] Given that a reduction of AMH in larger follicles is key to dominant follicle selection/development, elevated levels may contribute to failed growth and ovulation.[141] Indeed, altered cytokine profiles in PCOS have reproductive *sequelae*: Reduced cumulus cell GDF-9 expression causes premature luteinization, suboptimal oocyte maturation, and poor luteal function,[123,142] accounting for the higher miscarriage rates in these women.[143] Imbalances in oocyte GDF-9 and BMP-15 impair oocyte development, reduce embryo viability, and decrease pregnancy rates.[143–145] Similarly, lower follicular fluid LIF concentrations may lower implantation rates,[146] while elevated intrafollicular TNF-α levels and their inverse correlation with oestradiol levels are indicators of poor oocyte/embryo quality.[147,148]

Ovarian Hyperstimulation Syndrome

Ovarian hyperstimulation syndrome (OHSS) is an iatrogenic potentially fatal complication of controlled ovarian hyperstimulation in assisted reproduction featuring cystic ovarian enlargement, increased vascular permeability, hemoconcentration/thrombosis, oliguria, and extravascular fluid redistribution. Moderate and severe forms of OHSS affect 3%–6% and 0.1%–3% of treatment cycles, respectively[149] Intrafollicular IL-8, angiopoietin, TNF-α, the TGF-β family, IGF-1, endothelial cadherin, VEGF, and CXC chemokines have all been implicated in OHSS, although their roles are often unclear.[122,132,150–153] Others, such as IL-6, mediate increases in vascular permeability and suppress hepatic albumin production in favor of acute phase reactants.[151,154]

Premature Ovarian Insufficiency

Premature ovarian insufficiency (POI) affects 0.9%–1.0% of women <40 years and features a premature depletion of the primordial follicle pool which exhausts ovarian reserve.[155] Normally, pool preservation is orchestrated by the coordinated suppression of follicular activation by gonadotrophins, estrogen, growth hormone, BMP-15, IGF-1, EGF, TGF-α/β, bFGF, IL-1β, IL-6, gonadal GnRH-like peptide, and nitric oxide.[156,157] Disruptions in this configuration may trigger the premature and irreversible activation of the primordial pool, resulting in accelerated follicular atresia.[158] Inflammatory/immune perturbations in POI, partly involving abnormal infiltrating lymphocyte function, may induce a continuous rather than cyclical cytokine production, including a loss of the episodic IFN-γ-initiated cytokine cascade which normally governs follicular development/atresia and GC steroidogenesis.[5,159,160] Since AMH is produced by the GCs of growing preantral/small antral follicles recruited from the primary follicle pool but not selected for dominance, its falling concentrations are diagnostic of secondary hypogonadism.[161]

Endometriosis

Endometriosis affects 3%–10% of women of childbearing age and features extrauterine endometrial tissue growth and variable reproductive failure.[162] TGF-β1, IGF-1, and their receptors have been identified in the stroma and epithelium of ovarian endometriotic tissues and endometriomata.[163,164] Moreover, elevated concentrations of IFN-γ characterize both the serum and follicular fluid of women with endometriosis, which may reflect increased follicular IFN-γ production and the immune system's efforts to overcome apoptotic inhibition and decrease cell proliferation.[126,165] Similarly, GCs of women with endometriosis produce more IL-6,[166] while its mRNA levels are higher in ovaries containing endometriomata, possibly reflecting the contribution of infiltrating macrophages.[167] Higher ovarian TNF-α, IL-15, GM-CSF, VEGF,

RANTES, and epithelial neutrophil-activating peptide 78 (an angiogenic chemokine) levels are also found in women with endometriosis, although their precise involvement remains unclear.[166,168–172] Nonetheless, such an aberrant ovarian inflammatory microenvironment causes disordered folliculogenesis and adversely affects oocyte and embryo viability.[166,173]

Ovarian Cancer

Epithelial ovarian cancers are the most lethal gynecological malignancies, with circa 7300 new diagnoses and 4300 deaths/year in the United Kingdom alone.[174] These morphologically heterogeneous entities vary in their clinical course, mutational status, and precursor lesions.[175] Their histological origin is an enigma, since the embryological derivation of the ovarian surface epithelium (mesothelium) from which these tumors were thought to arise is distinct from the Müllerian system. It now recognized that most ovarian epithelial malignancies in the main do not originate in the ovary,[175] but rather from endometriotic cysts,[176] distal Fallopian tubes,[177] benign tubal epithelial inclusion cysts, or the tuboperitoneal junction.[178] The ovary seemingly offers an unusually favorable environment for malignant transformation and/or rapid progression of ovarian cancers, and compelling epidemiological data link the number of lifetime ovulations with cancer risk.[179] This suggests that ovulation releases chemotactic cytokines (e.g., GC-secreted SDF-1α and TNF-α) which may recruit ectopic premalignant/malignant cells to the surface epithelium, with subsequent episodes promoting malignant transformation/progression through inflammatory mechanisms.[180] Inflammation's involvement in ovarian carcinogenesis is supported by the observations that pelvic inflammatory disease and endometriosis increase ovarian cancer risk, while aspirin has a protective effect.[181–184] Variably synergistic ovarian cytokines such as IL-6, SDF-1, TNF-α, and VEGF also contribute to cancer cells' escape from immunosurveillance and the recruitment/subversion of accessory stromal/immune effector cells as well as favoring tumor cell proliferation/survival, angiogenesis, ECM remodeling, invasion and metastasis.[185,186] Various cells maintain this functional microenvironment, including cancer-associated fibroblasts, which secrete MMPs, versican, IL-8, SDF-1α, and IFN-γ-inducible protein (IP)-9 in response to ovarian cancer cell-secreted IL-1β, IL-6, IL-8, GRO-α, TGF-β, and TNF-β.[187–189] Similarly, cancer cells create an immunopermissive, pro-angiogenic microenvironmental niche which favors ECM remodeling and invasion by recruiting and polarizing macrophages toward the M2 phenotype through MCP-1, M-CSF, IL-6, macrophage migration inhibitory factor (MIF), and the TNF network.[186] In turn, tumor-associated macrophages (TAMs) secrete immunosuppressive TGF-β, IL-10, MIP-4, and macrophage-derived chemokine (MDC).[190] MDC (from ovarian cancer cells/TAMs) recruits regulatory T cells to tumor sites and malignant ascites where they suppress T-cell activation and IL-2/IFN-γ production, thus reducing antitumour activity.[191] The bidirectionality of the cancer cell: TAM dialog is underscored in vitro, where TAMs stimulate ovarian cancer cell IL-8 production,[192] while the latter promote TAM TNF-α and MIF-dependent MMP secretion.[193] TAMs also govern ascites formation and peritoneal metastasis through inflammatory mechanisms, possibly involving VEGF.[194] Understanding these interactions is critical to developing new treatments[195] such as antiangiogenic modalities,[196] recombinant IL-2,[197] immune checkpoint PD-1 receptor targeting,[195] and anti-IL-6 therapy.[198]

UNDERSTANDING OVARIAN CYTOKINE NETWORKS

The development of conceptual models to capture patho-/physiological cytokine interactions is challenging given cytokines' *modus operandi* as complex networks wherein these pleiotropic mediators exhibit relationships based on synergy, antagonism, and functional redundancy.[199] Bayesian networks (BNs) have been employed to characterize hierarchical cytokine interactions in both mice[200] and women where, in the latter, they model dynamic ovarian cytokine signatures allied to oocyte maturation.[201] These probabilistic models, which represent variables linked by directed interactions according to their respective conditional dependencies, offer a number of advantages. First, the so-called "edges" connecting nodes (variables) imply causal directionality of effect, with node-specific conditional probabilities offering a qualitative measure of their concentration. Second, BNs can be perturbed in silico perturbation, such that the status (concentration) of any node can be altered and the impact monitored throughout the network. This allows "virtual" experiments to be validated in vivo but also enables investigators to capture synergistic/antagonistic interactions. Third, the confidence threshold applied to the network edges ensures that only "robust" relationships are displayed. Although BNs appear to be a panacea for cytokine network interpretative problems, they have their idiosyncrasies. "Seeding" network core structures using literature-mining outputs alleviate computational demands, but potentially at the expense of introducing bias, especially where structural comparison between networks is required. Moreover, the number of samples analyzed can determine apparent network complexity. Finally, BNs provide a snapshot in time and their acyclicity precludes the inclusion of structural feedback loops, which fails to capture homeostatic mechanisms or the evolution of inflammatory/growth factor responses.

Dynamic BNs may offer a workable solution by linking sequential networks, although they pose the challenge of changing their structure and connections over time. BNs may stand the test of time as tools to illustrate cytokine interactions, but increased complexity will be one of their features given the need to integrate other mediators (e.g., eicosanoids) and multisource data from various analytical platforms and tissue constituents in future models.

CONCLUSIONS

Complex cytokine interplay governs ovarian development, folliculogenesis, oocyte maturation, and luteal function and deregulations of these networks are associated with a spectrum of ovarian disorders. Our understanding of these networks' structure remains poor: Our interpretative difficulties stem from our inability to conceptualize these in an accessible format and, in turn, limits our appreciation of ovarian physiology and pathology as well as limiting the inception of novel treatments. Future advances in this setting require holistic modeling of cytokine networks using machine learning-based approaches, and to expand these to include homeostatic regulatory/control systems.

References

1. Knight PG, Glister C. TGF-beta superfamily members and ovarian follicle development. *Reproduction* 2006;**132**:191–206.
2. Orsi NM, Tribe RM. Cytokine networks and the regulation of uterine function in pregnancy and parturition. *J Neuroendocrinol* 2008;**20**:462–9.
3. Bornstein SR, Rutkowski H, Vrezas I. Cytokines and steroidogenesis. *Mol Cell Endocrinol* 2004;**215**:135–41.
4. Shimasaki S, Moore RK, Erickson GF, Otsuka F. The role of bone morphogenetic proteins in ovarian function. *Reprod Suppl* 2003;**61**:323–37.
5. Wu R, Van der Hoek KH, Ryan NK, Norman RJ, Robker RL. Macrophage contributions to ovarian function. *Hum Reprod Update* 2004;**10**:119–33.
6. Vinatier D, Dufour P, Tordjeman-Rizzi N, Prolongeau JF, Depret-Moser S, Monnier JC. Immunological aspects of ovarian function: role of the cytokines. *Eur J Obstet Gynecol Reprod Biol* 1995;**63**:155–68.
7. Brannstrom M, Enskog A. Leukocyte networks and ovulation. *J Reprod Immunol* 2002; **57**:47–60.
8. Bukulmez O, Arici A. Leukocytes in ovarian function. *Hum Reprod Update* 2000;**6**: 1–15.
9. Hoyer PE, Byskov AG, Mollgard K. Stem cell factor and c-Kit in human primordial germ cells and fetal ovaries. *Mol Cell Endocrinol* 2005;**234**:1–10.
10. Ying Y, Qi X, Zhao GQ. Induction of primordial germ cells from pluripotent epiblast. *Scientific World J* 2002;**2**:801–10.
11. Ying Y, Liu XM, Marble A, Lawson KA, Zhao GQ. Requirement of Bmp8b for the generation of primordial germ cells in the mouse. *Mol Endocrinol* 2000;**14**:1053–63.

12. Childs AJ, Kinnell HL, Collins CS, Hogg K, Bayne RA, Green SJ, et al. BMP signaling in the human fetal ovary is developmentally regulated and promotes primordial germ cell apoptosis. *Stem Cells* 2010;**28**:1368–78.
13. Eddie SL, Childs AJ, Jabbour HN, Anderson RA. Developmentally regulated IL6-type cytokines signal to germ cells in the human fetal ovary. *Mol Hum Reprod* 2012;**18**: 88–95.
14. Matsui Y, Takehara A, Tokitake Y, Ikeda M, Obara Y, Morita-Fujimura Y, et al. The majority of early primordial germ cells acquire pluripotency by AKT activation. *Development* 2014;**141**:4457–67.
15. Morita Y, Manganaro TF, Tao XJ, Martimbeau S, Donahoe PK, Tilly JL. Requirement for phosphatidylinositol-3′-kinase in cytokine-mediated germ cell survival during fetal oogenesis in the mouse. *Endocrinology* 1999;**140**:941–9.
16. De Biasi SN, Apfelbaum LI, Apfelbaum ME. In vitro effect of leptin on LH release by anterior pituitary glands from female rats at the time of spontaneous and steroid-induced LH surge. *Eur J Endocrinol* 2001;**145**:659–65.
17. Palmert MR, Radovick S, Boepple PA. Leptin levels in children with central precocious puberty. *J Clin Endocrinol Metab* 1998;**83**:2260–5.
18. Strobel A, Issad T, Camoin L, Ozata M, Strosberg AD. A leptin missense mutation associated with hypogonadism and morbid obesity. *Nat Genet* 1998;**18**:213–15.
19. Serke H, Nowicki M, Kosacka J, Schroder T, Kloting N, Bluher M, et al. Leptin-deficient (ob/ob) mouse ovaries show fatty degeneration, enhanced apoptosis and decreased expression of steroidogenic acute regulatory enzyme. *Int J Obes (Lond)* 2012; **36**:1047–53.
20. Cortes ME, Carrera B, Rioseco H, Pablo del Rio J, Vigil P. The role of kisspeptin in the onset of puberty and in the ovulatory mechanism: a mini-review. *J Pediatr Adolesc Gynecol* 2015;**28**:286–91.
21. McLaughlin EA, McIver SC. Awakening the oocyte: controlling primordial follicle development. *Reproduction* 2009;**137**:1–11.
22. Richards JS, Hedin L. Molecular aspects of hormone action in ovarian follicular development, ovulation, and luteinization. *Annu Rev Physiol* 1988;**50**:441–63.
23. Eppig JJ. Oocyte control of ovarian follicular development and function in mammals. *Reproduction* 2001;**122**:829–38.
24. Eppig JJ. Oocyte-somatic cell interactions during oocyte growth and maturation in the mammal. *Dev Biol (NY 1985)* 1985;**1**:313–47.
25. Wigglesworth K, Lee KB, O'Brien MJ, Peng J, Matzuk MM, Eppig JJ. Bidirectional communication between oocytes and ovarian follicular somatic cells is required for meiotic arrest of mammalian oocytes. *Proc Natl Acad Sci USA* 2013;**110**:E3723–9.
26. Nilsson EE, Kezele P, Skinner MK. Leukemia inhibitory factor (LIF) promotes the primordial to primary follicle transition in rat ovaries. *Mol Cell Endocrinol* 2002;**188**:65–73.
27. Nilsson EE, Skinner MK. Bone morphogenetic protein-4 acts as an ovarian follicle survival factor and promotes primordial follicle development. *Biol Reprod* 2003;**69**:1265–72.
28. Nilsson EE, Skinner MK. Kit ligand and basic fibroblast growth factor interactions in the induction of ovarian primordial to primary follicle transition. *Mol Cell Endocrinol* 2004;**214**:19–25.
29. Wang J, Roy SK. Growth differentiation factor-9 and stem cell factor promote primordial follicle formation in the hamster: modulation by follicle-stimulating hormone. *Biol Reprod* 2004;**70**:577–85.
30. Gougeon A, Delangle A, Arouche N, Stridsberg M, Gotteland JP, Loumaye E. Kit ligand and the somatostatin receptor antagonist, BIM-23627, stimulate in vitro resting follicle growth in the neonatal mouse ovary. *Endocrinology* 2010;**151**:1299–309.
31. Parrott JA, Skinner MK. Kit-ligand/stem cell factor induces primordial follicle development and initiates folliculogenesis. *Endocrinology* 1999;**140**:4262–71.

32. da Nobrega Jr. JE, Goncalves PB, Chaves RN, Magalhaes Dde M, Rossetto R, Lima-Verde IB, et al. Leukemia inhibitory factor stimulates the transition of primordial to primary follicle and supports the goat primordial follicle viability in vitro. *Zygote* 2012;**20**:73–8.
33. Myers M, Pangas SA. Regulatory roles of transforming growth factor beta family members in folliculogenesis. *Wiley Interdiscip Rev Syst Biol Med* 2010;**2**:117–25.
34. Sun RZ, Lei L, Cheng L, Jin ZF, Zu SJ, Shan ZY, et al. Expression of GDF-9, BMP-15 and their receptors in mammalian ovary follicles. *J Mol Histol* 2010;**41**:325–32.
35. Carabatsos MJ, Elvin J, Matzuk MM, Albertini DF. Characterization of oocyte and follicle development in growth differentiation factor-9-deficient mice. *Dev Biol* 1998;**204**: 373–84.
36. Wang ZP, Mu XY, Guo M, Wang YJ, Teng Z, Mao GP, et al. Transforming growth factor-beta signaling participates in the maintenance of the primordial follicle pool in the mouse ovary. *J Biol Chem* 2014;**289**:8299–311.
37. Durlinger AL, Kramer P, Karels B, de Jong FH, Uilenbroek JT, Grootegoed JA, et al. Control of primordial follicle recruitment by anti-Mullerian hormone in the mouse ovary. *Endocrinology* 1999;**140**:5789–96.
38. Durlinger AL, Visser JA, Themmen AP. Regulation of ovarian function: the role of anti-Mullerian hormone. *Reproduction* 2002;**124**:601–9.
39. Grondahl ML, Nielsen ME, Dal Canto MB, Fadini R, Rasmussen IA, Westergaard LG, et al. Anti-Mullerian hormone remains highly expressed in human cumulus cells during the final stages of folliculogenesis. *Reprod Biomed Online* 2011;**22**:389–98.
40. McGee EA, Raj RS. Regulators of ovarian preantral follicle development. *Semin Reprod Med* 2015;**33**:179–84.
41. Otsuka F, Shimasaki S. A negative feedback system between oocyte bone morphogenetic protein 15 and granulosa cell kit ligand: its role in regulating granulosa cell mitosis. *Proc Natl Acad Sci USA* 2002;**99**:8060–5.
42. Otsuka F, Yamamoto S, Erickson GF, Shimasaki S. Bone morphogenetic protein-15 inhibits follicle-stimulating hormone (FSH) action by suppressing FSH receptor expression. *J Biol Chem* 2001;**276**:11387–92.
43. Shimizu T, Iijima K, Ogawa Y, Miyazaki H, Sasada H, Sato E. Gene injections of vascular endothelial growth factor and growth differentiation factor-9 stimulate ovarian follicular development in immature female rats. *Fertil Steril* 2008;**89**:1563–70.
44. Juengel JL, Bodensteiner KJ, Heath DA, Hudson NL, Moeller CL, Smith P, et al. Physiology of GDF9 and BMP15 signalling molecules. *Anim Reprod Sci* 2004;**82-83**:447–60.
45. Elvin JA, Yan C, Wang P, Nishimori K, Matzuk MM. Molecular characterization of the follicle defects in the growth differentiation factor 9-deficient ovary. *Mol Endocrinol* 1999;**13**:1018–34.
46. Elvin JA, Clark AT, Wang P, Wolfman NM, Matzuk MM. Paracrine actions of growth differentiation factor-9 in the mammalian ovary. *Mol Endocrinol* 1999;**13**:1035–48.
47. Spicer LJ, Aad PY, Allen DT, Mazerbourg S, Payne AH, Hsueh AJ. Growth differentiation factor 9 (GDF9) stimulates proliferation and inhibits steroidogenesis by bovine theca cells: influence of follicle size on responses to GDF9. *Biol Reprod* 2008;**78**:243–53.
48. Vitt UA, Hayashi M, Klein C, Hsueh AJ. Growth differentiation factor-9 stimulates proliferation but suppresses the follicle-stimulating hormone-induced differentiation of cultured granulosa cells from small antral and preovulatory rat follicles. *Biol Reprod* 2000;**62**:370–7.
49. Gospodarowicz D, Cheng J, Lui GM, Baird A, Esch F, Bohlen P. Corpus luteum angiogenic factor is related to fibroblast growth factor. *Endocrinology* 1985;**117**:2383–91.
50. Brown LF, Detmar M, Claffey K, Nagy JA, Feng D, Dvorak AM, et al. Vascular permeability factor/vascular endothelial growth factor: a multifunctional angiogenic cytokine. *EXS* 1997;**79**:233–69.

51. Chaves RN, de Matos MH, Buratini Jr. J, de Figueiredo JR. The fibroblast growth factor family: involvement in the regulation of folliculogenesis. *Reprod Fertil Dev* 2012;**24**:905–15.
52. Joo JK, Joo BS, Kim SC, Choi JR, Park SH, Lee KS. Role of leptin in improvement of oocyte quality by regulation of ovarian angiogenesis. *Anim Reprod Sci* 2010;**119**:329–34.
53. Araujo VR, Duarte AB, Bruno JB, Pinho Lopes CA, de Figueiredo JR. Importance of vascular endothelial growth factor (VEGF) in ovarian physiology of mammals. *Zygote* 2013;**21**:295–304.
54. Jin X, Han CS, Zhang XS, Yuan JX, Hu ZY, Liu YX. Signal transduction of stem cell factor in promoting early follicle development. *Mol Cell Endocrinol* 2005;**229**:3–10.
55. Parrott JA, Skinner MK. Direct actions of kit-ligand on theca cell growth and differentiation during follicle development. *Endocrinology* 1997;**138**:3819–27.
56. Yang P, Roy SK. Epidermal growth factor modulates transforming growth factor receptor messenger RNA and protein levels in hamster preantral follicles in vitro. *Biol Reprod* 2001;**65**:847–54.
57. McFee RM, Rozell TG, Cupp AS. The balance of proangiogenic and antiangiogenic VEGFA isoforms regulate follicle development. *Cell Tissue Res* 2012;**349**:635–47.
58. Gougeon A. Human ovarian follicular development: from activation of resting follicles to preovulatory maturation. *Ann Endocrinol (Paris)* 2010;**71**:132–43.
59. Eppig JJ, Chesnel F, Hirao Y, O'Brien MJ, Pendola FL, Watanabe S, et al. Oocyte control of granulosa cell development: how and why. *Hum Reprod* 1997;**12**:127–32.
60. Myllymaa S, Pasternack A, Mottershead DG, Poutanen M, Pulkki MM, Pelliniemi LJ, et al. Inhibition of oocyte growth factors in vivo modulates ovarian folliculogenesis in neonatal and immature mice. *Reproduction* 2010;**139**:587–98.
61. Dragovic RA, Ritter LJ, Schulz SJ, Amato F, Armstrong DT, Gilchrist RB. Role of oocyte-secreted growth differentiation factor 9 in the regulation of mouse cumulus expansion. *Endocrinology* 2005;**146**:2798–806.
62. Otsuka F, Yao Z, Lee T, Yamamoto S, Erickson GF, Shimasaki S. Bone morphogenetic protein-15. Identification of target cells and biological functions. *J Biol Chem* 2000;**275**: 39523–8.
63. Fatehi AN, van den Hurk R, Colenbrander B, Daemen AJ, van Tol HT, Monteiro RM, et al. Expression of bone morphogenetic protein2 (BMP2), BMP4 and BMP receptors in the bovine ovary but absence of effects of BMP2 and BMP4 during IVM on bovine oocyte nuclear maturation and subsequent embryo development. *Theriogenology* 2005;**63**:872–89.
64. Scheele F, Schoemaker J. The role of follicle-stimulating hormone in the selection of follicles in human ovaries: a survey of the literature and a proposed model. *Gynecol Endocrinol* 1996;**10**:55–66.
65. Spicer LJ, Francisco CC. The adipose obese gene product, leptin: evidence of a direct inhibitory role in ovarian function. *Endocrinology* 1997;**138**:3374–9.
66. Spicer LJ, Chamberlain CS, Francisco CC. Ovarian action of leptin: effects on insulin-like growth factor-I-stimulated function of granulosa and thecal cells. *Endocrine* 2000; **12**:53–9.
67. Agarwal SK, Vogel K, Weitsman SR, Magoffin DA. Leptin antagonizes the insulin-like growth factor-I augmentation of steroidogenesis in granulosa and theca cells of the human ovary. *J Clin Endocrinol Metab* 1999;**84**:1072–6.
68. Zachow RJ, Magoffin DA. Direct intraovarian effects of leptin: impairment of the synergistic action of insulin-like growth factor-I on follicle-stimulating hormone-dependent estradiol-17 beta production by rat ovarian granulosa cells. *Endocrinology* 1997;**138**:847–50.
69. Durlinger AL, Gruijters MJ, Kramer P, Karels B, Kumar TR, Matzuk MM, et al. Anti-Mullerian hormone attenuates the effects of FSH on follicle development in the mouse ovary. *Endocrinology* 2001;**142**:4891–9.
70. Juengel JL, McNatty KP. The role of proteins of the transforming growth factor-beta superfamily in the intraovarian regulation of follicular development. *Hum Reprod Update* 2005;**11**:143–60.

71. Dunkel L, Tilly JL, Shikone T, Nishimori K, Hsueh AJ. Follicle-stimulating hormone receptor expression in the rat ovary: increases during prepubertal development and regulation by the opposing actions of transforming growth factors beta and alpha. *Biol Reprod* 1994;**50**:940–8.
72. Ouellette Y, Price CA, Carriere PD. Follicular fluid concentration of transforming growth factor-beta1 is negatively correlated with estradiol and follicle size at the early stage of development of the first-wave cohort of bovine ovarian follicles. *Domest Anim Endocrinol* 2005;**29**:623–33.
73. Ye Y, Kawamura K, Sasaki M, Kawamura N, Groenen P, Gelpke MD, et al. Kit ligand promotes first polar body extrusion of mouse preovulatory oocytes. *Reprod Biol Endocrinol* 2009;**7**:26.
74. Miyoshi T, Otsuka F, Nakamura E, Inagaki K, Ogura-Ochi K, Tsukamoto N, et al. Regulatory role of kit ligand-c-kit interaction and oocyte factors in steroidogenesis by rat granulosa cells. *Mol Cell Endocrinol* 2012;**358**:18–26.
75. Ito M, Harada T, Tanikawa M, Fujii A, Shiota G, Terakawa N. Hepatocyte growth factor and stem cell factor involvement in paracrine interplays of theca and granulosa cells in the human ovary. *Fertil Steril* 2001;**75**:973–9.
76. Zackrisson U, Lofman CO, Janson PO, Wallin A, Mikuni M, Brannstrom M. Alterations of follicular microcirculation and apex structure during ovulation in the rat. *Eur J Obstet Gynecol Reprod Biol* 2011;**157**:169–74.
77. Oakley OR, Kim H, El-Amouri I, Lin PC, Cho J, Bani-Ahmad M, et al. Periovulatory leukocyte infiltration in the rat ovary. *Endocrinology* 2010;**151**:4551–9.
78. Richards JS, Liu Z, Shimada M. Immune-like mechanisms in ovulation. *Trends Endocrinol Metab* 2008;**19**:191–6.
79. Richards JS, Pangas SA. New insights into ovarian function. *Handb Exp Pharmacol* 2010;**198**:3–27.
80. Pate JL, Landis Keyes P. Immune cells in the corpus luteum: friends or foes? *Reproduction* 2001;**122**:665–76.
81. Szukiewicz D, Pyzlak M, Klimkiewicz J, Szewczyk G, Maslinska D. Mast cell-derived interleukin-8 may be involved in the ovarian mechanisms of follicle growth and ovulation. *Inflamm Res* 2007;**56**(Suppl. 1):S35–6.
82. Runesson E, Ivarsson K, Janson PO, Brannstrom M. Gonadotropin- and cytokine-regulated expression of the chemokine interleukin 8 in the human preovulatory follicle of the menstrual cycle. *J Clin Endocrinol Metab* 2000;**85**:4387–95.
83. Malizia BA, Wook YS, Penzias AS, Usheva A. The human ovarian follicular fluid level of interleukin-8 is associated with follicular size and patient age. *Fertil Steril* 2010;**93**: 537–43.
84. Brannstrom M, Norman RJ. Involvement of leukocytes and cytokines in the ovulatory process and corpus luteum function. *Hum Reprod* 1993;**8**:1762–75.
85. Schall TJ, Bacon K, Toy KJ, Goeddel DV. Selective attraction of monocytes and T lymphocytes of the memory phenotype by cytokine RANTES. *Nature* 1990;**347**:669–71.
86. Aust G, Simchen C, Heider U, Hmeidan FA, Blumenauer V, Spanel-Borowski K. Eosinophils in the human corpus luteum: the role of RANTES and eotaxin in eosinophil attraction into periovulatory structures. *Mol Hum Reprod* 2000;**6**:1085–91.
87. Aust G, Brylla E, Lehmann I, Kiessling S, Spanel-Borowski K. Cloning of bovine RANTES mRNA and its expression and regulation in ovaries in the periovulatory period. *FEBS Lett* 1999;**463**:160–4.
88. Machelon V, Nome F, Emilie D. Regulated on activation normal T expressed and secreted chemokine is induced by tumor necrosis factor-alpha in granulosa cells from human preovulatory follicle. *J Clin Endocrinol Metab* 2000;**85**:417–24.
89. Wong KH, Negishi H, Adashi EY. Expression, hormonal regulation, and cyclic variation of chemokines in the rat ovary: key determinants of the intraovarian residence of representatives of the white blood cell series. *Endocrinology* 2002;**143**:784–91.

90. Kuwabara Y, Katayama A, Igarashi T, Tomiyama R, Piao H, Kaneko R, et al. Rapid and transient upregulation of CCL11 (eotaxin-1) in mouse ovary during terminal stages of follicular development. *Am J Reprod Immunol* 2012;**67**:358–68.
91. Dahm-Kahler P, Ghahremani M, Lind AK, Sundfeldt K, Brannstrom M. Monocyte chemotactic protein-1 (MCP-1), its receptor, and macrophages in the perifollicular stroma during the human ovulatory process. *Fertil Steril* 2009;**91**:231–9.
92. Arici A, Oral E, Bukulmez O, Buradagunta S, Bahtiyar O, Jones EE. Monocyte chemotactic protein-1 expression in human preovulatory follicles and ovarian cells. *J Reprod Immunol* 1997;**32**:201–19.
93. Kawano Y, Fukuda J, Itoh H, Takai N, Nasu K, Miyakawa I. The effect of inflammatory cytokines on secretion of macrophage colony-stimulating factor and monocyte chemoattractant protein-1 in human granulosa cells. *Am J Reprod Immunol* 2004;**52**: 124–8.
94. Kawano Y, Kawasaki F, Nakamura S, Matsui N, Narahara H, Miyakawa I. The production and clinical evaluation of macrophage colony-stimulating factor and macrophage chemoattractant protein-1 in human follicular fluids. *Am J Reprod Immunol* 2001;**45**:1–5.
95. Nishimura K, Tanaka N, Ohshige A, Fukumatsu Y, Matsuura K, Okamura H. Effects of macrophage colony-stimulating factor on folliculogenesis in gonadotrophin-primed immature rats. *J Reprod Fertil* 1995;**104**:325–30.
96. Sugiura K, Su YQ, Eppig JJ. Does bone morphogenetic protein 6 (BMP6) affect female fertility in the mouse? *Biol Reprod* 2010;**83**:997–1004.
97. Shi J, Yoshino O, Osuga Y, Koga K, Hirota Y, Nose E, et al. Bone morphogenetic protein-2 (BMP-2) increases gene expression of FSH receptor and aromatase and decreases gene expression of LH receptor and StAR in human granulosa cells. *Am J Reprod Immunol* 2011;**65**:421–7.
98. Shi J, Yoshino O, Osuga Y, Akiyama I, Harada M, Koga K, et al. Growth differentiation factor 3 is induced by bone morphogenetic protein 6 (BMP-6) and BMP-7 and increases luteinizing hormone receptor messenger RNA expression in human granulosa cells. *Fertil Steril* 2012;**97**:979–83.
99. Ashkenazi H, Cao X, Motola S, Popliker M, Conti M, Tsafriri A. Epidermal growth factor family members: endogenous mediators of the ovulatory response. *Endocrinology* 2005;**146**:77–84.
100. Park JY, Su YQ, Ariga M, Law E, Jin SL, Conti M. EGF-like growth factors as mediators of LH action in the ovulatory follicle. *Science* 2004;**303**:682–4.
101. Shimada M, Hernandez-Gonzalez I, Gonzalez-Robayna I, Richards JS. Paracrine and autocrine regulation of epidermal growth factor-like factors in cumulus oocyte complexes and granulosa cells: key roles for prostaglandin synthase 2 and progesterone receptor. *Mol Endocrinol* 2006;**20**:1352–65.
102. Sriperumbudur R, Zorrilla L, Gadsby JE. Transforming growth factor-beta (TGFbeta) and its signaling components in peri-ovulatory pig follicles. *Anim Reprod Sci* 2010; **120**:84–94.
103. Gao H, Chen XL, Zhang ZH, Song XX, Hu ZY, Liu YX. IFN-gamma and TNF-alpha inhibit expression of TGF-beta-1, its receptors TBETAR-I and TBETAR-II in the corpus luteum of PMSG/hCG treated rhesus monkey. *Front Biosci* 2005;**10**:2496–503.
104. Su YQ, Sugiura K, Li Q, Wigglesworth K, Matzuk MM, Eppig JJ. Mouse oocytes enable LH-induced maturation of the cumulus-oocyte complex via promoting EGF receptor-dependent signaling. *Mol Endocrinol* 2010;**24**:1230–9.
105. Sugiura K, Su YQ, Li Q, Wigglesworth K, Matzuk MM, Eppig JJ. Estrogen promotes the development of mouse cumulus cells in coordination with oocyte-derived GDF9 and BMP15. *Mol Endocrinol* 2010;**24**:2303–14.
106. Adashi EY. The potential role of interleukin-1 in the ovulatory process: an evolving hypothesis. *Mol Cell Endocrinol* 1998;**140**:77–81.

107. Loret de Mola JR, Goldfarb JM, Hecht BR, Baumgardner GP, Babbo CJ, Friedlander MA. Gonadotropins induce the release of interleukin-1 beta, interleukin-6 and tumor necrosis factor-alpha from the human preovulatory follicle. *Am J Reprod Immunol* 1998;**39**:387–90.
108. Cerri PS, Pereira-Junior JA, Biselli NB, Sasso-Cerri E. Mast cells and MMP-9 in the lamina propria during eruption of rat molars: quantitative and immunohistochemical evaluation. *J Anat* 2010;**217**:116–25.
109. Oktem O, Urman B. Understanding follicle growth in vivo. *Hum Reprod* 2010;**25**:2944–54.
110. Guerra DM, Giometti IC, Price CA, Andrade PB, Castilho AC, Machado MF, et al. Expression of fibroblast growth factor receptors during development and regression of the bovine corpus luteum. *Reprod Fertil Dev* 2008;**20**:659–64.
111. Neuvians TP, Schams D, Berisha B, Pfaffl MW. Involvement of pro-inflammatory cytokines, mediators of inflammation, and basic fibroblast growth factor in prostaglandin F2alpha-induced luteolysis in bovine corpus luteum. *Biol Reprod* 2004;**70**: 473–80.
112. Neuvians TP, Berisha B, Schams D. Vascular endothelial growth factor (VEGF) and fibroblast growth factor (FGF) expression during induced luteolysis in the bovine corpus luteum. *Mol Reprod Dev* 2004;**67**:389–95.
113. Nishigaki A, Okada H, Okamoto R, Shimoi K, Miyashiro H, Yasuda K, et al. The concentration of human follicular fluid stromal cell-derived factor-1 is correlated with luteinization in follicles. *Gynecol Endocrinol* 2013;**29**:230–4.
114. Kitawaki J, Kusuki I, Koshiba H, Tsukamoto K, Honjo H. Leptin directly stimulates aromatase activity in human luteinized granulosa cells. *Mol Hum Reprod* 1999;**5**: 708–13.
115. Mitchell M, Armstrong DT, Robker RL, Norman RJ. Adipokines: implications for female fertility and obesity. *Reproduction* 2005;**130**:583–97.
116. Merhi Z, Buyuk E, Berger DS, Zapantis A, Israel DD, Chua Jr. S, et al. Leptin suppresses anti-Mullerian hormone gene expression through the JAK2/STAT3 pathway in luteinized granulosa cells of women undergoing IVF. *Hum Reprod* 2013;**28**:1661–9.
117. Shimizu T, Kaji A, Murayama C, Magata F, Shirasuna K, Wakamiya K, et al. Effects of interleukin-8 on estradiol and progesterone production by bovine granulosa cells from large follicles and progesterone production by luteinizing granulosa cells in culture. *Cytokine* 2012;**57**:175–81.
118. Acosta TJ, Hayashi KG, Ohtani M, Miyamoto A. Local changes in blood flow within the preovulatory follicle wall and early corpus luteum in cows. *Reproduction* 2003;**125**:759–67.
119. Niswender GD, Juengel JL, Silva PJ, Rollyson MK, McIntush EW. Mechanisms controlling the function and life span of the corpus luteum. *Physiol Rev* 2000;**80**:1–29.
120. Kwintkiewicz J, Giudice LC. The interplay of insulin-like growth factors, gonadotropins, and endocrine disruptors in ovarian follicular development and function. *Semin Reprod Med* 2009;**27**:43–51.
121. Bhatia V. Insulin resistance in polycystic ovarian disease. *South Med J* 2005;**98**:903–10 quiz 11-2, 23.
122. Lee A, Christenson LK, Patton PE, Burry KA, Stouffer RL. Vascular endothelial growth factor production by human luteinized granulosa cells in vitro. *Hum Reprod* 1997;**12**:2756–61.
123. Artini PG, Monteleone P, Toldin M, Matteucci C, Ruggiero M, Cela V. Growth factors and folliculogenesis in polycystic ovary patients. *Expert Rev Endocrinol Metab* 2007;**2**: 215–23.
124. Amer MI, Hassan TA, Awad EA, Fawzy OA, Fatah MA. Insulin-like growth factor-1 (IGF-1) and hormonal patterns in polycystic ovarian syndrome (PCOS). *Mid East Fert Soc J* 2002;**7**:115–23.

125. Dahm-Kahler P, Runesson E, Lind AK, Brannstrom M. Monocyte chemotactic protein-1 in the follicle of the menstrual and IVF cycle. *Mol Hum Reprod* 2006;**12**:1–6.
126. Sarapik A, Velthut A, Haller-Kikkatalo K, Faure GC, Bene MC, de Carvalho Bittencourt M, et al. Follicular proinflammatory cytokines and chemokines as markers of IVF success. *Clin Dev Immunol* 2012;**2012**:606459.
127. El Orabi H, Ghalia AA, Khalifa A, Mahfouz H, El Shalkani A, Shoieb N. Serum leptin as an additional possible pathogenic factor in polycystic ovary syndrome. *Clin Biochem* 1999;**32**:71–5.
128. Duggal PS, Van Der Hoek KH, Milner CR, Ryan NK, Armstrong DT, Magoffin DA, et al. The in vivo and in vitro effects of exogenous leptin on ovulation in the rat. *Endocrinology* 2000;**141**:1971–6.
129. Moschos S, Chan JL, Mantzoros CS. Leptin and reproduction: a review. *Fertil Steril* 2002;**77**:433–44.
130. Cao R, Brakenhielm E, Wahlestedt C, Thyberg J, Cao Y. Leptin induces vascular permeability and synergistically stimulates angiogenesis with FGF-2 and VEGF. *Proc Natl Acad Sci USA* 2001;**98**:6390–5.
131. Stanek MB, Borman SM, Molskness TA, Larson JM, Stouffer RL, Patton PE. Insulin and insulin-like growth factor stimulation of vascular endothelial growth factor production by luteinized granulosa cells: comparison between polycystic ovarian syndrome (PCOS) and non-PCOS women. *J Clin Endocrinol Metab* 2007;**92**:2726–33.
132. Agrawal R, Conway G, Sladkevicius P, Tan SL, Engmann L, Payne N, et al. Serum vascular endothelial growth factor and Doppler blood flow velocities in in vitro fertilization: relevance to ovarian hyperstimulation syndrome and polycystic ovaries. *Fertil Steril* 1998;**70**:651–8.
133. Peitsidis P, Agrawal R. Role of vascular endothelial growth factor in women with PCO and PCOS: a systematic review. *Reprod Biomed Online* 2010;**20**:444–52.
134. Mason HD, Margara R, Winston RM, Beard RW, Reed MJ, Franks S. Inhibition of oestradiol production by epidermal growth factor in human granulosa cells of normal and polycystic ovaries. *Clin Endocrinol (Oxf)* 1990;**33**:511–17.
135. Amato G, Conte M, Mazziotti G, Lalli E, Vitolo G, Tucker AT, et al. Serum and follicular fluid cytokines in polycystic ovary syndrome during stimulated cycles. *Obstet Gynecol* 2003;**101**:1177–82.
136. Bruno JB, Matos MH, Chaves RN, Celestino JJH, Saraiva MV, Lima-Verde IB. Angiogenic factors and ovarian follicle development. *Anim Reprod Sci* 2009;**6**:371–9.
137. Almahbobi G, Misajon A, Hutchinson P, Lolatgis N, Trounson AO. Hyperexpression of epidermal growth factor receptors in granulosa cells from women with polycystic ovary syndrome. *Fertil Steril* 1998;**70**:750–8.
138. Das M, Gillott DJ, Saridogan E, Djahanbakhch O. Anti-Mullerian hormone is increased in follicular fluid from unstimulated ovaries in women with polycystic ovary syndrome. *Hum Reprod* 2008;**23**:2122–6.
139. Mulders AG, Laven JS, Eijkemans MJ, de Jong FH, Themmen AP, Fauser BC. Changes in anti-Mullerian hormone serum concentrations over time suggest delayed ovarian ageing in normogonadotrophic anovulatory infertility. *Hum Reprod* 2004;**19**:2036–42.
140. Franks S, Stark J, Hardy K. Follicle dynamics and anovulation in polycystic ovary syndrome. *Hum Reprod Update* 2008;**14**:367–78.
141. Pellatt L, Hanna L, Brincat M, Galea R, Brain H, Whitehead S, et al. Granulosa cell production of anti-Mullerian hormone is increased in polycystic ovaries. *J Clin Endocrinol Metab* 2007;**92**:240–5.
142. Takebayashi K, Takakura K, Wang H, Kimura F, Kasahara K, Noda Y. Mutation analysis of the growth differentiation factor-9 and -9B genes in patients with premature ovarian failure and polycystic ovary syndrome. *Fertil Steril* 2000;**74**:976–9.

143. Zhao SY, Qiao J, Chen YJ, Liu P, Li J, Yan J. Expression of growth differentiation factor-9 and bone morphogenetic protein-15 in oocytes and cumulus granulosa cells of patients with polycystic ovary syndrome. *Fertil Steril* 2010;**94**:261–7.
144. Teixeira Filho FL, Baracat EC, Lee TH, Suh CS, Matsui M, Chang RJ, et al. Aberrant expression of growth differentiation factor-9 in oocytes of women with polycystic ovary syndrome. *J Clin Endocrinol Metab* 2002;**87**:1337–44.
145. Gilchrist RB, Lane M, Thompson JG. Oocyte-secreted factors: regulators of cumulus cell function and oocyte quality. *Hum Reprod Update* 2008;**14**:159–77.
146. Ledee-Bataille N, Lapree-Delage G, Taupin JL, Dubanchet S, Taieb J, Moreau JF, et al. Follicular fluid concentration of leukaemia inhibitory factor is decreased among women with polycystic ovarian syndrome during assisted reproduction cycles. *Hum Reprod* 2001;**16**:2073–8.
147. Wu R, Fujii S, Ryan NK, Van der Hoek KH, Jasper MJ, Sini I, et al. Ovarian leukocyte distribution and cytokine/chemokine mRNA expression in follicular fluid cells in women with polycystic ovary syndrome. *Hum Reprod* 2007;**22**:527–35.
148. Kim CH, Ahn JW, You RM, Kim SH, Chae HD, Kang BM. Pioglitazone treatment decreases follicular fluid levels of tumor necrosis factor-alpha and interleukin-6 in patients with polycystic ovary syndrome. *Clin Exp Reprod Med* 2011;**38**:98–102.
149. Kumar P, Sait SF, Sharma A, Kumar M. Ovarian hyperstimulation syndrome. *J Hum Reprod Sci* 2011;**4**:70–5.
150. Abramov Y, Schenker JG, Lewin A, Friedler S, Nisman B, Barak V. Plasma inflammatory cytokines correlate to the ovarian hyperstimulation syndrome. *Hum Reprod* 1996;**11**:1381–6.
151. Chen CD, Chen HF, Lu HF, Chen SU, Ho HN, Yang YS. Value of serum and follicular fluid cytokine profile in the prediction of moderate to severe ovarian hyperstimulation syndrome. *Hum Reprod* 2000;**15**:1037–42.
152. Villasante A, Pacheco A, Pau E, Ruiz A, Pellicer A, Garcia-Velasco JA. Soluble vascular endothelial-cadherin levels correlate with clinical and biological aspects of severe ovarian hyperstimulation syndrome. *Hum Reprod* 2008;**23**:662–7.
153. Chen SU, Chou CH, Lin CW, Lee H, Wu JC, Lu HF, et al. Signal mechanisms of vascular endothelial growth factor and interleukin-8 in ovarian hyperstimulation syndrome: dopamine targets their common pathways. *Hum Reprod* 2010;**25**:757–67.
154. Artini PG, Monti M, Fasciani A, Battaglia C, D'Ambrogio G, Genazzani AR. Vascular endothelial growth factor, interleukin-6 and interleukin-2 in serum and follicular fluid of patients with ovarian hyperstimulation syndrome. *Eur J Obstet Gynecol Reprod Biol* 2002;**101**:169–74.
155. Beck-Peccoz P, Persani L. Premature ovarian failure. *Orphanet J Rare Dis* 2006;**1**:9.
156. Hsueh AJ, Eisenhauer K, Chun SY, Hsu SY, Billig H. Gonadal cell apoptosis. *Recent Prog Horm Res* 1996;**51**:433–55 [discussion 455-6].
157. Tiotiu D, Alvaro Mercadal B, Imbert R, Verbist J, Demeestere I, De Leener A, et al. Variants of the BMP15 gene in a cohort of patients with premature ovarian failure. *Hum Reprod* 2010;**25**:1581–7.
158. Adhikari D, Liu K. Molecular mechanisms underlying the activation of mammalian primordial follicles. *Endocr Rev* 2009;**30**:438–64.
159. Coulam CB, Stern JJ. Immunology of ovarian failure. *Am J Reprod Immunol* 1991;**25**:169–74.
160. Best CL, Griffin PM, Hill JA. Interferon gamma inhibits luteinized human granulosa cell steroid production in vitro. *Am J Obstet Gynecol* 1995;**172**:1505–10.
161. van Rooij IA, Broekmans FJ, te Velde ER, Fauser BC, Bancsi LF, de Jong FH, et al. Serum anti-Mullerian hormone levels: a novel measure of ovarian reserve. *Hum Reprod* 2002;**17**:3065–71.

162. Iwabe T, Harada T, Terakawa N. Role of cytokines in endometriosis-associated infertility. *Gynecol Obstet Invest* 2002;**53**(Suppl. 1):19–25.
163. Tamura M, Fukaya T, Enomoto A, Murakami T, Uehara S, Yajima A. Transforming growth factor-beta isoforms and receptors in endometriotic cysts of the human ovary. *Am J Reprod Immunol* 1999;**42**:160–7.
164. Loverro G, Maiorano E, Napoli A, Selvaggi L, Marra E, Perlino E. Transforming growth factor-beta 1 and insulin-like growth factor-1 expression in ovarian endometriotic cysts: a preliminary study. *Int J Mol Med* 2001;**7**:423–9.
165. Nishida M, Nasu K, Ueda T, Fukuda J, Takai N, Miyakawa I. Endometriotic cells are resistant to interferon-gamma-induced cell growth inhibition and apoptosis: a possible mechanism involved in the pathogenesis of endometriosis. *Mol Hum Reprod* 2005;**11**:29–34.
166. Pellicer A, Albert C, Mercader A, Bonilla-Musoles F, Remohi J, Simon C. The follicular and endocrine environment in women with endometriosis: local and systemic cytokine production. *Fertil Steril* 1998;**70**:425–31.
167. Tsudo T, Harada T, Iwabe T, Tanikawa M, Nagano Y, Ito M, et al. Altered gene expression and secretion of interleukin-6 in stromal cells derived from endometriotic tissues. *Fertil Steril* 2000;**73**:205–11.
168. Xu H, Schultze-Mosgau A, Agic A, Diedrich K, Taylor RN, Hornung D. Regulated upon activation, normal T cell expressed and secreted (RANTES) and monocyte chemotactic protein 1 in follicular fluid accumulate differentially in patients with and without endometriosis undergoing in vitro fertilization. *Fertil Steril* 2006;**86**:1616–20.
169. Wunder DM, Mueller MD, Birkhauser MH, Bersinger NA. Increased ENA-78 in the follicular fluid of patients with endometriosis. *Acta Obstet Gynecol Scand* 2006;**85**:336–42.
170. Falconer H, Sundqvist J, Gemzell-Danielsson K, von Schoultz B, D'Hooghe TM, Fried G. IVF outcome in women with endometriosis in relation to tumour necrosis factor and anti-Mullerian hormone. *Reprod Biomed Online* 2009;**18**:582–8.
171. Garrido N, Navarro J, Remohi J, Simon C, Pellicer A. Follicular hormonal environment and embryo quality in women with endometriosis. *Hum Reprod Update* 2000;**6**: 67–74.
172. Fasciani A, D'Ambrogio G, Bocci G, Monti M, Genazzani AR, Artini PG. High concentrations of the vascular endothelial growth factor and interleukin-8 in ovarian endometriomata. *Mol Hum Reprod* 2000;**6**:50–4.
173. Garrido N, Navarro J, Garcia-Velasco J, Remoh J, Pellice A, Simon C. The endometrium versus embryonic quality in endometriosis-related infertility. *Hum Reprod Update* 2002;**8**:95–103.
174. Cancer Research UK. *Ovarian cancer statistics* <http://www.cancerresearchuk.org/health-professional/cancer-statistics/statistics-by-cancer-type/ovarian-cancer#heading-Zero>; 2016 [accessed 06/06/16].
175. Shih Ie M, Kurman RJ. Ovarian tumorigenesis: a proposed model based on morphological and molecular genetic analysis. *Am J Pathol* 2004;**164**:1511–18.
176. Pearce CL, Templeman C, Rossing MA, Lee A, Near AM, Webb PM, et al. Association between endometriosis and risk of histological subtypes of ovarian cancer: a pooled analysis of case-control studies. *Lancet Oncol* 2012;**13**:385–94.
177. Kindelberger DW, Lee Y, Miron A, Hirsch MS, Feltmate C, Medeiros F, et al. Intraepithelial carcinoma of the fimbria and pelvic serous carcinoma: Evidence for a causal relationship. *Am J Surg Pathol* 2007;**31**:161–9.
178. Kurman RJ, Shih Ie M. The dualistic model of ovarian carcinogenesis: revisited, revised, and expanded. *Am J Pathol* 2016;**186**:733–47.
179. Beral V, Doll R, Hermon C, Peto R, Reeves G. Ovarian cancer and oral contraceptives: collaborative reanalysis of data from 45 epidemiological studies including 23,257 women with ovarian cancer and 87,303 controls. *Lancet* 2008;**371**:303–14.

180. Yang-Hartwich Y, Gurrea-Soteras M, Sumi N, Joo WD, Holmberg JC, Craveiro V, et al. Ovulation and extra-ovarian origin of ovarian cancer. *Sci Rep* 2014;**4**:6116.
181. Nezhat F, Datta MS, Hanson V, Pejovic T, Nezhat C, Nezhat C. The relationship of endometriosis and ovarian malignancy: a review. *Fertil Steril* 2008;**90**:1559–70.
182. Lin HW, Tu YY, Lin SY, Su WJ, Lin WL, Lin WZ, et al. Risk of ovarian cancer in women with pelvic inflammatory disease: a population-based study. *Lancet Oncol* 2011;**12**:900–4.
183. Munksgaard PS, Blaakaer J. The association between endometriosis and ovarian cancer: a review of histological, genetic and molecular alterations. *Gynecol Oncol* 2012;**124**:164–9.
184. Trabert B, Ness RB, Lo-Ciganic WH, Murphy MA, Goode EL, Poole EM, et al. Aspirin, nonaspirin nonsteroidal anti-inflammatory drug, and acetaminophen use and risk of invasive epithelial ovarian cancer: a pooled analysis in the Ovarian Cancer Association Consortium. *J Natl Cancer Inst* 2014;**106**: djt431.
185. Kryczek I, Lange A, Mottram P, Alvarez X, Cheng P, Hogan M, et al. CXCL12 and vascular endothelial growth factor synergistically induce neoangiogenesis in human ovarian cancers. *Cancer Res* 2005;**65**:465–72.
186. Kulbe H, Chakravarty P, Leinster DA, Charles KA, Kwong J, Thompson RG, et al. A dynamic inflammatory cytokine network in the human ovarian cancer microenvironment. *Cancer Res* 2012;**72**:66–75.
187. Schauer IG, Sood AK, Mok S, Liu J. Cancer-associated fibroblasts and their putative role in potentiating the initiation and development of epithelial ovarian cancer. *Neoplasia* 2011;**13**:393–405.
188. Yeung TL, Leung CS, Wong KK, Samimi G, Thompson MS, Liu J, et al. TGF-beta modulates ovarian cancer invasion by upregulating CAF-derived versican in the tumor microenvironment. *Cancer Res* 2013;**73**:5016–28.
189. Lau TS, Chung TK, Cheung TH, Chan LK, Cheung LW, Yim SF, et al. Cancer cell-derived lymphotoxin mediates reciprocal tumour-stromal interactions in human ovarian cancer by inducing CXCL11 in fibroblasts. *J Pathol* 2014;**232**:43–56.
190. Colvin EK. Tumor-associated macrophages contribute to tumor progression in ovarian cancer. *Front Oncol* 2014;**4**:137.
191. Curiel TJ, Coukos G, Zou L, Alvarez X, Cheng P, Mottram P, et al. Specific recruitment of regulatory T cells in ovarian carcinoma fosters immune privilege and predicts reduced survival. *Nat Med* 2004;**10**:942–9.
192. Wang X, Zhao X, Wang K, Wu L, Duan T. Interaction of monocytes/macrophages with ovarian cancer cells promotes angiogenesis in vitro. *Cancer Sci* 2013;**104**:516–23.
193. Hagemann T, Wilson J, Kulbe H, Li NF, Leinster DA, Charles K, et al. Macrophages induce invasiveness of epithelial cancer cells via NF-kappa B and JNK. *J Immunol* 2005;**175**:1197–205.
194. Robinson-Smith TM, Isaacsohn I, Mercer CA, Zhou M, Van Rooijen N, Husseinzadeh N, et al. Macrophages mediate inflammation-enhanced metastasis of ovarian tumors in mice. *Cancer Res* 2007;**67**:5708–16.
195. Hansen JM, Coleman RL, Sood AK. Targeting the tumour microenvironment in ovarian cancer. *Eur J Cancer* 2016;**56**:131–43.
196. Korkmaz T, Seber S, Basaran G. Review of the current role of targeted therapies as maintenance therapies in first and second line treatment of epithelial ovarian cancer. In the light of completed trials. *Crit Rev Oncol Hematol* 2016;**98**:180–8.
197. Vlad AM, Budiu RA, Lenzner DE, Wang Y, Thaller JA, Colonello K, et al. A phase II trial of intraperitoneal interleukin-2 in patients with platinum-resistant or platinum-refractory ovarian cancer. *Cancer Immunol Immunother* 2010;**59**:293–301.
198. Coward J, Kulbe H, Chakravarty P, Leader D, Vassileva V, Leinster DA, et al. Interleukin-6 as a therapeutic target in human ovarian cancer. *Clin Cancer Res* 2011;**17**:6083–96.

199. Orsi NM. Cytokine networks in the establishment and maintenance of pregnancy. *Hum Fertil (Camb)* 2008;**11**:222–30.
200. Field SL, Dasgupta T, Cummings M, Savage RS, Adebayo J, McSara H, et al. Bayesian modeling suggests that IL-12 (p40), IL-13 and MCP-1 drive murine cytokine networks in vivo. *BMC Syst Biol* 2015;**9**:76.
201. Orsi N, Dasgupta T, Cummings M, Adebayo J, Sharma V, Gunawardena J, et al. Predicting oocyte fertilisability in intracytoplasmic sperm injection cycles: a retrospective observational study. *Lancet* 2017;**389**:75.

CHAPTER

4

Cytokines in the Liver

Cytokine Mechanisms in Liver Health and Disease

Hani S. Mousa[1], *Francesca Bernuzzi*[2], *and Pietro Invernizzi*[2,3]

[1]University of Cambridge, Cambridge, United Kingdom

[2]University of Milan-Bicocca, Milan, Italy

[3]University of California Davis, Davis, CA, United States

OUTLINE

Cytokine Effector Functions in Tissues.

DOI: http://dx.doi.org/10.1016/B978-0-12-804214-4.00003-8

CYTOKINE MECHANISMS UNDERLYING ACUTE LIVER INJURY

Insult to the liver starts manifold signaling cascades which are mediated mainly by cytokines. Depending on the etiology, these cascades are ignited by the release of damage-associated molecular pattern molecules (DAMPs) or pathogen-associated molecular pattern molecules (PAMPs) to which the parenchymal and the nonparenchymal cellular constituents of the liver respond (Fig. 4.1).

DAMPs encompass different cellular components which are released upon hepatocyte necrosis to induce sterile inflammation.[1] This group includes fragments of DNA, ATP, the high mobility group box-1 (HMGB1) protein as well as IL-33 released from intracellular stores upon injury.[2]

This results in activation of Kupffer cells and the release of cytokines that recruit neutrophils and monocytes which, beside their cytotoxic capabilities, become themselves a source of proinflammatory cytokines such as TNF-α, IL-1β, IL-12, and IL-18.

TNF-α released from Kupffer cells, NKT cells, and intrahepatic macrophages act on TNF receptor on hepatocytes to induce apoptosis and concurrently activates the NF-κB signaling cascade in these cells.[3] Activation of NF-κB by itself has antiapoptotic effects and may provide a molecular basis for the link between recurrent insult and carcinogenesis (Fig. 4.1, Panel C).[4]

In fulminant hepatic failure (FHF), TNF-α is significantly elevated, apoptotic hepatocytes are increased, and TNF-α levels correlate with disease prognosis. Moreover, hepatocytes from explanted livers of FHF patients show overexpression of the TNF-R1 receptor paralleled with overexpression of TNF-α by the infiltrating mononuclear cells.[5]

IL-6 is another pivotal cytokine mediating inflammation. It is produced by Kupffer cells, dendritic cells (DCs), and biliary epithelial cells (BECs) and participate in skewing naive T-cells into Th17 phenotype.[6]

Hepatocytes express different glycoproteins (gp80 and gp130) which interact with IL-6 to induce a Stat3-mediated signaling cascade that results in the acute phase response (APR).[7] In animal models, disruption of IL-6 signaling (e.g., through gp130 knockout) impairs APR and renders the animals more susceptible to LPS challenge.[8]

As a general inducer of APR, IL-6 is often used clinically as a general marker for inflammation.[9] In patients with acute and chronic liver diseases, IL-6 levels are elevated and they are inversely correlated with biochemical markers of liver function.[10]

IL-6 signaling is tightly controlled; it induces the hepatic expression of suppressor of cytokine signaling (SOCS) 3, a potent feedback inhibitor which in turn terminates IL-6 signal transduction (Fig. 4.1, Panel C).[11]

IL1β is another important cytokine in propagating inflammation, providing feed-forward stimulus for several proinflammatory cytokines.

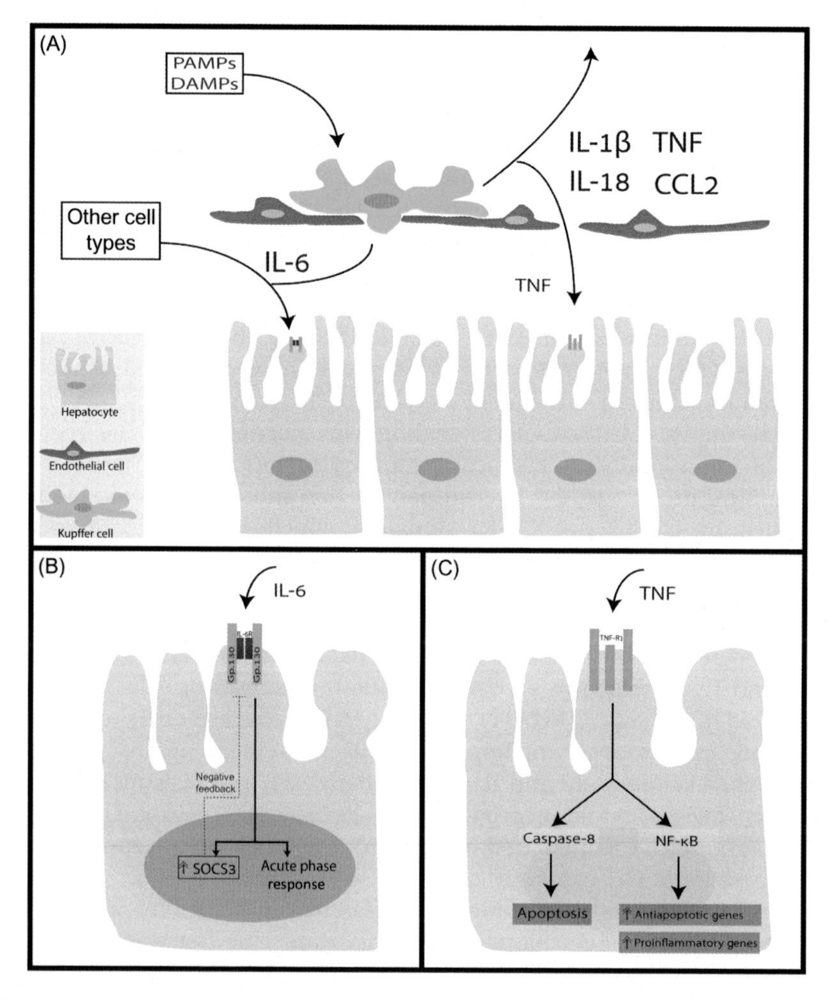

FIGURE 4.1 **Key cytokine mechanisms in liver inflammation.** (A) Upon activation by damage-associated molecular pattern molecules or pathogen-associated molecular pattern molecules, Kupffer cells are activated to release a myriad of cytokines including IL-6 and TNF. Both of these cytokines exert direct effect on hepatocytes to induce acute phase response, negative feedback inhibition, apoptosis, and NF-κB activation as shown in (B) and (C).

Similar to IL-18, it is proteolytically activated by caspase-1 upon inflammasome assembly.[12] DAMPs, such as ATP signaling through purinergic receptors, can activate the Nalp3 inflammasome in Kupffer cells which results in the activation of caspase-1.[2]

IL1β signaling has been implicated in liver injury of various etiologies—including acetaminophen-induced, immune-mediated, alcoholic, and nonalcoholic hepatic injury.[13–16]

Liver injury and the inflammatory response that ensues are also marked by the chemokine-mediated recruitment of different cellular subsets that further amplify inflammation and the release of proinflammatory cytokines.

CCL2 is released by hepatocytes and Kupffer cells in response to insult and signaling through the CCL2–CCR2 axis plays a key role in the recruitment of monocytes/macrophages, which contribute to the inflammation, angiogenesis, and fibrosis and inhibition of this axis has been shown to decrease the number of intrahepatic macrophages both in acute and chronic liver injury.[17,18] Another chemokine which has been shown to play a role in monocyte/macrophage recruitment and differentiation is CCL1 which signals through its cognate receptor, CCR8 expressed on macrophages. In the absence of CCL1–CCR8 signaling, hepatic macrophages had an altered phenotype with overexpression of the T-cell attracting chemokine CCL3 and subsequent increase in Th1 polarization.[19]

Interestingly, CCR8–CCL1 signaling was necessary for the recruitment of inflammatory ("classical") monocytes but not for the CD16+ nonclassical monocyte.[19,20] Recruitment of this subset of macrophages is mediated by CX3CL1–CX3CR1 signaling which also attenuate monocytic proinflammatory activity.[21,22] In the absence of CX3CL1–CX3CR1 signaling, intrahepatic monocytes are more prone to develop into proinflammatory macrophages.[21]

Monocytes, when stimulated by toll-like receptor (TLR)-8, produce IL-12 and IL-18 which in turn activate another subset of cells: Mucosal-associated invariant T (MAIT) cells.[23] MAIT cells are enriched in the liver and seem to play an important role in defense against bacterial infections.[24] While IL-12 and IL-18 activate MAIT in a TCR-independent manner, IL-7—a cytokine produced by hepatocytes during inflammation—regulates TCR-mediated activation of MAIT cells, licensing them to increase their Th1 cytokines and IL-17 production.[25]

Neutrophils are also recruited in different types of liver injury and their presence in a liver biopsy is an important clue that can guide the differential diagnosis.[26] CXCR2-active chemokines, such as CXCL1, CXCL2, and CXCL8, guide their recruitment and play a role in their subsequent activation.[27–29]

However, cytokines also take part in different checks that are meant to curtail the inflammatory response in homeostatic conditions; one of these cytokines is IL-10, produced by DCs upon TLR signaling and resulting in reduced production of TNF, IL-6, and reactive oxygen species (ROS) by monocytes (for further discussion see *Cytokine Mechanisms Mediating Liver Tolerance* section).[30] IL-10 is also secreted by regulatory T-cells (Tregs) and Kupffer cells.[31]

Another example of such a negative feedback mechanism is mediated by NKT cells. NKT cells are recruited to liver inflammatory foci through CXCR6 signaling, where they become a source of IL-4, promoting neutrophil survival and hepatitis.[32] Yet, they also sequentially produce IFN-γ which induces neutrophil apoptosis and thus ameliorating inflammation.[33]

NKT cells also produce IL-17 which has a protective role against liver inflammation and against neutrophil and monocyte recruitment IL-17.[34]

CYTOKINE MECHANISMS UNDERLYING LIVER REGENERATION

Inflammation and regeneration are intertwined in the liver and many of the cytokines that drive inflammation and their downstream signaling cascades, including NF-κB and Stat3, have been shown to be crucial at the initial priming phase of liver regeneration and contribute also to the subsequent expansion and termination phases of the regenerative process.[35]

TNF-α and IL-6 are both increased after partial hepatectomy and antibodies against TNF-α inhibit liver regeneration and abolish the rise normally seen in IL-6 levels.[36]

While TNF-α might not have direct mitogenic effects on hepatocytes, it exerts its effect through IL-6 mediated activation of protective mechanisms that facilitate hepatocyte proliferation.[10] It has also been shown that TNF-α signaling renders hepatocytes more responsive to key growth factors in liver regeneration including HGF and TGF-α which drive the expansion phase.[37–39] TNF-α derived from recruited monocytes in liver injury also induces the proliferation of the liver progenitor cells (LPCs) subset.[40]

Other than TNF-α, signaling through the TNF-R1 receptor can be initiated by lymphotoxin-α (LT-α) which is increasingly expressed by lymphocytes after hepatectomy and has been shown to exert similar facilitative effects on liver regeneration.[41,42] LT-α can promote liver regeneration signaling through the lymphotoxin β receptor (LTβ-R).[41,43]

At the termination phase of liver regeneration, TGF-β seems to be important and mice deficient for the type II TGF-β receptor show increased proliferation upon partial hepatectomy. While it has been shown that TGF-β1 expression is increased after partial hepatectomy, the responsiveness of hepatocytes to TGF-β signaling is transiently dampened to allow for their regeneration.[44,45]

IL-22 is another cytokine that is increased after partial hepatectomy and it has been shown to induce Stat3 activation, act on LPCs to promote their proliferation and evidence from animal models suggests possible advantageous effects of IL-22 treatment for recovery after liver injury.[46,47] This effect is reversed by SOCS3 overexpression.[48]

These shared cytokine networks in inflammation and regeneration provide a mechanistic link between these two processes and supports increasing body of evidence about the involvement of inflammation in liver carcinogenesis.

Different studies have indeed linked the increased production of IL-6 in males with the observed higher risk of hepatocellular carcinoma (HCC).[49,50] Similar studies have also linked the increased risk of HCC in obese patients to the chronic inflammatory response and the enhanced production of TNF-α, IL-6, and the subsequent activation of Stat3.[51] The prosurvival effect mediated by Stat3 signaling is also at the base of the link between IL-22 and the development of HCC.[52]

Dysregulation in the NLRP3 inflammasome and its associated cytokines (e.g., IL-1β) has also been reported in HCC.[53]

The LTα and LTβ and their LTβ-R have been shown to be upregulated in HBV and HCV hepatitis, causally linking viral hepatitis with HCC development.[54]

CHRONIC LIVER INJURY, FIBROSIS, AND CIRRHOSIS

Hepatic stellate cells (HSCs) are the main producer of collagen in liver fibrosis, and profibrotic and antifibrotic effects of cytokines converge on their activation and transdifferentiation into myofibroblasts; profibrotic cytokines can activate HSCs, induce their transdifferentiation into myofibroblasts, and exert a protective role on their survival. On the other hand, antifibrotic cytokines interfere with HSC activation and induce senescence or apoptosis (e.g., through the well-described NK-induced HSC apoptosis).[55]

Gene expression studies suggest that the HSC activation pattern is largely independent of the mechanism of liver injury and involves NF-κB mediated secretion of different inflammatory mediators including the chemokines CCL2, CCL3, CXCL2, and CXCL5.[56] These chemokines bring into the scene other key cellular constituents of hepatic fibrosis.

Resident Kupffer cells and different subsets of recruited monocytes seem to be of paramount importance for the development of fibrosis and cirrhosis. The expansion of the above mentioned "nonclassical" CD14+CD16+ monocytes both in the circulation and liver of patients with chronic liver injury seems to correlate with intrahepatic inflammation and HSC activation in cirrhosis.[57]

Yet, released chemokines may also recruit circulating cells that can limit fibrosis and induce resolution. CCL20 and its receptor, CCR6, are both overexpressed in patients with chronic liver injury and this axis has been shown to mediate recruitment of γδ T-cells to the injured liver, inducing HSC apoptosis through cell-to-cell contact. γδ T-cells also produce IL-22 which has some antifibrotic effects (see below).[58] Yet, recent evidence suggests that CCL20 expression can also have deleterious effects in alcoholic hepatitis-related inflammation and fibrosis as well as in autoimmune hepatitis (see below).[59,60]

The balance between quiescent and activated HSCs is largely governed by different T-cell responses. While Th2 and Th17 cytokines such as TGF-β, IL-13, and MMP9 all promote this activation, Tregs and Th1 cells limit it through IFN-γ and IL-12 production and T-cell induced apoptosis.[55,61]

IL-13 is a master promoter of fibrosis, acting mainly through the activation of HSC. It has historically been studied in the context of Schistosomiasis-associated fibrosis. However, accumulating evidence has implicated it in fibrosis of other etiologies—including viral hepatitis and steatohepatitis of alcoholic and nonalcoholic origin.[62] Beside direct effects on HSCs, IL-13 also activates macrophages which further contribute to the developing fibrosis.[63]

Other than IL-13, IL-4, and IL-5 are both Th2 cytokines that have been shown to contribute to fibrosis—the former by direct effects on HSC activation and collagen production and the latter indirectly by regulating IL-13 activity.[64,65] IL-33 is another cytokine that contributes to fibrosis indirectly through the induction of type 2 cytokines. It is a member of the IL-1 family and a classic alarmin (DAMP) released by activated HSCs and hepatocytes to induce the release of IL-13 and IL-5 in part by the expansion of group 2 innate lymphoid cells.[66,67]

Another master promoter of fibrosis is IL-17 which is overexpressed, along with its receptor, following liver injury and can potently activate HSCs to induce the production of type I collagen in a Stat3 dependent manner.[68] Kupffer cells are also responsive to IL-17 and it can promote the expression of IL-6, IL-1, TNF-α, and TGF-β.[69] Signaling through the cannabinoid receptor 2 is capable of reducing fibrosis through the decreased production of IL-17 by Th17 cells and by attenuating its effects on macrophages and myofibroblasts.[70]

TGF-β is a key profibrogenic cytokine that has long been shown to be overexpressed in patients with liver fibrosis.[71] Moreover, TGF-β polymorphism seems to be predictive of fibrosis in certain patients.[72–74] Hepatic injury induces paracrine release of TGF-β from hepatocytes, Kupffer cells, and platelets with its main target being the HSC/myofibroblast subset.[75] Once activated, myfibroblasts become themselves an important source of TGF-β and they also upregulate its receptor in an autocrine loop. Interestingly, it has been shown that the response of HSCs to TGF-β is modulated by direct TLR-4 activation which induces downregulation of its pseudoreceptor, Bambi, thus rendering them more sensitive to TGF-β and enhancing Kupffer cell activation. This sheds light on the link between intestinal bacterial microflora and hepatic fibrogenesis.[76]

Different genes have been identified as direct targets of TGF-β in transdifferentiated myofibroblasts including procollagens, tissue inhibitors of maetaloproteinases (TIMPs), and plasminogen activator inhibitors 1.[77–81] TGF-β also induces the activity of NADPH oxidase, contributing to the oxidative stress in liver fibrosis.[82,83] Interestingly, the effect of TGF-β seems to change according to the phenotypic state of the HSC/

myofibroblast cells. For example, in quiescent HSCs, TGF-β initiates a negative feedback loop that results in the production of Smad7—an inhibitor of TGF-β signaling. In myofriblasts, this upregulation of Smad7 is absent allowing a more potent stimulation.[77,84] TGF-β-mediated myofibroblast activation has also been shown to be dependent upon Galectin-3, a lectin released by macrophages in response to LPS as well as TNF-α and IFN-γ.[85,86] This further supports the pivotal role that macrophages play in liver fibrosis. Gal-3 also exerts autocrine effects on macrophages themselves, promoting the release of proinflammatory cytokines such as TNF-α, IL-12, and chemokines (e.g., CCL3 and CCL4).[87–89]

PPARδ agonists, which have been shown to have some antifibrotic effect in liver fibrosis, exert their effect at least partially by intefering with downstream TGF-β signaling.[90,91] Other agents interfering with TGF-β signaling have also been suggested as possible therapy for liver fibrosis.[92–94]

Yet, not only hepatic resident cells (such as HSC) contribute to fibrosis. Other cellular constituents, such as recruited nonclassical macrophages (CD14 + CD16 +) play a major role in fibrosis—through the release of proinflammatory cytokines and possible direct activation of HSCs. Targeting their chemokine-mediated recruitment could be an approach to treat fibrosis.[57] DCs, which make up to 25% of the leukocytes in the fibrotic liver, also contribute to the altered hepatic immunity in fibrosis and this is contingent upon their production of TNF-α.[95]

Other than TNF-α, IL-1 is another inflammatory cytokine that participates in the progression from hepatic injurt to fibrosis, directly activating HSCs in an autocrine manner and stimulating their production of matrix metalloproteinases.[96]

Some cytokines can also limit fibrosis, protect against inflammation, and hamper activation of HSCs; IL-10 and IL-22 both belong to the IL-10 family of cytokines and they both share the IL-10R2 receptor subunit. IL-10 limits collagen production, TGF-β secretion, and DNA synthesis in HSCs and has been shown to reverse fibrosis in animal models.[97–99] IL-10 also interferes with Kupffer cell activation and knockout mice for IL-10 are more prone to develop severe liver fibrosis in experimental models.[100,101] Other than HSCs, IL-10 is also produced by hepatocytes, sinusoidal endothelial cells, lymphocytes monocytes, and Kupffer cells.[102,103]

IL-22 has been also been shown to have a protective antiapoptotic and mitogenic effects on hepatocytes, signaling through Stat3 (see above).[104] Moreover, IL-22 acts directly on HSCs, through interaction with the IL-10R2 and IL-22R1 receptor subunits. This results in HSC senescence and limits fibrosis through different mechanisms—including SOC3 activation and protective effects against oxidative stress.[105,106] Moreover, the balance between IL-22 and the soluble IL-22 binding protein (IL-22BP)—which competitively inhibits its binding to its cognate receptor—determines the extent of fibrosis in animal models.[107] The therapeutic potential of these IL-22 mediated effects is still to be explored.[108]

In contrast, it is noteworthy to mention that some studies have attributed pathological role to IL-22 in chronic injury and fibrosis through the induction of CXCL10 and CCL20 expression and the subsequent recruitment of Th17 cells in a transgenic mouse model.[109]

Inteferons generally play an antifibrotic role—acting through different Stats: IFN-α, signaling through Stat1, downregulates collagen gene transcription.[110] IFN-β suppresses hepatocellular injury and downregulates the expressions of TGF-β, bFGF, collagen, and the TIMP-1 an important mediator of fibrosis in HSC.[111,112] Stat1-induced abrogation of TGF-β is also induced by IFN-γ.[113] Indeed, NKT-cell secreted IFN-γ has been shown to induce Smad7, an inhibitor of TGF-β signaling.[74,84,114,115] Some evidence shows that Stat1 signaling in HSC can also be induced by IL-27, which might result in antifibrotic effects.[116] IL-6 signaling—in nonparenchymal liver cells rather than hepatocytes—has also been suggested to have a protective role during fibrosis Fig. 4.2.[10]

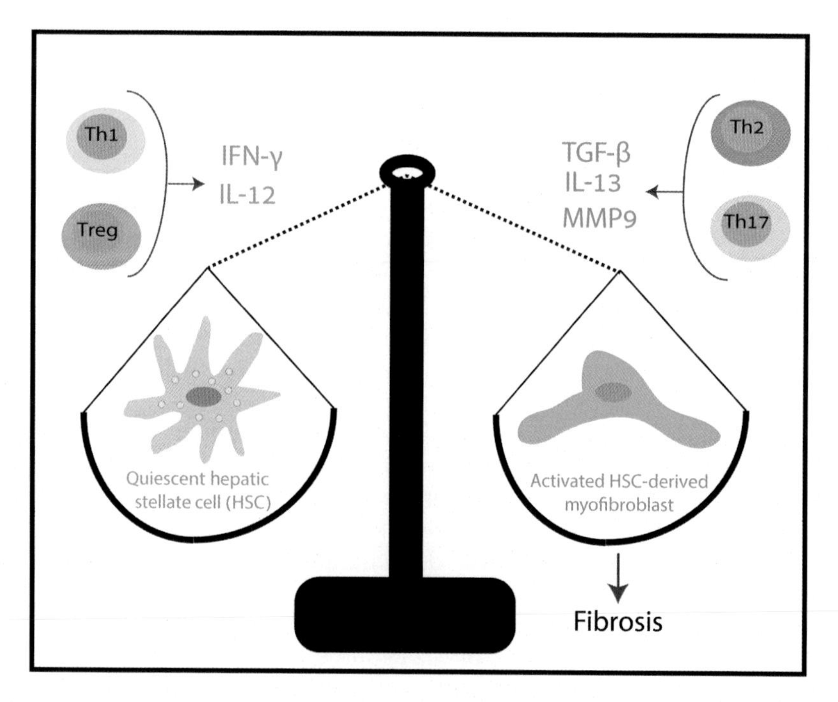

FIGURE 4.2 **T-cell cytokines govern the balance between quiescent and activated hepatic stellate cells (HSCs).** Th2 and Th17 cytokines tip the balance toward HSC activation and as such promote fibrosis. Th1 and Treg cytokines exert an antifibrotic effect by maintaining HSCs in their quiescent state.

CYTOKINE MECHANISMS MEDIATING LIVER TOLERANCE

In addition to peripheral and central tolerance mechanisms, liver-specific mechanisms of tolerance have been described in what is often collectively referred to as the "liver tolerance effect." These mechanisms are mediated by resident cellular components such as Kupffer cells and liver sinusoidal endothelial cells (LSECs) as well as nonparenchymal cells such as DCs.

The liver is constantly exposed to bacterial residues from the gut through the portal circulation which do not mount an immune response despite the expression of TLRs by hepatocytes and other parenchymal liver cells.[117] This tolerogenic effect in constant exposure to low levels of endotoxin is attributed to the release of IL-10 and TGF-β from Kupffer cells and LSECs and the downregulation of MHC class II molecules which subsequently downregulates the activation of Th2 cells.[118,119] The activation of Kupffer cells subsequently affects NK cell activation. Interestingly, it has been shown that the phenotype of the activated NK cells depends on the pathways through which Kupffer cells were stimulated. Signaling through the myeloid differentiation factor 88 (MyD88) induces IL-10, resulting in attentuated activation of NK cells. On the other hand, signaling through the Toll/IL-1 receptor domain-containing adapter inducing IFN-β induces less IL-10 and concurrently potentiates the stimulatory effects of IL-18 on NK cells enhancing their activation.[120]

IL-10 is also expressed by liver DCs in homeostatic condition and this has been shown to generate more suppressive CD4+ CD25+ FoxP3+ Tregs and IL-4-producing Th2 cells[121] as well as attenuate proinflammatory functions of monocytes.[30] Exposure of DCs themselves to IL-10 promotes this immature immunosuppressive phenotype and may also prevent their homing to secondary lymphoid tissue by upregulating CCR5 and downregulating CCR7.[122]

However, DCs are also highly sensitive to danger signals (e.g., HMGB1) which can switch them for this tolerogenic IL-10 producing state to an immunogenic phenotype which subsequently assumes inflammatory functions (such as TNF secretion).[61]

Tregs also release IL-10 and dysregulation of their function has been associated with different autoimmune liver diseases including primary sclerosing cholangitis (PSC) and primary biliary cholangitis (PBC).[123,124] Different subsets of Tregs have been studied in this context, but the CD4 + CD25 + subset is probably the one that is better characterized.[125]

CD25, used to characterize these cells, constitutes the α chain of the IL-2 receptor (IL-2Rα). Mice deficient for IL-2Rα exhibit PBC-like features and a case of a child with congenital IL-2R deficinecy presenting with PBC-like features has also been reported.[126,127] IL-2 therapy has been suggested to have advantageous effects on the Treg population in a variety of autoimmune conditions including HCV-induced vasculitis and graft-versus-host disease.[128,129]

CYTOKINE MECHANISMS IN SELECT LIVER PATHOLOGIES

Nonalcoholic Steatohepatitis and Nonalcoholic Fatty Liver Disease

Nonalcoholic fatty liver disease (NAFLD) and nonalcoholic steatohepatitis (NASH) are both becoming increasingly common in developed countries. Fibrosis distinguishes the two entities and before specific imaging techniques were widely available (e.g., elastography) serum levels of TGF-β have been suggested as a possible surrogate marker to make this distinction.[130]

Lipid accumulation in Kupffer cells primes them to exhibit a proinflammatory phenotype with the subsequent secretion of IL-1β, TNF, CCL2, and CCL5.[131] In this metabolically driven inflammation, TNF-α plays a significant role both in NAFLD and in NASH and a link between TNF-α signaling and insulin resistance—a key feature of NASH—has been suggested more than two decades ago.[132] Indeed, overexpression of TNF along with its receptors in both adipose and liver tissue have been reported in NASH patients and their levels seem to correlate with disease stage and level of fibrosis.[133,134] Certain polymorphs of the TNF gene are presented in higher prevalence among NAFLD patients and the accumullating evidence of the advantageous effect of pentoxifyline treatment—a potent inhibitor of TNF-α—in patients with NASH and NAFLD suggests that TNF-α signaling is actively involved in disease progression.[135–137]

IL-6 is also overexpressed in NAFLD, NASH, and generally insulin resistance and its levels correlate with disease severity and decrease upon Vitamin E treatment.[138–141] Hepatic steatosis in NAFLD patients has also been associated with IL-17, which also facilitates the transition from simple steatosis to steatohepatitis in this subset of patients.[142] CCL2 has also been shown to be elevated in patients with NASH and NAFLD and this occurs even before significant liver inflammation develops.[140,143,144] This systemic CCL2 increase could originate from release at the level of the liver tissue but also from visceral adipose tissue which reaches the liver through the portal circulation and plays a role in the progression from simple steatosis to steatohepatitis.[145] This increase in proinflammatory cytokines is paralleled with decreased levels of antiinflammatory cytokines (such as IL-10) observed in metabolic syndrome.[146]

Viral Hepatitis

The response to an infection with a hepatotropic virus involves a network of cells that orchestrates the acute reaction to the pathogen and at later stages mediates the progressive injury that occurs in chronic viral hepatitis. Viral infection of hepatocytes induces the release of type I interferons (IFN-α and IFN-β) activating NK and NKT cells—which constitute the initial line of defense against hepatotropic viruses.[147] Once activated, these cells become an important source of IFN-γ and TNF-α which—beside their classical role as cytokines—also directly interfere with viral replication. IFN-γ augments the antigen presentation capabilities of DCs, which are of paramount importance in defense against viral infections and a source of crucial cytokines in the antiviral immune response.[148,149] DCs interact with viral particles through different TLRs and upon activation, become a major source of IL-12, TNF-α, and IFN-α but also of IL-10.[150] Some of these cytokines (e.g., IL-12) in turn regulate the interaction between the NK cells and the DCs.[151,152] The balance between the different cytokines secreted relies on the specific TLR activated: TLR-3, a receptor for double stranded RNA which is of great importance in viral infection, is capable of inducing IL-12 release without the coproduction of the immunosuppressive IL-10—seen upon TLR-2 or TLR-4 activation.[120] IL-12 is a key cytokine in defense against viral hepatitis; it promotes noncytotoxic antiviral defense through the induction of IFN-γ secretion by T and NK cells but also promote cytotoxic Th1 and NK responses.[153] TLR-7 activation also contributes to the defense against viral infection and TLR-7 mediated induction of type 1 interferons results in viral clearance.[154]

NKT cells also participate in defense against viral hepatitis, probably responding to self-lipids that have been altered by the viral infection, to produce IL-4 and IFN-γ.[155] NKT cells can also play a protective role against hepatocyte damage in viral hepatitis through the release of IL-17 and IL-22.[34] IFN-γ from NKT cells also triggers the production of CXCL10 which attracts Tregs to the liver in a cytokine-to-chemokine pathway.[156]

Some cytokines mediating innate immune responses may play adverse effects in viral hepatitis, rendering hepatocytes more susceptible to viral infection. Recent evidence suggests that TNF-α from LPS-stimulated macrophages facilitates the infection of hepatocytes with HCV.[157] Interestingly, this is in line with previous reports linking TNF polymorphism and susceptibility to infections.[158] In chronic viral hepatitis, viral persistence is enabled by a general intrahepatic immunosuppressive milieu facilitated by IL-10, TGF-β, and also SOCS3.[159–161]

Autoimmune Liver Diseases

Imbalance between immunogenic and Treg subsets implicated the great majority of autoimmune pathologies affecting the liver, with enhanced Th1 and Th17 activity and diminished Treg activity.[162] In mouse models, signaling through the IL-12/Th1 axis has been shown essential for the development of autoimmune cholangitis and this has been later confirmed by genetic studies in PBC patients which also implicated the IL-23/Th17 axis.[163–165] TGF-β signaling induces the differentiation of naïve T-cells into Tregs or—in the presence of IL-6—into the proinflammatory Th17 phenotype.[6] BECs can actively promote the Th17 phenotype, with the production of IL-6, IL-1b, and IL-23 are themselves responsive to IL-17 signaling.[166] Moreover, BECs can present antigens to NKT cells which are also implicated in liver autoimmunity.[167,168] Genetic studies have implicated the IL-2/IL-21 locus in the pathogenesis of PSC. As mentioned above, dysfunctional IL-2 signaling can impair the Treg cell activity, while IL-21 is an important regulator in the development of Th17—inducing the expression of IL-23 receptor on naïve T-cells.[169] Several chemokines have also been implicated in the pathogenesis of liver autoimmune diseases. CXCL10 plays a role in T-cell homing to the portal tract in both PBC and PSC and has been shown to be elevated in patients with PBC.[170,171]

In autoimmune hepatitis, TNF-α stimulation enhances CCL20 expression by hepatocytes which contributes to the pathogenesis through promigratory effect on dysregulated splenic T-cells.[59]

The great therapeutic potential these studies offer is yet to be realized (Table 4.1).[172]

TABLE 4.1 Cytokine in the Liver

Cytokine	Role in liver	Reference
TNF-α	Induction of liver regeneration and hepatocytes proliferation through IL-6. It renders hepatocytes more responsive to key growth factors (HGF, TGF-α, and TGF-β) in liver injury. Antiapoptotic.	4,5
IFN-α	Antifibrotic effect, downregulation of collagen gene transcription.	110
IFN-β	Suppression of hepatocellular injury and downregulation of TGF-β, collagen, and TIMP-1.	111,112
IFN-γ	Induces HSC apoptosis, depletes NKT cells and decreases liver damage and fibrosis. It also induced neutrophil apoptosis and ameliorates inflammation.	33
NF-kB	Induction of fibrosis. Secretion of CCL2, CCL3, CXCL2, and CXCL5.	56
TGF-β	Key role in fibrosis and in myofibroblast and HSC activation.	71,76
CCL20/CCR6	Antifibrotic effect. Induction of HSCs apoptosis and recruitment of γδ T-cells. Production of IL-22.	59,60
IL-13	Profibrotic. HSC and macrophages activation.	63
IL-33, IL-17, IL-4, IL-5	Profibrotic. HSC activation.	66,67,68,69
IL-10, IL-22	Protect against inflammation and hamper activation of HSCs.	97,99
IL-6	Protective role in fibrosis. Stat3-mediated signaling cascade and activation of APR.	7
IL-22	Induction of Stat3 activation and liver progenitor cell proliferation.	46,47
CCL2/CCR2	Role in recruitment of monocytes/macrophages. Induction of inflammation, angiogenesis, and fibrosis.	17,18
IL-10	Role in tolerance. Downregulation of TNF, IL-6, and ROS.	31

ABBREVIATIONS

CCR CC chemokine receptor
CCL CC chemokine ligand
CX3CL chemokine (C-X3-C motif) ligand
CX3CR chemokine (C-X3-C motif) receptor
CXCL chemokine (C-X-C motif) ligand
CXCR chemokine (C-X-C motif) receptor
Gal-3 galectin 3
HGF hepatocyte growth factor

IFN	interferon
IL1β	interleukin 1 beta
IL12	interleukin 12
IL18	interleukin 18
IL-33	interleukin 33
LPS	lipopolysaccharide
MMP-9	matrix metallopeptidase 9
NF-κB	nuclear factor kappa-light-chain-enhancer of activated B cells
NKT	natural killer T cells
NLRP3	nucleotide-binding domain and leucine-rich repeat protein 3
PPARδ	peroxisome proliferator-activated receptor delta
Smad7	mothers against decapentaplegic homolog 7
STAT3	signal transducer and activator of transcription 3
TCR	T cell receptor
TGF	transforming growth factor
TNF	tumor necrosis factor
TNF-α	tumor necrosis factor alpha

References

1. Woolbright BL, Jaeschke H. Sterile inflammation in acute liver injury: myth or mystery? *Exp Rev Gastroenterol Hepatol* 2015;**9**(8):1027–9.
2. Kubes P, Mehal WZ. Sterile inflammation in the liver. *Gastroenterology* 2012; **143**(5):1158–72.
3. Leist M, et al. Activation of the 55 kDa TNF receptor is necessary and sufficient for TNF-induced liver failure, hepatocyte apoptosis, and nitrite release. *J Immunol* 1995;**154**(3):1307–16.
4. Luedde T, Schwabe RF. NF-κB in the liver—linking injury, fibrosis and hepatocellular carcinoma. *Nat Rev Gastroenterol Hepatol* 2011;**8**(2):108–18.
5. Streetz K, et al. Tumor necrosis factor α in the pathogenesis of human and murine fulminant hepatic failure. *Gastroenterology* 2000;**119**(2):446–60.
6. Mucida D, et al. Reciprocal TH17 and regulatory T cell differentiation mediated by retinoic acid. *Science* 2007;**317**(5835):256–60.
7. Moshage H. Cytokines and the hepatic acute phase response. *J Pathol* 1997; **181**(3):257–66.
8. StreetZ KL, et al. Lack of gp130 expression in hepatocytes promotes liver injury. *Gastroenterology* 2003;**125**(2):532–43.
9. Rincon M. Interleukin-6: from an inflammatory marker to a target for inflammatory diseases. *Trends Immunol* 2012;**33**(11):571–7.
10. Streetz KL, et al. Interleukin 6/gp130-dependent pathways are protective during chronic liver diseases. *Hepatology* 2003;**38**(1):218–29.
11. Yang X-P, et al. Interleukin-6 plays a crucial role in the hepatic expression of SOCS3 during acute inflammatory processes in vivo. *J Hepatol* 2005;**43**(4):704–10.
12. Szabo G, Petrasek J. Inflammasome activation and function in liver disease. *Nat Rev Gastroenterol Hepatol* 2015;**12**(7):387–400.
13. Imaeda AB, et al. Acetaminophen-induced hepatotoxicity in mice is dependent on Tlr9 and the Nalp3 inflammasome. *J Clin Invest* 2009;**119**(2):305–14.
14. Miura K, et al. Toll-like receptor 9 promotes steatohepatitis by induction of interleukin-1β in mice. *Gastroenterology* 2010;**139**(1):323–34 [e7].
15. Petrasek J, et al. Type I interferons protect from toll-like receptor 9-associated liver injury and regulate IL-1 receptor antagonist in mice. *Gastroenterology* 2011; **140**(2):697–708 [e4].

16. Petrasek J, et al. IL-1 receptor antagonist ameliorates inflammasome-dependent alcoholic steatohepatitis in mice. *J Clin Invest* 2012;**122**(10):3476–89.
17. Ehling J, et al. CCL2-dependent infiltrating macrophages promote angiogenesis in progressive liver fibrosis. *Gut* 2014;**63**(12):1960–71.
18. Baeck C, et al. Pharmacological inhibition of the chemokine CCL2 (MCP-1) diminishes liver macrophage infiltration and steatohepatitis in chronic hepatic injury. *Gut 61.3* 2012;416.
19. Heymann F, et al. Hepatic macrophage migration and differentiation critical for liver fibrosis is mediated by the chemokine receptor C–C motif chemokine receptor 8 in mice. *Hepatology* 2012;**55**(3):898–909.
20. Liaskou E, et al. Monocyte subsets in human liver disease show distinct phenotypic and functional characteristics. *Hepatology* 2013;**57**(1):385–98.
21. Karlmark KR, et al. The fractalkine receptor CX3CR1 protects against liver fibrosis by controlling differentiation and survival of infiltrating hepatic monocytes. *Hepatology* 2010;**52**(5):1769–82.
22. Aspinall AI, et al. CX3CR1 and vascular adhesion protein-1-dependent recruitment of CD16 + monocytes across human liver sinusoidal endothelium. *Hepatology* 2010; **51**(6):2030–9.
23. Jo J, et al. Toll-like receptor 8 agonist and bacteria trigger potent activation of innate immune cells in human liver. *PLoS Pathog* 2014;**10**(6) [e1004210].
24. Napier RJ, et al. The role of mucosal associated invariant T cells in antimicrobial immunity. *Front Immunol* 2015;**6**:344.
25. Tang X-Z, et al. IL-7 licenses activation of human liver intrasinusoidal mucosal-associated invariant T cells. *J Immunol* 2013;**190**(7):3142–52.
26. Spahr L, et al. Early liver biopsy, intraparenchymal cholestasis, and prognosis in patients with alcoholic steatohepatitis. *BMC Gastroenterol* 2011;**11**(1):115.
27. Moles A, et al. A TLR2/S100A9/CXCL-2 signaling network is necessary for neutrophil recruitment in acute and chronic liver injury in the mouse. *J Hepatol* 2014; **60**(4):782–91.
28. Marques PE, et al. Chemokines and mitochondrial products activate neutrophils to amplify organ injury during mouse acute liver failure. *Hepatology* 2012;**56**(5):1971–82.
29. McDonald B, et al. Intravascular danger signals guide neutrophils to sites of sterile inflammation. *Science* 2010;**330**(6002):362–6.
30. Bamboat ZM, et al. Conventional DCs reduce liver ischemia/reperfusion injury in mice via IL-10 secretion. *J Clin Invest* 2010;**120**(2):559–69.
31. Erhardt A, et al. IL-10, regulatory T cells, and Kupffer cells mediate tolerance in concanavalin A–induced liver injury in mice. *Hepatology* 2007;**45**(2):475–85.
32. Wehr A, et al. Chemokine receptor CXCR6-dependent hepatic NK T Cell accumulation promotes inflammation and liver fibrosis. *J Immunol* 2013;**190**(10):5226–36.
33. Wang H, et al. Invariant NKT cell activation induces neutrophil accumulation and hepatitis: opposite regulation by IL-4 and IFN-γ. *Hepatology* 2013;**58**(4):1474–85.
34. Wondimu Z, et al. Protective role of interleukin-17 in murine NKT cell-driven acute experimental hepatitis. *Am J Pathol* 2010;**177**(5):2334–46.
35. Böhm F, et al. Regulation of liver regeneration by growth factors and cytokines. *EMBO Mol Med* 2010;**2**(8):294–305.
36. Akerman P, et al. Antibodies to tumor necrosis factor-alpha inhibit liver regeneration after partial hepatectomy. *Am J Physiol—Gastrointest Liver Physiol* 1992;**263**(4):G579–85.
37. Webber EM, et al. Tumor necrosis factor primes hepatocytes for DNA replication in the rat. *Hepatology* 1998;**28**(5):1226–34.
38. Phaneuf D, Chen S-J, Wilson JM. Intravenous injection of an adenovirus encoding hepatocyte growth factor results in liver growth and has a protective effect against apoptosis. *Mol Med* 2000;**6**(2):96.

39. Fausto N. Liver regeneration and repair: hepatocytes, progenitor cells, and stem cells. *Hepatology* 2004;**39**(6):1477–87.
40. Elsegood CL, et al. Kupffer cell-monocyte communication is essential for initiating murine liver progenitor cell-mediated liver regeneration. *Hepatology* 2015; **62**(4):1272–84.
41. Anders RA, et al. Contribution of the lymphotoxin β receptor to liver regeneration. *J Immunol* 2005;**175**(2):1295–300.
42. Knight B, Yeoh GC. TNF/LTα double knockout mice display abnormal inflammatory and regenerative responses to acute and chronic liver injury. *Cell Tissue Res* 2005; **319**(1):61–70.
43. Tumanov AV, et al. T cell-derived lymphotoxin regulates liver regeneration. *Gastroenterology* 2009;**136**(2):694–704 [e4].
44. Chart RS, et al. Down-regulation of transforming growth factor beta receptor type I, II, and III during liver regeneration. *Am J Surg* 1995;**169**(1):126–32.
45. Houck KA, Michalopoulos GK. Altered responses of regenerating hepatocytes to norepinephrine and transforming growth factor type β. *J Cell Physiol* 1989;**141** (3):503–9.
46. Feng D, et al. Interleukin-22 promotes proliferation of liver stem/progenitor cells in mice and patients with chronic hepatitis B virus infection. *Gastroenterology* 2012;**143** (1):188–98 [e7].
47. Ren X, Hu B, Colletti LM. IL-22 is involved in liver regeneration after hepatectomy. *Am J Physiol—Gastrointest Liver Physiol* 2010;**298**(1):G74–80.
48. Brand S, et al. IL-22-mediated liver cell regeneration is abrogated by SOCS-1/3 overexpression in vitro. *Am J Physiol—Gastrointest Liver Physiol* 2007;**292**(4): G1019–28.
49. Naugler WE, et al. Gender disparity in liver cancer due to sex differences in MyD88-dependent IL-6 production. *Science* 2007;**317**(5834):121–4.
50. Yeh S-H, Chen P-J. Gender disparity of hepatocellular carcinoma: the roles of sex hormones. *Oncology* 2010;**78**(Suppl. 1):172–9.
51. Park EJ, et al. Dietary and genetic obesity promote liver inflammation and tumorigenesis by enhancing IL-6 and TNF expression. *Cell* 2010;**140**(2):197–208.
52. Jiang R, et al. Interleukin-22 promotes human hepatocellular carcinoma by activation of STAT3. *Hepatology* 2011;**54**(3):900–9.
53. Wei Q, et al. Deregulation of the NLRP3 inflammasome in hepatic parenchymal cells during liver cancer progression. *Lab Invest* 2014;**94**(1):52–62.
54. Haybaeck J, et al. A lymphotoxin-driven pathway to hepatocellular carcinoma. *Cancer Cell* 2009;**16**(4):295–308.
55. Pellicoro A, et al. Liver fibrosis and repair: immune regulation of wound healing in a solid organ. *Nat Rev Immunol* 2014;**14**(3):181–94.
56. De Minicis S, et al. Gene expression profiles during hepatic stellate cell activation in culture and in vivo. *Gastroenterology* 2007;**132**(5):1937–46.
57. Zimmermann HW, et al. Functional contribution of elevated circulating and hepatic non-classical CD14+ CD16+ monocytes to inflammation and human liver fibrosis. *PLoS One* 2010;**5**(6):e11049.
58. Hammerich L, et al. Chemokine receptor CCR6-dependent accumulation of γδ T cells in injured liver restricts hepatic inflammation and fibrosis. *Hepatology* 2014;**59** (2):630–42.
59. Iwamoto S, et al. TNF-α is essential in the induction of fatal autoimmune hepatitis in mice through upregulation of hepatic CCL20 expression. *Clin Immunol* 2013;**146** (1):15–25.
60. Affò S, et al. CCL20 mediates lipopolysaccharide induced liver injury and is a potential driver of inflammation and fibrosis in alcoholic hepatitis. *Gut 63.11* 2014;1782.

61. Heymann F, Tacke F. Immunology in the liver—from homeostasis to disease. *Nat Rev Gastroenterol Hepatol* 2016;**13**:88–110.
62. Liu Y, et al. IL-13 signaling in liver fibrogenesis. *Front Immunol* 2012;**3**:116.
63. Barron L, Wynn TA. Fibrosis is regulated by Th2 and Th17 responses and by dynamic interactions between fibroblasts and macrophages. *Am J Physiol Gastrointest Liver Physiol* 2011;**300**(5):G723–8.
64. Aoudjehane L, et al. Interleukin-4 induces the activation and collagen production of cultured human intrahepatic fibroblasts via the STAT-6 pathway. *Lab Invest* 2008;**88**(9):973–85.
65. Reiman RM, et al. Interleukin-5 (IL-5) augments the progression of liver fibrosis by regulating IL-13 activity. *Infect Immun* 2006;**74**(3):1471–9.
66. Marvie P, et al. Interleukin-33 overexpression is associated with liver fibrosis in mice and humans. *J Cell Mol Med* 2010;**14**(6b):1726–39.
67. Mchedlidze T, et al. Interleukin-33-dependent innate lymphoid cells mediate hepatic fibrosis. *Immunity* 2013;**39**(2):357–71.
68. Tan Z, et al. IL-17A plays a critical role in the pathogenesis of liver fibrosis through hepatic stellate cell activation. *J Immunol* 2013;**191**(4):1835–44.
69. Meng F, et al. Interleukin-17 signaling in inflammatory, Kupffer cells, and hepatic stellate cells exacerbates liver fibrosis in mice. *Gastroenterol* 2012;**143**(3):765–76 [e3].
70. Guillot A, et al. Cannabinoid receptor 2 counteracts interleukin-17-induced immune and fibrogenic responses in mouse liver. *Hepatology* 2014;**59**(1):296–306.
71. Mann DA, Marra F. Fibrogenic signalling in hepatic stellate cells. *J Hepatol* 2010;**52**(6):949–50.
72. Dixon JB, et al. Pro-fibrotic polymorphisms predictive of advanced liver fibrosis in the severely obese. *J Hepatol* 2003;**39**(6):967–71.
73. Milani S, et al. Transforming growth factors beta 1 and beta 2 are differentially expressed in fibrotic liver disease. *Am J Pathol* 1991;**139**(6):1221.
74. Castilla A, Prieto J, Fausto N. Transforming growth factors β1 and α in chronic liver disease effects of interferon alfa therapy. *N Engl J Med* 1991;**324**(14):933–40.
75. Gressner A, Weiskirchen R. Modern pathogenetic concepts of liver fibrosis suggest stellate cells and TGF-β as major players and therapeutic targets. *J Cell Mol Med* 2006;**10**(1):76–99.
76. Seki E, et al. TLR4 enhances TGF-β signaling and hepatic fibrosis. *Nat Med* 2007;**13**(11):1324–32.
77. Dooley S, et al. Modulation of transforming growth factor β response and signaling during transdifferentiation of rat hepatic stellate cells to myofibroblasts. *Hepatology* 2000;**31**(5):1094–106.
78. Nieto N, et al. Rat hepatic stellate cells contribute to the acute-phase response with increased expression of α1 (I) and α1 (IV) collagens, tissue inhibitor of metalloproteinase-1, and matrix-metalloproteinase-2 messenger RNAs. *Hepatology* 2001;**33**(3):597–607.
79. Inagaki Y, et al. Regulation of the α2 (I) collagen gene transcription in fat-storing cells derived from a cirrhotic liver. *Hepatology* 1995;**22**(2):573–9.
80. Herbst H, et al. Tissue inhibitor of metalloproteinase-1 and-2 RNA expression in rat and human liver fibrosis. *Am J Pathol* 1997;**150**(5):1647.
81. Knittel T, Fellmer P, Ramadori G. Gene expression and regulation of plasminogen activator inhibitor type I in hepatic stellate cells of rat liver. *Gastroenterology* 1996;**111**(3):745–54.
82. Proell V, et al. TGF-β dependent regulation of oxygen radicals during transdifferentiation of activated hepatic stellate cells to myofibroblastoid cells. *Comp Hepatol* 2007;**6**(1):1.

83. Sánchez A, et al. Apoptosis induced by transforming growth factor-in fetal hepatocyte primary cultures involvement of reactive oxygen intermediates. *J Biol Chem* 1996;**271**(13):7416–22.
84. Dooley S, Dijke P Ten. TGF-β in progression of liver disease. *Cell Tissue Res* 2012; **347**(1):245–56.
85. Henderson NC, et al. Galectin-3 regulates myofibroblast activation and hepatic fibrosis. *Proc Natl Acad Sci USA* 2006;**103**(13):5060–5.
86. Liu F-T, et al. Expression and function of galectin-3, a beta-galactoside-binding lectin, in human monocytes and macrophages. *Am J Pathol* 1995;**147**(4):1016.
87. Papaspyridonos M, et al. Galectin-3 is an amplifier of inflammation in atherosclerotic plaque progression through macrophage activation and monocyte chemoattraction. *Arterioscler Thromb Vasc Biol* 2008;**28**(3):433–40.
88. Dragomir A-CD, et al. Role of galectin-3 in classical and alternative macrophage activation in the liver following acetaminophen intoxication. *J Immunol* 2012;**189** (12):5934–41.
89. Nishi Y, et al. Role of galectin-3 in human pulmonary fibrosis. *Allergol Int* 2007; **56**(1):57–65.
90. Zhao C, et al. PPARγ agonists prevent TGFβ1/Smad3-signaling in human hepatic stellate cells. *Biochem Biophys Res Commun* 2006;**350**(2):385–91.
91. Iwaisako K, et al. Protection from liver fibrosis by a peroxisome proliferator-activated receptor δ agonist. *Proc Natl Acad Sci* 2012;**109**(21):E1369–76.
92. Liu X, Hu H, Yin JQ. Therapeutic strategies against TGF-β signaling pathway in hepatic fibrosis. *Liver Int* 2006;**26**(1):8–22.
93. Ohyama T, et al. Azelnidipine is a calcium blocker that attenuates liver fibrosis and may increase antioxidant defence. *Br J Pharmacol* 2012;**165**(4b):1173–87.
94. Park S-A, et al. EW-7197 inhibits hepatic, renal, and pulmonary fibrosis by blocking TGF-β/Smad and ROS signaling. *Cell Mol Life Sci* 2015;**72**(10):2023–39.
95. Connolly MK, et al. In liver fibrosis, dendritic cells govern hepatic inflammation in mice via TNF-α. *J Clin Invest* 2009;**119**(11):3213–25.
96. Gieling RG, Wallace K, Han Y-P. Interleukin-1 participates in the progression from liver injury to fibrosis. *Am J Physiol Gastrointest Liver Physiol* 2009;**296**(6):G1324.
97. Hung K-S, et al. Interleukin-10 gene therapy reverses thioacetamide-induced liver fibrosis in mice. *Biochem Biophys Res Commun* 2005;**336**(1):324–31.
98. Mathurin P, et al. IL-10 receptor and coreceptor expression in quiescent and activated hepatic stellate cells. *Am J Physiol Gastrointest Liver Physiol* 2002;**282**(6):G981–90.
99. Reitamo S, et al. Interleukin-10 modulates type I collagen and matrix metalloprotease gene expression in cultured human skin fibroblasts. *J Clin Invest* 1994;**94** (6):2489.
100. Louis H, et al. Production and role of interleukin-10 in concanavalin A-induced hepatitis in mice. *Hepatology* 1997;**25**(6):1382–9.
101. Louis H, et al. Interleukin-10 controls neutrophilic infiltration, hepatocyte proliferation, and liver fibrosis induced by carbon tetrachloride in mice. *Hepatology* 1998;**28** (6):1607–15.
102. Wan S, et al. Hepatic release of interleukin-10 during cardiopulmonary bypass in steroid-pretreated patients. *Am Heart J* 1997;**133**(3):335–9.
103. Del Prete G, et al. Human IL-10 is produced by both type 1 helper (Th1) and type 2 helper (Th2) T cell clones and inhibits their antigen-specific proliferation and cytokine production. *J Immunol* 1993;**150**(2):353–60.
104. Radaeva S, et al. Interleukin 22 (IL-22) plays a protective role in T cell-mediated murine hepatitis: IL-22 is a survival factor for hepatocytes via STAT3 activation. *Hepatology* 2004;**39**(5):1332–42.
105. Kong X, et al. Interleukin-22 induces hepatic stellate cell senescence and restricts liver fibrosis in mice. *Hepatology* 2012;**56**(3):1150–9.

106. Ki SH, et al. Interleukin-22 treatment ameliorates alcoholic liver injury in a murine model of chronic-binge ethanol feeding: role of signal transducer and activator of transcription 3. *Hepatology* 2010;**52**(4):1291–300.
107. Sertorio M, et al. IL-22 and IL-22 binding protein (IL-22BP) regulate fibrosis and cirrhosis in hepatitis C virus and schistosome infections. *Hepatology* 2015;**61**(4):1321–31.
108. Sabat R, Ouyang W, Wolk K. Therapeutic opportunities of the IL-22-IL-22R1 system. *Nat Rev Drug Discovery* 2014;**13**(1):21–38.
109. Zhao J, et al. Pathological functions of interleukin-22 in chronic liver inflammation and fibrosis with hepatitis B virus infection by promoting T helper 17 cell recruitment. *Hepatology* 2014;**59**(4):1331–42.
110. Inagaki Y, et al. Interferon alfa down-regulates collagen gene transcription and suppresses experimental hepatic fibrosis in mice. *Hepatology* 2003;**38**(4):890–9.
111. Tanabe J, et al. Interferon-β reduces the mouse liver fibrosis induced by repeated administration of concanavalin A via the direct and indirect effects. *Immunology* 2007;**122**(4):562–70.
112. Arthur M, Mann DA, Iredale JP. Tissue inhibitors of metalloproteinases, hepatic stellate cells and liver fibrosis. *J Gastroenterol Hepatol* 1998;**13**:S33–8.
113. Weng H, et al. IFN-γ abrogates profibrogenic TGF-β signaling in liver by targeting expression of inhibitory and receptor Smads. *J Hepatol* 2007;**46**(2):295–303.
114. Ulloa L, Doody J, Massagué J. Inhibition of transforming growth factor-β/SMAD signalling by the interferon-γ/STAT pathway. *Nature* 1999;**397**(6721):710–13.
115. Weng H-L, et al. Effect of interferon-gamma on hepatic fibrosis in chronic hepatitis B virus infection: a randomized controlled study. *Clin Gastroenterol Hepatol* 2005;**3**(8):819–28.
116. Schoenherr C, Weiskirchen R, Haan S. Interleukin-27 acts on hepatic stellate cells and induces signal transducer and activator of transcription 1-dependent responses. *Cell Commun Signal* 2010;**8**(1):1.
117. Crispe IN. The liver as a lymphoid organ. *Ann Rev Immunol* 2009;**27**:147–63.
118. Knolle PA, et al. Endotoxin down-regulates T cell activation by antigen-presenting liver sinusoidal endothelial cells. *J Immunol* 1999;**162**(3):1401–7.
119. Knoll P, et al. Human Kupffer cells secrete IL-10 in response to lipopolysaccharide (LPS) challenge. *J Hepatol* 1995;**22**(2):226–9.
120. Tu Z, et al. TLR-dependent cross talk between human Kupffer cells and NK cells. *J Exp Med* 2008;**205**(1):233–44.
121. Bamboat ZM, et al. Human liver dendritic cells promote T cell hyporesponsiveness. *J Immunol* 2009;**182**(4):1901–11.
122. Takayama T, et al. Mammalian and viral IL-10 enhance CC chemokine receptor 5 but down-regulate CC chemokine receptor 7 expression by myeloid dendritic cells: impact on chemotactic responses and in vivo homing ability. *J Immunol* 2001;**166**(12):7136–43.
123. Sebode M, et al. Reduced FOXP3+ regulatory T cells in patients with primary sclerosing cholangitis are associated with IL2RA gene polymorphisms. *J Hepatol* 2014;**60**(5):1010–16.
124. Lan RY, et al. Liver-targeted and peripheral blood alterations of regulatory T cells in primary biliary cirrhosis. *Hepatology* 2006;**43**(4):729–37.
125. Bernuzzi F, et al. Phenotypical and functional alterations of CD8 regulatory T cells in primary biliary cirrhosis. *J Autoimmunity* 2010;**35**(3):176–80.
126. Aoki CA, et al. IL-2 receptor alpha deficiency and features of primary biliary cirrhosis. *J Autoimmunity* 2006;**27**(1):50–3.
127. Wakabayashi K, et al. IL-2 receptor α−/− mice and the development of primary biliary cirrhosis. *Hepatology* 2006;**44**(5):1240–9.

128. Saadoun D, et al. Regulatory T-cell responses to low-dose interleukin-2 in HCV-induced vasculitis. *N Engl J Med* 2011;**365**(22):2067–77.
129. Matsuoka K-i, et al. Low-dose interleukin-2 therapy restores regulatory T cell homeostasis in patients with chronic graft-versus-host disease. *Sci Transl Med* 2013;**5**(179) [179ra43].
130. Hasegawa T, et al. Plasma transforming growth factor-β1 level and efficacy of α-tocopherol in patients with non-alcoholic steatohepatitis: a pilot study. *Alimentary Pharmacol Ther* 2001;**15**(10):1667–72.
131. Leroux A, et al. Toxic lipids stored by Kupffer cells correlates with their pro-inflammatory phenotype at an early stage of steatohepatitis. *J Hepatol* 2012; **57**(1):141–9.
132. Hotamisligil GS, et al. Increased adipose tissue expression of tumor necrosis factor-alpha in human obesity and insulin resistance. *J Clin Invest* 1995;**95**(5):2409.
133. Crespo J, et al. Gene expression of tumor necrosis factor α and TNF-receptors, p55 and p75, in nonalcoholic steatohepatitis patients. *Hepatology* 2001;**34**(6):1158–63.
134. Lesmana CRA, et al. Diagnostic value of a group of biochemical markers of liver fibrosis in patients with non-alcoholic steatohepatitis. *J Dig Dis* 2009;**10**(3):201–6.
135. Valenti L, et al. Tumor necrosis factor α promoter polymorphisms and insulin resistance in nonalcoholic fatty liver disease. *Gastroenterology* 2002;**122**(2):274–80.
136. Lee Y-M, et al. A randomized controlled pilot study of pentoxifylline in patients with non-alcoholic steatohepatitis (NASH). *Hepatol Int* 2008;**2**(2):196–201.
137. Li W, et al. Systematic review on the treatment of pentoxifylline in patients with non-alcoholic fatty liver disease. *Lipids Health Dis* 2011;**10**(1):1.
138. Kopp H-P, et al. Impact of weight loss on inflammatory proteins and their association with the insulin resistance syndrome in morbidly obese patients. *Arterioscler Thromb Vasc Biol* 2003;**23**(6):1042–7.
139. Kugelmas M, et al. Cytokines and NASH: a pilot study of the effects of lifestyle modification and vitamin E. *Hepatology* 2003;**38**(2):413–19.
140. Haukeland JW, et al. Systemic inflammation in nonalcoholic fatty liver disease is characterized by elevated levels of CCL2. *J Hepatol* 2006;**44**(6):1167–74.
141. Wieckowska A, et al. Increased hepatic and circulating interleukin-6 levels in human nonalcoholic steatohepatitis. *Am J Gastroenterol* 2008;**103**(6):1372–9.
142. Tang Y, et al. Interleukin-17 exacerbates hepatic steatosis and inflammation in non-alcoholic fatty liver disease. *Clin Exp Immunol* 2011;**166**(2):281–90.
143. Westerbacka J, et al. Genes involved in fatty acid partitioning and binding, lipolysis, monocyte/macrophage recruitment, and inflammation are overexpressed in the human fatty liver of insulin-resistant subjects. *Diabetes* 2007;**56**(11):2759–65.
144. Greco D, et al. Gene expression in human NAFLD. *Am J Physiol Gastrointest Liver Physiol* 2008;**294**(5):G1281–7.
145. Kirovski G, et al. Elevated systemic monocyte chemoattractrant protein-1 in hepatic steatosis without significant hepatic inflammation. *Exp Mol Pathol* 2011;**91**(3):780–3.
146. Esposito K, et al. Association of low interleukin-10 levels with the metabolic syndrome in obese women. *J Clin Endocrinol Metab* 2003;**88**(3):1055–8.
147. Biron CA, Brossay L. NK cells and NKT cells in innate defense against viral infections. *Curr Opin Immunol* 2001;**13**(4):458–64.
148. Akbar S, Inaba K, Onji M. Upregulation of MHC class II antigen on dendritic cells from hepatitis B virus transgenic mice by interferon-gamma: abrogation of immune response defect to a T-cell-dependent antigen. *Immunology* 1996;**87**(4):519.
149. Akbar S, et al. Low responsiveness of hepatitis B virus-transgenic mice in antibody response to T-cell-dependent antigen: defect in antigen-presenting activity of dendritic cells. *Immunology* 1993;**78**(3):468.

150. Banchereau J, et al. Immunobiology of dendritic cells. *Ann Rev Immunol* 2000; **18**(1):767–811.
151. Moretta A. Natural killer cells and dendritic cells: rendezvous in abused tissues. *Nat Rev Immunol* 2002;**2**(12):957–65.
152. Marcenaro E, et al. IL-12 or IL-4 prime human NK cells to mediate functionally divergent interactions with dendritic cells or tumors. *J Immunol* 2005;**174**(7):3992–8.
153. Cavanaugh VJ, Guidotti LG, Chisari FV. Interleukin-12 inhibits hepatitis B virus replication in transgenic mice. *J Virol* 1997;**71**(4):3236–43.
154. Lanford RE, et al. GS-9620, an oral agonist of Toll-like receptor-7, induces prolonged suppression of hepatitis B virus in chronically infected chimpanzees. *Gastroenterology* 2013;**144**(7):1508–17 [e10].
155. Zeissig S, et al. Hepatitis B virus-induced lipid alterations contribute to natural killer T cell-dependent protective immunity. *Nat Med* 2012;**18**(7):1060–8.
156. Santodomingo-Garzon T, et al. Natural killer T cells regulate the homing of chemokine CXC receptor 3-positive regulatory T cells to the liver in mice. *Hepatology* 2009;**49**(4):1267–76.
157. Fletcher NF, et al. Activated macrophages promote hepatitis C virus entry in a tumor necrosis factor-dependent manner. *Hepatology* 2014;**59**(4):1320–30.
158. Qidwai T, Khan F. Tumour necrosis factor gene polymorphism and disease prevalence. *Scand J Immunol* 2011;**74**(6):522–47.
159. Fletcher SP, et al. Transcriptomic analysis of the woodchuck model of chronic hepatitis B. *Hepatology* 2012;**56**(3):820–30.
160. Peppa D, et al. Blockade of immunosuppressive cytokines restores NK cell antiviral function in chronic hepatitis B virus infection. *PLoS Pathog* 2010;**6**(12):e1001227.
161. Sun C, et al. TGF-β1 down-regulation of NKG2D/DAP10 and 2B4/SAP expression on human NK cells contributes to HBV persistence. *PLoS Pathog* 2012;**8**(3):e1002594.
162. Rong G, et al. Imbalance between T helper type 17 and T regulatory cells in patients with primary biliary cirrhosis: the serum cytokine profile and peripheral cell population. *Clin Exp Immunol* 2009;**156**(2):217–25.
163. Yoshida K, et al. Deletion of interleukin-12p40 suppresses autoimmune cholangitis in dominant negative transforming growth factor β receptor type II mice. *Hepatology* 2009;**50**(5):1494–500.
164. Hirschfield GM, et al. Primary biliary cirrhosis associated with HLA, IL12A, and IL12RB2 variants. *N Engl J Med* 2009;**360**(24):2544–55.
165. Liu X, et al. Genome-wide meta-analyses identify three loci associated with primary biliary cirrhosis. *Nat Genet* 2010;**42**(8):658–60.
166. Harada K, et al. Periductal interleukin-17 production in association with biliary innate immunity contributes to the pathogenesis of cholangiopathy in primary biliary cirrhosis. *Clin Exp Immunol* 2009;**157**(2):261–70.
167. Mattner J, et al. Liver autoimmunity triggered by microbial activation of natural killer T cells. *Cell Host Microbe* 2008;**3**(5):304–15.
168. Schrumpf E, et al. The biliary epithelium presents antigens to and activates natural killer T cells. *Hepatology* 2015;**62**(4):1249–59.
169. Huber M, et al. IRF4 is essential for IL-21-mediated induction, amplification, and stabilization of the Th17 phenotype. *Proc Natl Acad Sci* 2008;**105**(52):20846–51.
170. Borchers AT, et al. *Lymphocyte recruitment and homing to the liver in primary biliary cirrhosis and primary sclerosing cholangitis. Seminars in immunopathology,* vol. 31. Springer-Verlag: Berlin; 2009, No. 3.
171. Chuang Y-H, et al. Increased levels of chemokine receptor CXCR3 and chemokines IP-10 and MIG in patients with primary biliary cirrhosis and their first degree relatives. *J Autoimmunity* 2005;**25**(2):126–32.
172. Mousa HS, et al. Advances in pharmacotherapy for primary biliary cirrhosis. *Exp Opin Pharmacother* 2015;**16**(5):633–43

PART III

CYTOKINE REGULATION IN DISEASE

CHAPTER

5

The Role of Adipokines and Adipogenesis in the Pathogenesis of Osteoarthritis

Poulami Datta[1], *Anirudh Sharma*[1,2], *Bandna Pal*[1], *and Kapoor Mohit*[1,2,3]

[1]Krembil Research Institute, Toronto, ON, Canada [2]Toronto Western Hospital, Toronto, ON, Canada [3]University of Toronto, Toronto, ON, Canada

OUTLINE

Cytokine Effector Functions in Tissues.
DOI: http://dx.doi.org/10.1016/B978-0-12-804214-4.00004-X

INTRODUCTION

Osteoarthritis (OA) is a common degenerative disease of the joints associated with the process of aging. It is mainly characterized by gradual destruction of the articular cartilage, changes in the subchondral bone, and osteophyte formation—leading to pain and severe disability.[1] The etiology of OA is not well understood but recent evidence suggests that obesity is a critical player in its pathogenesis.[2]

The World Health Organization (WHO) has defined obesity as "abnormal or excessive fat accumulation presenting a risk to one's health." It is measured using body mass index (BMI, weight in kg divided by height in m^2), with obesity being defined as having a BMI of greater than or equal to 30 kg/m^2. It has been estimated by WHO that the prevalence of obesity has more than doubled since 1980, with more than 10% of the world's population now being defined as obese.[3] The risk of developing OA is threefold higher in an obese or overweight population.[4] While this may explain the occurrence of OA in weight-bearing joints, the co-existence of obesity and OA in nonweight bearing joints cannot be explained by mechanical loads alone.

Today, adipose tissue is believed to play an important role in the regulation of different pathological processes, along with acting as a storehouse for energy. In the human body, two forms of adipose tissue have been observed: White adipose tissue (WAT) and brown adipose tissue (BAT). The WAT is more prevalent, associated with energy storage, and has been reported to be very active in secreting adipokines such as leptin, adiponectin, resistin, and visfatin; as well as certain classic proinflammatory cytokines like interleukin 6 (IL-6), IL-1β, and tumor necrosis factor α (TNFα).[5]

Researchers have recently become interested in the role of these adipokines in the pathophysiology of OA, as evidence suggests that they are directly involved in causing cartilage degradation and synovial inflammation.[6] However, the exact molecular mechanisms linking obesity and OA are not yet fully understood. In this chapter, we will explore the current literature regarding the association of these adipokines in the pathogenesis of OA in relation to obesity and adipogenesis.

LEPTIN

Leptin is a 16 kDa nonglycosylated adipokine, secreted mostly by the adipocytes present in WAT. It is mainly encoded by the obese (ob) gene and activates the Ob-Rb long form receptor isoform.[7] By acting on specific hypothalamic nuclei and inducing anorexigenic factors such as cocaine and amphetamine-related transcript, and by suppressing orexigenic neuropeptides such as neuropeptide Y, leptin has been found to decrease food intake and increase energy consumption.[8,9] Since its presence is mostly related to body fat while its synthesis is regulated by inflammatory mediators, higher levels of leptin have been observed in obese individuals.

LEPTIN AND OSTEOARTHRITIS

Leptin has been observed to play a critical role in the pathogenesis of OA. While its concentration in plasma is positively correlated with BMI in both healthy and OA patients, the circulating levels have been found to be higher in obese than nonobese individuals. It also increases the cartilage degradation process by inducing monocyte infiltration and vascular cell adhesion molecule 1 (VCAM-1) expression.[10] Many authors have reported that joint tissues like chondrocytes, osteophytes, synovium, as well as the infrapatellar fat pad are able to secrete leptin and express its receptor.[11,12] This expression has been shown to change during the process of OA. Conde et al. reported that the expression of leptin is higher in the infrapatellar fat pad and synovial tissues obtained from OA patients, and its production increases with disease progression.[13]

Studies report a dual role of leptin with both anabolic and catabolic activities. Griffin et al. reported[14] that although impaired leptin signaling was associated with extreme obesity and subchondral bone changes, it did not increase OA incidence.[14] Similarly, intraarticular injections of leptin increase the secretion and expression of insulin-like growth factors and transforming growth factors (TGFs), both of which exert anabolic activities in cartilage metabolism.[15] Leptin has also been reported to act with collagen, osteoblast proliferation, bone mineralization, and endochondral ossification.[16] An increase in the levels of alkaline phosphatase, osteocalcin, type I collagen, and TGF-β1 in the subchondral osteoblasts of OA is related to increased leptin secretion, moreover recent immunohistochemical studies demonstrate a higher leptin expression in osteophytes.[17,18]

On the contrary, leptin has also shown catabolic effects in OA by increasing the secretion of cysteine proteases as well as matrix metalloproteinases (MMPs) such as MMP-1, MMP-3, MMP-9, and MMP-13.[19] In combination with IL-1β, leptin upregulates MMP-1 and MMP-3 production through nuclear factor kappa beta (NFκβ), protein kinase C, and MAP kinase pathways in human OA cartilage, positively correlating with synovial fluid levels of MMP-1 and MMP-3 in OA patients.[20] Cartilage breakdown markers like urinary C-terminal telopeptide of type II collagen (uCTX-II), serum oligomeric matrix protein, and serum hyaluronic acid have been found to be significantly associated with serum leptin levels.[21] Studies in human and murine chondrocytes have also shown that leptin not only activates type-2 nitric oxide synthase (NOS), it also increases NOS activation (mediated by IL-1β) via the JAK2, PI3K, and mitogen activated kinases.[22]

In various models of inflammation, a leptin deficiency exhibited a protective effect by decreasing the production of proinflammatory T_H1 cytokines and shifting the immune response toward a T_H2 response.[23] Leptin levels in the serum and adipose tissue have been found to increase in response to proinflammatory stimuli, increase the production of T_H1 type cytokines, and supresses T_H2 type cytokines.[24] Thus, these findings suggest that the role of leptin is still not clearly understood,[25] elucidating that its mechanism requires further work.

ADIPONECTIN

Adipose tissue has been known to produce antiinflammatory adipokines such as adiponectin. Adiponectin, also known as AdipoQ, is a 244 protein residue with structural homology to type VIII collagen, type X collagen, and complement factor CIq.[10] It exists in three main isoforms: Trimer, hexamer, and high molecular weight.[26] It exerts its effects via two receptors, AdipoR1 and AdipoR2. AdipoR1 is mainly present in skeletal muscle while AdipoR2 is predominant in the liver.[27] Adiponectin levels tend to be low in obese patients and increase as weight is lost.[28] In OA patients adiponectin levels are reported to be lower than in healthy individuals. It has also been reported that adiponectin levels increase in the infrapatellar fat pad during OA, which may play an important role in its pathophysiology.[29] Matsuzawa reported that it plays an important role in several inflammatory diseases like type II diabetes, OA, rheumatoid arthritis (RA), and cardiovascular diseases.[30]

ADIPONECTIN AND OSTEOARTHRITIS

Like leptin, adiponectin has also been shown to induce both anabolic and catabolic effects in joint tissues. In chondrocytes and cartilage, adiponectin is able to induce several proinflammatory mediators like NO, IL-6, IL-8, PGE2, MMP-1, MMP-3, MMP-9, VGEF, and MCP-1.[12] By inducing IL-8 and VCAM-1 secretion in human primary chondrocytes, adiponectin contributes not only to leukocyte and monocyte infiltration, but also to the chemotactic gradient in inflamed joints.[31] Kang et al.[32] observed that after incubating the OA cartilage explants with adiponectin, levels of type II collagen degradation products increased in the supernatants of the cartilage explants.[32] Aggrecan degradation has also been observed in the synovium due to adiponectin secretion.[33] Adiponectin treated chondrocytes lead to induction of NOS2 and increased secretion of inflammatory mediators like IL-6, MMP-3, MMP-9, and MCP-1 through a signaling pathway involving PI3 kinase. Moreover, it also acts through the P38, AMPK, and NFκβ pathway for the secretion of MMP-3 in chondrocytes.[34]

Adiponectin has been observed to stimulate the release of antiinflammatory mediators such as IL-10 and IL-1 receptor antagonists, showing protection against OA. Although a positive correlation has been observed with adiponectin levels and serum cartilage breakdown markers like IL-1β, CTX- II, and synovial fluid high sensitivity C-reactive protein (CRP); a negative correlation has also been reported with serum high sensitivity CRP.[6,35] Adiponectin also inhibits IL-1β induced MMP-13 expression and upregulation of TIMP-2 production in chondrocytes.[36] Furthermore, Yusuf et al. recently reported that high adiponectin levels decrease the risk of progression of hand OA.[37] It has been found that adiponectin mainly suppresses the proinflammatory cytokine production, thus increasing the antiinflammatory cytokine production and depicting its protective role in inflammatory diseases. Globular adiponectin has been reported to suppress TLR induced NFκβ activation, indicating that it negatively regulates macrophage responses to TLR ligands showing an important innate immune response.[38] High and low molecular weight adipokines have been found to induce apoptosis by activation of AMP activated protein kinases and suppress scavenger receptor expression by macrophages.[39] Since it plays both an antiinflammatory and proinflammatory role in inflammatory diseases, the question of whether adiponectin should be considered a proinflammatory or antiinflammatory adipokine still remains unanswered. Fig. 5.1 provides a quick overview on how these adipokines can influence the pathology of OA.

ROLE OF ADIPOGENESIS IN PATHOGENESIS OF OSTEOARTHRITIS

Adipogenesis is mostly defined as the process by which preadipocytes get differentiated into adipocytes, whereas osteoblastogenesis is defined as the process of maturation of osteoblasts. Both adipocytes and osteoblasts share mesenchymal stem cells (MSCs) as a common precursor for their differentiation and are inversely related.[40] Various clinical studies depicted that bone volume loss is associated with osteoporosis, and age-related osteopenia is associated with an increase in bone marrow adipose tissue.[41–43]

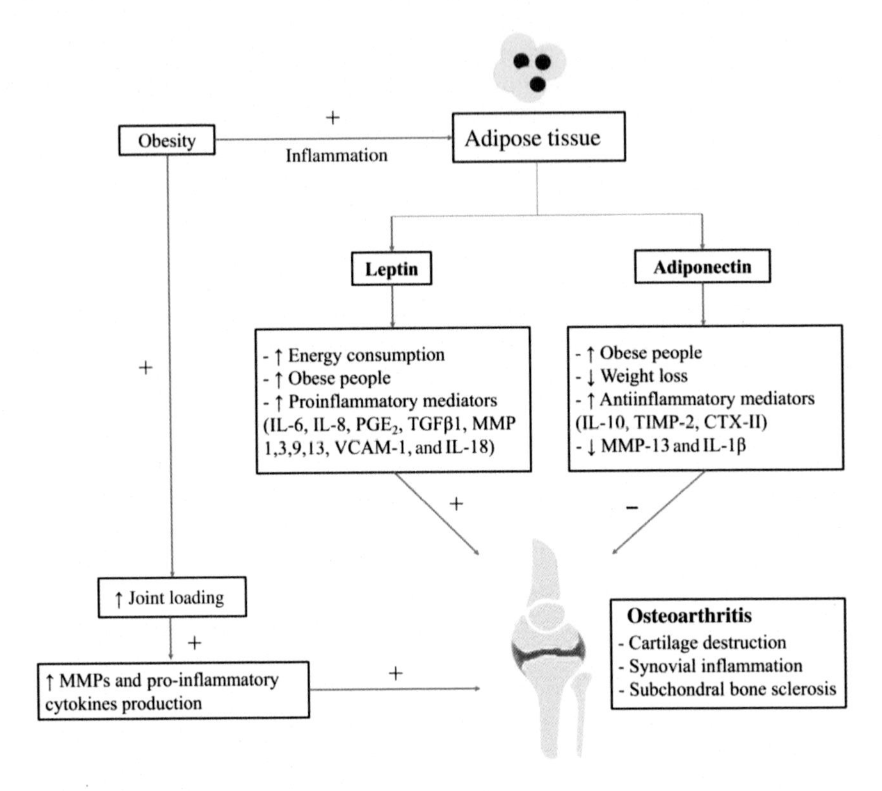

FIGURE 5.1 The following schematic provides an overview of the relationship of adipokines (adiponectin and leptin) in the pathogenesis of osteoarthritis (OA). Adipose tissue secrets adiponectin and leptin which in turn can cause imbalance between the catabolic and anabolic factors in the joint via activation of various cytokines thus leading to an increase in the pathogenesis of OA.

Several transcription factors have been observed to play important roles in osteoblast differentiation from MSCs, such as runt-related transcription factor 2 (RUNX2), osteoblastic specific transcription factor osterix (acts downstream to RUNX2), and peroxisome proliferator-activated receptor (PPAR). PPAR appears in three isoforms: PPARα, PPARδ, and PPARγ. PPARγ plays the most important role in adipogenesis during MSC differentiation to adipocytes and has also been found to limit inflammation, inhibit production of inflammatory cytokines, and inhibit macrophage activation. Its agonists also inhibit the production of MMP-13 in human chondrocytes and MMP-1 in human synovial fibroblasts, thus showing its antiinflammatory capacity.[44,45]

Today, increased resources in research are being used to study the notch/delta factors and the receptors associated with epidermal growth factor (EGF). In vitro studies have shown that overexpression of the notch inhibits osteogenesis while promoting adipogenesis in cell models, thereby downregulating important factors in adipogenesis such as Wnt receptor function and inhibiting β-catenin.[43,46,47] It has been observed that the Wnt peptides interact with the low density receptor-related proteins (LRPS), with mutations of these proteins causing loss of bone mass. Preadipocyte factor 1 has shown to be expressed in undifferentiated stromal cells and its involvement in the Wnt/LRP signaling pathway, causing inhibition in adipogenesis and leading to abnormalities in skeletal tissue.[48] Nuttall et al. reported that the dlk protein, a member of the EGF family, plays an important role in both adipogenesis and bone regeneration by inhibiting adipogenesis, hematopoiesis, and neuroendocrine cell differentiation. This is accomplished by interrupting the Notch 1 signaling pathway, leading to bone regeneration. Thus, several studies showed that the balance between adipogenesis and osteoblastogenesis is key since disturbances can cause obesity, diabetes, and joint/bone disorders.

CONCLUSION

Various research studies have shown that there is a pathophysiological relationship between obesity and OA. Adipokines are being postulated as likely mediators, and their dysregulation initiates the progression of cartilage destruction via a metabolic link. Further studies are warranted to fully elucidate the role of adipokines and their use as biomarkers of disease, which may open new dimensions in preventing obesity-induced OA.

References

1. Griffin TM, Fermor B, Huebner JL, et al. Diet-induced obesity differentially regulates behavioral, biomechanical, and molecular risk factors for osteoarthritis in mice. *Arthritis Res Ther* 2010;**12**:R130.
2. Conrozier T, Chappuis-Cellier C, Richard M, Mathieu P, Richard S, Vignon E. Increased serum C-reactive protein levels by immunonephelometry in patients with rapidly destructive hip osteoarthritis. *Rev Rhum* 1998;**65**:759–65.
3. Salih S, Sutton P. Obesity, knee osteoarthritis and knee arthroplasty: a review. *BMC Sports Sci Med Rehab* 2013;**5**:25.
4. Blagojevic M, Jinks C, Jeffery A, Jordan KP. Risk factors for onset of osteoarthritis of the knee in older adults: a systematic review and meta-analysis. *Osteoarthr Cartil/OARS Osteoarthr Res Soc* 2010;**18**:24–33.
5. Conde J, Scotece M, Gomez R, Lopez V, Gomez-Reino JJ, Gualillo O. Adipokines and osteoarthritis: novel molecules involved in the pathogenesis and progression of disease. *Arthritis* 2011;**2011**:203901.
6. de Boer TN, van Spil WE, Huisman AM, et al. Serum adipokines in osteoarthritis; comparison with controls and relationship with local parameters of synovial inflammation and cartilage damage. *Osteoarthr Cartil/OARS Osteoarthr Res Soc* 2012;**20**:846–53.

7. Zhang Y, Proenca R, Maffei M, Barone M, Leopold L, Friedman JM. Positional cloning of the mouse obese gene and its human homologue. *Nature* 1994;**372**:425–32.
8. Ahima RS, Prabakaran D, Mantzoros C, et al. Role of leptin in the neuroendocrine response to fasting. *Nature* 1996;**382**:250–2.
9. Ahima RS, Flier JS. Leptin. *Ann Rev Physiol* 2000;**62**:413–37.
10. Abella V, Scotece M, Conde J. Adipokines, metabolic syndrome and rheumatic diseases. *J Immunol Res* 2014;**2014**:343746.
11. Gegout PP, Francin PJ, Mainard D, Presle N. Adipokines in osteoarthritis: friends or foes of cartilage homeostasis? *Joint Bone Spine: Rev Rhum* 2008;**75**:669–71.
12. Presle N, Pottie P, Dumond H, et al. Differential distribution of adipokines between serum and synovial fluid in patients with osteoarthritis. Contribution of joint tissues to their articular production. *Osteoarthr Cartil/OARS Osteoarthr Res Soc* 2006;**14**:690–5.
13. Conde J, Scotece M, Lopez V, et al. Adiponectin and leptin induce VCAM-1 expression in human and murine chondrocytes. *PloS One* 2012;**7**:e52533.
14. Griffin TM, Huebner JL, Kraus VB, Guilak F. Extreme obesity due to impaired leptin signaling in mice does not cause knee osteoarthritis. *Arthritis Rheum* 2009;**60**:2935–44.
15. Conde J, Gomez R, Bianco G, et al. Expanding the adipokine network in cartilage: identification and regulation of novel factors in human and murine chondrocytes. *Ann Rheum Dis* 2011;**70**:551–9.
16. Kume K, Satomura K, Nishisho S, et al. Potential role of leptin in endochondral ossification. *J Histochem Cytochem: Off J Histochem Soc* 2002;**50**:159–69.
17. Gordeladze JO, Drevon CA, Syversen U, Reseland JE. Leptin stimulates human osteoblastic cell proliferation, de novo collagen synthesis, and mineralization: Impact on differentiation markers, apoptosis, and osteoclastic signaling. *J Cell Biochem* 2002;**85**: 825–36.
18. Mutabaruka MS, Aoulad Aissa M, Delalandre A, Lavigne M, Lajeunesse D. Local leptin production in osteoarthritis subchondral osteoblasts may be responsible for their abnormal phenotypic expression. *Arthritis Res Ther* 2010;**12**:R20.
19. Hu PF, Bao JP, Wu LD. The emerging role of adipokines in osteoarthritis: a narrative review. *Mol Biol Rep* 2011;**38**:873–8.
20. Koskinen A, Vuolteenaho K, Nieminen R, Moilanen T, Moilanen E. Leptin enhances MMP-1, MMP-3 and MMP-13 production in human osteoarthritic cartilage and correlates with MMP-1 and MMP-3 in synovial fluid from OA patients. *Clin Exp Rheum* 2011;**29**:57–64.
21. Berry PA, Jones SW, Cicuttini FM, Wluka AE, Maciewicz RA. Temporal relationship between serum adipokines, biomarkers of bone and cartilage turnover, and cartilage volume loss in a population with clinical knee osteoarthritis. *Arthritis Rheum* 2011;**63**: 700–7.
22. Otero M, Gomez Reino JJ, Gualillo O. Synergistic induction of nitric oxide synthase type II: in vitro effect of leptin and interferon-gamma in human chondrocytes and ATDC5 chondrogenic cells. *Arthritis Rheum* 2003;**48**:404–9.
23. Faggioni R, Jones-Carson J, Reed DA, et al. Leptin-deficient (ob/ob) mice are protected from T cell-mediated hepatotoxicity: role of tumor necrosis factor alpha and IL-18. *Proc Natl Acad Sci USA* 2000;**97**:2367–72.
24. Ouchi N, Parker JL, Lugus JJ, Walsh K. Adipokines in inflammation and metabolic disease. *Nat Rev Immunol* 2011;**11**:85–97.
25. Tilg H, Moschen AR. Adipocytokines: mediators linking adipose tissue, inflammation and immunity. *Nat Rev Immunol* 2006;**6**:772–83.
26. Antuna-Puente B, Feve B, Fellahi S, Bastard JP. Adipokines: the missing link between insulin resistance and obesity. *Diabetes Metab* 2008;**34**:2–11.
27. Kadowaki T, Yamauchi T. Adiponectin and adiponectin receptors. *Endocr Rev* 2005;**26**: 439–51.

28. Oh DK, Ciaraldi T, Henry RR. Adiponectin in health and disease. *Diabetes Obes Metab* 2007;**9**:282–9.
29. Distel E, Cadoudal T, Durant S, Poignard A, Chevalier X, Benelli C. The infrapatellar fat pad in knee osteoarthritis: an important source of interleukin-6 and its soluble receptor. *Arthritis Rheum* 2009;**60**:3374–7.
30. Matsuzawa Y. Adiponectin: identification, physiology and clinical relevance in metabolic and vascular disease. *Atheroscler Suppl* 2005;**6**:7–14.
31. Gomez R, Scotece M, Conde J, Gomez-Reino JJ, Lago F, Gualillo O. Adiponectin and leptin increase IL-8 production in human chondrocytes. *Ann Rheum Dis* 2011;**70**: 2052–4.
32. Kang EH, Lee YJ, Kim TK, et al. Adiponectin is a potential catabolic mediator in osteoarthritis cartilage. *Arthritis Res Ther* 2010;**12**:R231.
33. Hao D, Li M, Wu Z, Duan Y, Li D, Qiu G. Synovial fluid level of adiponectin correlated with levels of aggrecan degradation markers in osteoarthritis. *Rheumatol Int* 2011;**31**:1433–7.
34. Tong KM, Chen CP, Huang KC, et al. Adiponectin increases MMP-3 expression in human chondrocytes through AdipoR1 signaling pathway. *J Cell Biochem* 2011;**112**: 1431–40.
35. Ehling A, Schaffler A, Herfarth H, et al. The potential of adiponectin in driving arthritis. *J Immunol (Baltimore, MD: 1950)* 2006;**176**:4468–78.
36. Chen TH, Chen L, Hsieh MS, Chang CP, Chou DT, Tsai SH. Evidence for a protective role for adiponectin in osteoarthritis. *Biochim Biophys Acta* 2006;**1762**:711–18.
37. Yusuf E, Ioan-Facsinay A, Bijsterbosch J, et al. Association between leptin, adiponectin and resistin and long-term progression of hand osteoarthritis. *Ann Rheum Dis* 2011;**70**:1282–4.
38. Yamauchi T, Nio Y, Maki T, et al. Targeted disruption of AdipoR1 and AdipoR2 causes abrogation of adiponectin binding and metabolic actions. *Nat Med* 2007;**13**:332–9.
39. Neumeier M, Weigert J, Schaffler A, et al. Different effects of adiponectin isoforms in human monocytic cells. *J Leukoc Biol* 2006;**79**:803–8.
40. Dragojevic J, Logar DB, Komadina R, Marc J. Osteoblastogenesis and adipogenesis are higher in osteoarthritic than in osteoporotic bone tissue. *Arch Med Res* 2011;**42**:392–7.
41. Justesen J, Pedersen SB, Stenderup K, Kassem M. Subcutaneous adipocytes can differentiate into bone-forming cells in vitro and in vivo. *Tissue Eng* 2004;**10**:381–91.
42. Gimble JM, Robinson CE, Wu X, Kelly KA. The function of adipocytes in the bone marrow stroma: an update. *Bone* 1996;**19**:421–8.
43. Nuttall ME, Gimble JM. Controlling the balance between osteoblastogenesis and adipogenesis and the consequent therapeutic implications. *Curr Opin Pharmacol* 2004;**4**:290–4.
44. Fahmi H, Pelletier JP, Di Battista JA, Cheung HS, Fernandes JC, Martel-Pelletier J. Peroxisome proliferator-activated receptor gamma activators inhibit MMP-1 production in human synovial fibroblasts likely by reducing the binding of the activator protein 1. *Osteoarthr Cartil/OARS Osteoarthr Res Soc* 2002;**10**:100–8.
45. Li X, Afif H, Cheng S, et al. Expression and regulation of microsomal prostaglandin E synthase-1 in human osteoarthritic cartilage and chondrocytes. *J Rheumatol* 2005;**32**:887–95.
46. Day TF, Guo X, Garrett-Beal L, Yang Y. Wnt/beta-catenin signaling in mesenchymal progenitors controls osteoblast and chondrocyte differentiation during vertebrate skeletogenesis. *Dev Cell* 2005;**8**:739–50.
47. Kennell JA, MacDougald OA. Wnt signaling inhibits adipogenesis through beta-catenin-dependent and -independent mechanisms. *J Biol Chem* 2005;**280**:24004–10.
48. Smas CM, Sul HS. Pref-1, a protein containing EGF-like repeats, inhibits adipocyte differentiation. *Cell* 1993;**73**:725–34.

CHAPTER

6

Cytokines in Atherosclerosis

Joe W.E. Moss and Dipak P. Ramji

Cardiff School of Biosciences, Cardiff University, Cardiff, United Kingdom

OUTLINE

INTRODUCTION

Due to global increases in the prevalence of obesity and diabetes, in combination with the diets of developing countries becoming more Westernized, cardiovascular disease (CVD)-related events are set to remain the world's leading cause of death.[1] CVD-related events such as heart attacks and strokes are responsible for approximately one in three global deaths annually and an estimated 17.5 million people died in 2012 due to a CVD-related event.[1]

Atherosclerosis is an inflammatory disease of medium and large arteries and is the primary cause of CVD.[2] The disease is triggered when an inflammatory response is generated in endothelial cells (ECs) within the artery walls in response to a build-up of oxidized low-density lipoprotein (oxLDL) and other risk factors.[2] Cytokines and chemokines are then

Cytokine Effector Functions in Tissues.
DOI: http://dx.doi.org/10.1016/B978-0-12-804214-4.00005-1

secreted from the activated ECs, which also express cell adhesion molecules on their surface, in order to recruit circulating monocytes and other immune cells from the bloodstream to the site of oxLDL accumulation.[2–4] The recruited monocytes then migrate into the wall of the artery where they become exposed to macrophage colony-stimulating factor and differentiate into macrophages.[2–4] Macrophages express scavenger receptors (SRs) on their surface which allows them to start taking up oxLDL and form foam cells.[2–4] Unlike the uptake of low-density lipoprotein (LDL) by the LDL receptor (LDLr), which is highly regulated by a negative feedback loop, the uptake of oxLDL by SRs is unregulated and leads to its continuous uptake.[4] Due to the unmanageable build-up of oxLDL leading to apoptosis/necrosis of foam cells, the inadequate clearance of such apoptotic/necrotic cell debris and the continuous recruitment of macrophages to the affected area, a sustained inflammatory response is triggered that then contributes to the development of an atherosclerotic plaque.[2–4] As the plaque grows in size, vascular smooth muscle cells (VSMCs) undergo proliferation and migration following stimuli from ECs, macrophages, and T cells.[2] These VSMCs migrate to the top of the necrotic core and deposit extracellular matrix (ECM), which forms a fibrous cap and stabilizes the plaque.[2] However, some proinflammatory cytokines are able to stimulate the expression of matrix metalloproteinases (MMPs), which are a class of ECM degrading enzymes, leading to the plaque becoming unstable and rupturing, which in turn can trigger a thrombotic reaction and cause a heart attack or stroke.[2–4] During atherosclerosis development, cytokines can either contribute to disease progression or reduce lesion formation. However, the role of some cytokines is not as well defined and they can exert both pro- or anti-atherogenic effects depending on their surrounding microenvironment. The role of key cytokines during atherosclerosis development is summarized in Fig. 6.1 and some of the major ones are discussed below in detail.

PROATHEROGENIC CYTOKINES

Interferon (IFN)-γ is a proinflammatory cytokine which is often considered to be a key orchestrator of atherosclerosis development due to its ability to influence every stage of disease progression.[5] The importance of IFN-γ is further demonstrated by its ability to regulate the expression of 25% of the genes in macrophages.[5] While on a high fat diet, mouse models of atherosclerosis that are also deficient for the IFN-γ receptor develop smaller plaques compared to the control mice.[6] IFN-γ plays an important role in the recruitment of immune cells to the site

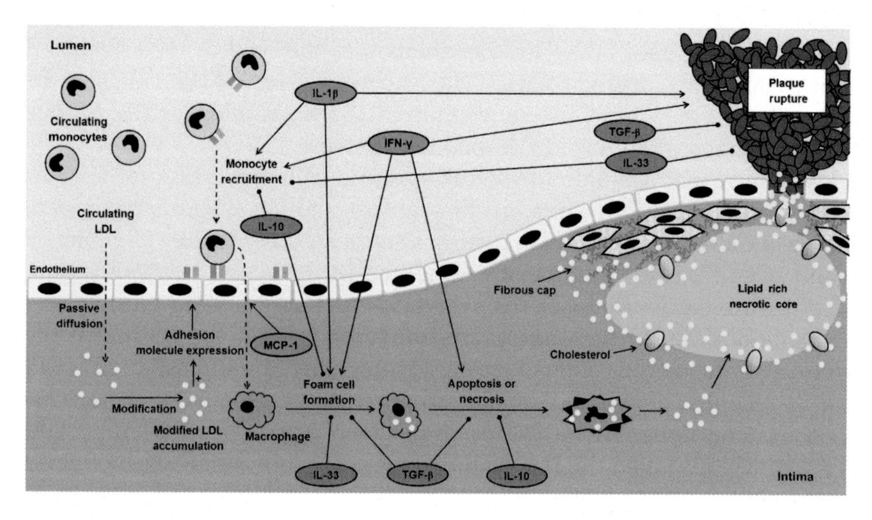

FIGURE 6.1 **The role of cytokines during atherosclerosis development.** Circulating monocytes are recruited from the bloodstream following the expression of adhesion molecules on the surface of activated endothelial cells. Once the monocytes differentiate into macrophages within the walls of the artery, they begin to express scavenger receptors on their surface which allows them to uncontrollably take up modified low-density lipoprotein (LDL). Foam cells eventually go through apoptosis or necrosis, causing them to "spill" their lipid contents into the wall of the artery that accumulates as a lipid-rich necrotic core. Over time a fibrous cap forms over the lipid-rich necrotic core due to the migration of smooth muscle cells from the media into the intima and the secretion of extracellular matrix. Constant foam cell formation and an inflammatory response leads to the growth of the atherosclerotic plaque, which can eventually become unstable and rupture. Several proinflammatory (*blue*) and antiinflammatory (*purple*) cytokines are able to influence several stages of plaque development. *IFN*, interferon; *IL*, interleukin; *LDL*, low-density lipoprotein; *MCP*, monocyte chemoattractant protein; and *TGF*, transforming growth factor.

of oxLDL accumulation and is therefore important for the growth of lesions. For example, IFN-γ is capable of inducing the expression of monocyte chemoattractant protein (MCP)-1, an important chemokine involved in the initial recruitment of monocytes during atherosclerosis development.[5,6] The expression of cell adhesion molecules, including intercellular adhesion molecule (ICAM)-1 and vascular cellular adhesion molecule (VCAM)-1, on the surface of ECs can also be induced by IFN-γ.[5] The formation of foam cells is also influenced by IFN-γ, as it is capable of attenuating the expression of key genes implicated in the efflux of cholesterol, such as ATP-binding cassette transporter A1 (ABCA1), and increasing the expression of those involved in the uptake and storage of this sterol, including acyl-CoA acyltransferase 1.[5,6] Altering the expression of these genes affects cholesterol homeostasis within macrophages to favor the uptake and intracellular accumulation of cholesterol, meaning they increase their uptake of oxLDL and form foam cells.[5,6]

Furthermore, IFN-γ is capable of inducing foam cell apoptosis, which causes them to "spill" their lipid contents into the wall of the artery and form a lipid-rich necrotic core.[5] The expression of MMPs can also be induced by IFN-γ and this contributes to the weakening of the fibrous cap and increases the risk of the plaque rupturing.[5,6]

Cytokines that are able to recruit immune cells including monocytes to a specific location, such as the site of oxLDL accumulation, are known as chemokines. MCP-1, also known as chemokine C–C motif ligand (CCL)2, is an important chemokine in the recruitment of monocytes during early atherosclerosis development as it able to guide circulating monocytes to the oxLDL accumulation within the intima of the artery. Atherosclerotic mouse models with reduced expression of MCP-1 or its receptor developed smaller atherosclerotic lesions.[7,8] Chemokine C–X3–C motif ligand (CX3CL)1 and CCL5 are two other important chemokines involved in the rolling and adhering stage of monocyte recruitment.[3,9] They work together with proteoglycans and P-selectins on the surface of activated ECs to aid the receptors on circulating monocytes to interact with their target ligands and slow their speed.[3,9] Once the monocytes have a firm interaction with ICAM-1 and VCAM-1 on the surface of the ECs and come to a stop, they are able to migrate into the wall of the artery. Deficiency of CCL2, CCL5, and CX3CL1 or their corresponding receptors in mouse models of atherosclerosis virtually stops the development of atherosclerosis by preventing the initial recruitment of monocytes.[3,10]

Other important proinflammatory cytokines during the progression of atherosclerosis belong to the interleukin (IL) family and include IL-1α and IL-1β. Smaller atherosclerotic lesions have been found in mouse models of this disease that also lacked either IL-1α or IL-1β compared to control mice.[3] Additionally, irradiated atherosclerotic mouse models which had either IL-1$\alpha^{-/-}$ or IL-1$\beta^{-/-}$ cells transplanted into them by a bone marrow transplantation approach developed smaller plaques by approximately 52% and 32% respectively when compared to mice which were transplanted with IL-$1^{+/+}$ cells.[11] Furthermore, the importance of IL-1α and IL-1β during atherosclerosis development has been observed in mice which lacked the IL-1 receptor 1.[3,12] Not only did these mice develop smaller atherosclerotic plaques but the expression of MCP-1, ICAM-1, and VCAM-1 was reduced in the vascular wall, highlighting the monocyte recruitment properties of IL-1 during atherosclerosis progression.[12] Other proatherogenic cytokines that have been comparatively less studied than IFN-γ or IL-1 are tumor necrosis factor (TNF)-α, IL-2, IL-18, and IFN-α. Extensive analysis of these proinflammatory cytokines and their role during atherosclerosis disease development is beyond the scope of this chapter and can be found in two recent reviews on this topic.[3,13]

ANTIATHEROGENIC CYTOKINES

Despite proinflammatory cytokines being the main driving force of atherosclerosis development, there are several antiinflammatory cytokines which are able to attenuate atherosclerosis progression and accelerate the resolution of inflammation. Therefore, if the levels of antiinflammatory cytokines could be safely raised in an individual, they may represent a promising therapeutic avenue to reduce atherosclerosis development. One of the main antiinflammatory cytokines capable of attenuating atherosclerosis development is transforming growth factor (TGF)-β.[3] Atherosclerotic mouse models which overexpressed TGF-β while on a high fat diet for 24 weeks developed smaller and more stable plaques when compared to the control mice.[14] Furthermore, the size of atherosclerotic plaques were found to have doubled following the abolishment of TGF-β signaling by expressing a functional mutation in the TGF-β receptor within mice.[15] The levels of proinflammatory markers, including IFN-γ, were also found to be significantly increased in the mice lacking a functional TGF-β receptor.[15] Foam cell formation is also inhibited by TGF-β due to its ability to increase the expression of genes implicated in cholesterol efflux in addition to decreasing the expression of those involved in lipoprotein uptake.[16] It should however be noted that a recent study found larger atherosclerotic lesions in mice which were lacking an inhibitor of TGF-β signaling (Smad7) in T cells.[17] However, the fibrous cap of the lesions had increased levels of collagen and were therefore more stable.[17]

Another antiinflammatory cytokine capable of attenuating foam cell formation is IL-33.[3] The expression of key genes implicated in atherosclerosis, such as apolipoprotein (Apo)E and ABCA1, was induced by IL-33 in macrophages.[18] Additionally the levels of SR genes such as CD36 were reduced in these cells, indicating that IL-33 is capable of attenuating foam cell formation.[18] Furthermore, these changes in the expression of cholesterol uptake and efflux genes correlated with reduced modified LDL uptake and increase cholesterol efflux in vitro, resulting in reduced intracellular cholesterol accumulation and foam cell formation.[18] Indeed the same study found that the number of foam cells found within atherosclerotic lesions were reduced in a mouse model receiving IL-33 injections when compared to the control mice.[18] Furthermore, larger plaques were found in atherosclerotic mice injected with soluble ST2 (a decoy receptor for IL-33),[19] indicating that the cell-surface ST2 receptor may be a potential therapeutic target capable of exerting antiatherogenic effects. Antiatherogenic effects have also been observed with other ILs, including IL-10 and IL-13.[3] Atherosclerotic mouse models which overexpressed IL-10 were found to have the expression of proatherogenic cytokines, such as MCP-1 and TNF-α, reduced compared to the control mice.[20] The same study also observed a decrease in the size of the atherosclerotic plaques, which correlated to an increase in cholesterol efflux and a decrease in macrophage apoptosis.[20] This study highlights the antiinflammatory properties of IL-10 and its potential to attenuate atherosclerosis disease progression. Mice deficient for IL-13 developed bigger lesions due to an increase in the size of the necrotic core without any changes in plasma cholesterol levels.[21] The atherosclerotic plaques which developed in mice receiving an IL-13 injection once every 2 weeks while receiving a high fat diet for 16 weeks were more stable due to an increase in the collagen levels within the fibrous cap and a decrease in the number of macrophages in the plaques.[21]

CYTOKINES WHOSE ROLE IN ATHEROSCLEROSIS IS LESS CLEAR

Unlike the previously mentioned cytokines that can be clearly defined as both pro- or anti-atherogenic due to their effects on disease progression, at least in vivo, there are some cytokines whose actions are not always clear-cut and their effects are often dependent on the microenvironment.[3,13] For example, two cytokines that have been found to exert both pro- and antiinflammatory effects are IL-4 and IL-18. Atherosclerotic mice that were lacking IL-4 were found to develop smaller plaques compared to the control mice indicating a proatherogenic role.[22] However, a later study by the same group found that IL-4 deficiency in atherosclerotic mice had no significant effect on the progression of the disease.[23] Another study has potentially shown a proatherogenic effect of IL-4 in later stage plaque maturation.[24] IL-4 deficiency in atherosclerotic mice resulted in a reduction in the size of the plaque by 27% compared to the control mice.[24] Further studies are required to fully understand the role of IL-4 during atherosclerosis disease development. Atherosclerotic mice lacking IL-18 on a high fat diet for 12 weeks developed more plaques compared to the control mice, furthermore, these plaques were larger and less stable, emphasizing the antiatherogenic properties of IL-18.[25] However, another study found that daily injections of recombinant IL-18 in atherosclerotic mice increased the expression of MMP-9 and the SR CD36, potentially via nuclear factor kappa-light-chain-enhancer of activated B cells, which resulted in larger and more unstable plaques.[26]

CONCLUSIONS

Several of the key steps in atherosclerosis disease progression, including monocyte recruitment, foam cell formation, and plaque stability are influenced by cytokines. Furthermore, the majority of current antiatherogenic therapies target cholesterol homeostasis in order to lower an individual's cholesterol level rather than preventing the proinflammatory effects regulated by cytokines. Therefore, due to their significant role in atherosclerosis, cytokines and their receptors are possible therapeutic targets to enhance inflammation resolution and disease regression. Indeed novel therapeutics are currently being developed which can block cytokine signaling by either using decoy receptors[27] or antibodies.[28] The canakinumab antiinflammatory thrombosis outcomes study (CANTOS)[29] and the cardiovascular inflammation reduction trial (CIRT)[30] are two large, robust clinical trials which will hopefully provide the first evidence of attenuating the inflammatory response in human patients in order to reduce their risk of suffering a CVD-related event. The CANTOS trial involves using the neutralizing antibody canakinumab to target IL-1β,[29] whereas the CIRT trial involves treating patients with methotrexate as a broad range antiinflammatory therapy.[30] However, one disadvantage of current cytokine therapies is that they exert their effects systemically. Due to the role of cytokines in the immune response, attenuating their effects systemically may leave individuals immunocompromised. Therefore, cytokine-targeted therapies should have their usage limited to those who are at the greatest risk of a CVD-related event until the development of drug delivery systems, such as nanoparticles, which are capable of directly targeting atherosclerotic plaques.[31] Additionally the acceleration of inflammation resolution by using nutraceuticals may be another potential therapeutic avenue that needs to be thoroughly explored. Many nutraceuticals including omega-3 polyunsaturated fatty acids,[32] polyphenols,[33] phytosterols,[34] and flavanols[35] have been shown to possess antiatherogenic properties. Additionally, several antiinflammatory effects of a unique combination of these nutraceuticals have been observed in vitro.[36] As nutraceuticals do not exert the potentially harmful effects of systemic inhibition associated with cytokine-targeted therapies, they are a promising therapeutic strategy for atherosclerosis treatment possibly in conjunction with current lipid lowering pharmaceutical drugs[37]. Further studies are required to fully understand the roles of cytokines and how to target them therapeutically in atherosclerosis disease development. Once a more comprehensive understanding of the role of cytokines has been established, it will lead to the discovery of novel antiatherogenic therapeutics.

References

1. WHO. World Health Organisation fact sheet 317 2015. <http://www.who.int/mediacentre/factsheets/fs317/en/>.
2. McLaren JE, Michael DR, Ashlin TG, Ramji DP. Cytokines, macrophage lipid metabolism and foam cells: implications for cardiovascular disease therapy. *Prog Lipid Res* 2011;**50**:331–47. Available from: http://dx.doi.org/10.1016/j.plipres.2011.04.002.

3. Ramji DP, Davies TS. Cytokines in atherosclerosis: key players in all stages of disease and promising therapeutic targets. *Cytokine Growth Factor Rev* 2015;**26**:673–85. Available from: http://dx.doi.org/10.1016/j.cytogfr.2015.04.003.
4. Buckley ML, Ramji DP. The influence of dysfunctional signaling and lipid homeostasis in mediating the inflammatory responses during atherosclerosis. *Biochim Biophys Acta—Mol Basis Dis* 2015;**1852**:1498–510. Available from: http://dx.doi.org/10.1016/j.bbadis.2015.04.011.
5. Moss JWE, Ramji DP. Interferon-γ: promising therapeutic target in atherosclerosis. *World J Exp Med* 2015;**5**:154–9. Available from: http://dx.doi.org/10.5493/wjem.v5.i3.154.
6. McLaren JE, Ramji DP. Interferon gamma: a master regulator of atherosclerosis. *Cytokine Growth Factor Rev* 2009;**20**:125–35. Available from: http://dx.doi.org/10.1016/j.cytogfr.2008.11.003.
7. Öhman MK, Wright AP, Wickenheiser KJ, Luo W, Russo HM, Eitzman DT. Monocyte chemoattractant protein-1 deficiency protects against visceral fat-induced atherosclerosis. *Arterioscler Thromb Vasc Biol* 2010;**30**:1151–8. Available from: http://dx.doi.org/10.1161/ATVBAHA.110.205914.
8. Liu XL, Zhang PF, Ding SF, Wang Y, Zhang M, Zhao YX, et al. Local gene silencing of monocyte chemoattractant protein-1 prevents vulnerable plaque disruption in apolipoprotein E-knockout mice. *PLoS One* 2012;**7**:e33497. Available from: http://dx.doi.org/10.1371/journal.pone.0033497.
9. Moore KJ, Sheedy FJ, Fisher EA. Macrophages in atherosclerosis: a dynamic balance. *Nat Rev Immunol* 2013;**13**:709–21. Available from: http://dx.doi.org/10.1038/nri3520.
10. Combadière C, Potteaux S, Rodero M, Simon T, Pezard A, Esposito B, et al. Combined inhibition of CCL2, CX3CR1, and CCR5 abrogates Ly6C(hi) and Ly6C(lo) monocytosis and almost abolishes atherosclerosis in hypercholesterolemic mice. *Circulation* 2008;**117**:1649–57. Available from: http://dx.doi.org/10.1161/CIRCULATIONAHA.107.745091.
11. Kamari Y, Shaish A, Shemesh S, Vax E, Grosskopf I, Dotan S, et al. Reduced atherosclerosis and inflammatory cytokines in apolipoprotein-E-deficient mice lacking bone marrow-derived interleukin-1α. *Biochem Biophys Res Commun* 2011;**405**:197–203. Available from: http://dx.doi.org/10.1016/j.bbrc.2011.01.008.
12. Shemesh S, Kamari Y, Shaish A, Olteanu S, Kandel-Kfir M, Almog T, et al. Interleukin-1 receptor type-1 in non-hematopoietic cells is the target for the pro-atherogenic effects of interleukin-1 in apoE-deficient mice. *Atherosclerosis* 2012;**222**:329–36. Available from: http://dx.doi.org/10.1016/j.atherosclerosis.2011.12.010.
13. Ait-Oufella H, Taleb S, Mallat Z, Tedgui A. Recent advances on the role of cytokines in atherosclerosis. *Arterioscler Thromb Vasc Biol* 2011;**31**:969–79. Available from: http://dx.doi.org/10.1161/ATVBAHA.110.207415.
14. Reifenberg K, Cheng F, Orning C, Crain J, Küpper I, Wiese E, et al. Overexpression of TGF-ß1 in macrophages reduces and stabilizes atherosclerotic plaques in ApoE-deficient mice. *PLoS One* 2012;**7**:e40990. Available from: http://dx.doi.org/10.1371/journal.pone.0040990.
15. Lievens D, Habets KL, Robertson A-K, Laouar Y, Winkels H, Rademakers T, et al. Abrogated transforming growth factor beta receptor II (TGFβRII) signalling in dendritic cells promotes immune reactivity of T cells resulting in enhanced atherosclerosis. *Eur Heart J* 2013;**34**:3717–27. Available from: http://dx.doi.org/10.1093/eurheartj/ehs106.
16. Michael DR, Salter RC, Ramji DP. TGF-β inhibits the uptake of modified low density lipoprotein by human macrophages through a Smad-dependent pathway: a dominant role for Smad-2. *Biochim Biophys Acta* 2012;**1822**:1608–16. Available from: http://dx.doi.org/10.1016/j.bbadis.2012.06.002.
17. Gistera A, Robertson AK, Andersson J, Ketelhuth DFJ, Ovchinnikova O, Nilsson SK, et al. Transforming growth factor-beta signaling in t cells promotes stabilization of

atherosclerotic plaques through an interleukin-17 dependent pathway. *Sci Transl Med* 2013;**5**:196a–100. Available from: http://dx.doi.org/10.1126/scitranslmed.3006133.
18. McLaren JE, Michael DR, Salter RC, Ashlin TG, Calder CJ, Miller AM, et al. IL-33 reduces macrophage foam cell formation. *J Immunol* 2010;**185**:1222–9.
19. Miller AM, Xu D, Asquith DL, Denby L, Li Y, Sattar N, et al. IL-33 reduces the development of atherosclerosis. *J Exp Med* 2008;**205**:339–46. Available from: http://dx.doi.org/10.1084/jem.20071868.
20. Han X, Kitamoto S, Wang H, Boisvert WA. Interleukin-10 overexpression in macrophages suppresses atherosclerosis in hyperlipidemic mice. *FASEB J* 2010;**24**:2869–80. Available from: http://dx.doi.org/10.1096/fj.09-148155.
21. Cardilo-Reis L, Gruber S, Schreier SM, Drechsler M, Papac-Milicevic N, Weber C, et al. Interleukin-13 protects from atherosclerosis and modulates plaque composition by skewing the macrophage phenotype. *EMBO Mol Med* 2012;**4**:1072–86. Available from: http://dx.doi.org/10.1002/emmm.201201374.
22. King VL. Interleukin-4 deficiency decreases atherosclerotic lesion formation in a site-specific manner in female LDL receptor-/- mice. *Arterioscler Thromb Vasc Biol* 2002;**22**:456–61. Available from: http://dx.doi.org/10.1161/hq0302.104905.
23. King VL, Cassis LA, Daugherty A. Interleukin-4 does not influence development of hypercholesterolemia or angiotensin II-induced atherosclerotic lesions in mice. *Am J Pathol* 2007;**171**:2040–7. Available from: http://dx.doi.org/10.2353/ajpath.2007.060857.
24. Davenport P, Tipping PG. The role of interleukin-4 and interleukin-12 in the progression of atherosclerosis in apolipoprotein E-deficient mice. *Am J Pathol* 2003;**163**: 1117–25. Available from: http://dx.doi.org/10.1016/S0002-9440(10)63471-2.
25. Pejnovic N, Vratimos A, Lee SH, Popadic D, Takeda K, Akira S, et al. Increased atherosclerotic lesions and Th17 in interleukin-18 deficient apolipoprotein E-knockout mice fed high-fat diet. *Mol Immunol* 2009;**47**:37–45. Available from: http://dx.doi.org/10.1016/j.molimm.2008.12.032.
26. Bhat OM, Kumar PU, Giridharan NV, Kaul D, Kumar MJM, Dhawan V. Interleukin-18-induced atherosclerosis involves CD36 and NF-κB crosstalk in Apo E-/- mice. *J Cardiol* 2015;**66**:28–35. Available from: http://dx.doi.org/10.1016/j.jjcc.2014.10.012.
27. Koga M, Kai H, Yasukawa H, Kato S, Yamamoto T, Kawai Y, et al. Postnatal blocking of interferon-γ function prevented atherosclerotic plaque formation in apolipoprotein E-knockout mice. *Hypertens Res* 2007;**30**:259–67. Available from: http://dx.doi.org/10.1291/hypres.30.259.
28. Weber C, Noels H. Atherosclerosis: current pathogenesis and therapeutic options. *Nat Med* 2011;**17**:1410–22. Available from: http://dx.doi.org/10.1038/nm.2538.
29. Ridker PM. From C-reactive protein to interleukin-6 to interleukin-1: moving upstream to identify novel targets for atheroprotection. *Circ Res* 2016;**118**:145–56. Available from: http://dx.doi.org/10.1161/circresaha.115.306656.
30. Everett BM, Pradhan AD, Solomon DH, Paynter N, Macfadyen J, Zaharris E, et al. Rationale and design of the cardiovascular inflammation reduction trial: a test of the inflammatory hypothesis of atherothrombosis. *Am Heart J* 2013;**166** 199–207. e15. Available from: http://dx.doi.org/10.1016/j.ahj.2013.03.018.
31. Fredman G, Kamaly N, Spolitu S, Milton J, Ghorpade D, Chiasson R, et al. Targeted nanoparticles containing the proresolving peptide Ac2-26 protect against advanced atherosclerosis in hypercholesterolemic mice. *Sci Transl Med* 2015;**7**:275ra20. Available from: http://dx.doi.org/10.1126/scitranslmed.aaa1065.
32. Brown AL, Zhu X, Rong S, Shewale S, Seo J, Boudyguina E, et al. Omega-3 fatty acids ameliorate atherosclerosis by favorably altering monocyte subsets and limiting monocyte recruitment to aortic lesions. *Arterioscler Thromb Vasc Biol* 2012;**32**:2122–30. Available from: http://dx.doi.org/10.1161/ATVBAHA.112.253435.

33. Rosignoli P, Fuccelli R, Fabiani R, Servili M, Morozzi G. Effect of olive oil phenols on the production of inflammatory mediators in freshly isolated human monocytes. *J Nutr Biochem* 2013;**24**:1513–19. Available from: http://dx.doi.org/10.1016/j.jnutbio.2012.12.011.
34. Sabeva NS, McPhaul CM, Li X, Cory TJ, Feola DJ, Graf GA. Phytosterols differentially influence ABC transporter expression, cholesterol efflux and inflammatory cytokine secretion in macrophage foam cells. *J Nutr Biochem* 2011;**22**:777–83. Available from: http://dx.doi.org/10.1016/j.jnutbio.2010.07.002.
35. Sansone R, Rodriguez-Mateos A, Heuel J, Falk D, Schuler D, Wagstaff R, et al. Cocoa flavanol intake improves endothelial function and Framingham risk score in healthy men and women: a randomised, controlled, double-masked trial: the Flaviola Health Study. *Br J Nutr* 2015;**114**:1246–55. Available from: http://dx.doi.org/10.1017/S0007114515002822.
36. Moss JWE, Davies TS, Garaiova I, Plummer SF, Michael DR, Ramji DP. A unique combination of nutritionally active ingredients can prevent several key processes associated with atherosclerosis in vitro. *PLoS One* 2016;**11**:e0151057. Available from: http://dx.doi.org/10.1371/journal.pone.0151057.
37. Moss JWE, Ramji DP. Nutraceutical therapies for atherosclerosis. *Nat Rev Cardiol* 2016;**13**:513–32.

CHAPTER

7

Cytokines in Diabetes and Diabetic Complications

Antonio Rivero-González, Edduin Martín-Izquierdo, Carlos Marín-Delgado, Anabel Rodríguez-Muñoz, and Juan F. Navarro-González

University Hospital of the Nuestra Señora de Candelaria, Santa Cruz de Tenerife, Spain

OUTLINE

INTRODUCTION

The incidence of diabetes mellitus and obesity is increasing and approaching to epidemic proportions. Nowadays, diabetes affects more than 380 million people and obesity around 500 million people

Cytokine Effector Functions in Tissues.
DOI: http://dx.doi.org/10.1016/B978-0-12-804214-4.00006-3

worldwide.[1] From a pathogenic perspective, inflammation is now recognized as a significant factor in the development of both type 1 (T1D) and type 2 diabetes (T2D),[2,3] resulting in a complex network of inflammatory–immune system interrelationships. The accumulation of activated innate immune cells in metabolic tissues results in release of inflammatory mediators, including inflammatory cytokines, which promote systemic insulin resistance (IR) and β-cell damage.[4,5] In addition to its role in the pathogenesis of diabetes mellitus, inflammation is also an underlying mechanism in the development of diabetic complications, as well as in the increased risk of cardiovascular disease in diabetic population (Fig. 7.1).[6–8]

Although it is necessary to recognize the specific characteristics of the two main forms of diabetes regarding different aspects, such as clinical presentation, rate of β-cell destruction, profiles of autoantibodies, histology of islets, etc., diverse epidemiological, basic and clinical studies show that differences regarding the pathogenesis of T1D and T2D are blurred.[9] Thus, a recent systematic review of nine clinical studies supports the association of obesity and subsequent T1D development,[10] whereas accumulating evidence supports the involvement of innate

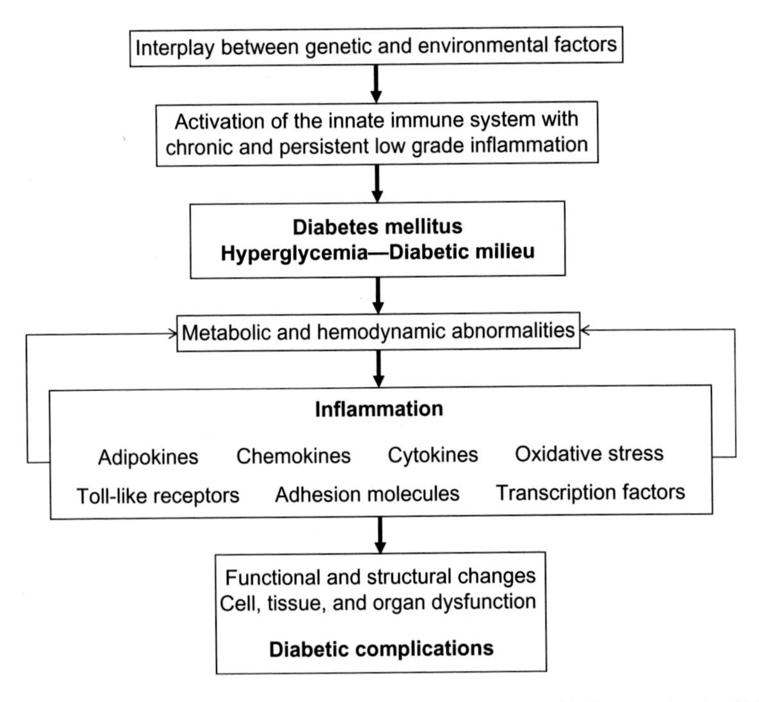

FIGURE 7.1 Schematic representation of the participation of inflammation in diabetes and diabetic complications.

TABLE 7.1 Potential Effects of Inflammatory Cytokines Related to Development and Progression of Diabetes Mellitus and Diabetic Complications

Direct cytotoxicity
Induction of apoptotis and necrosis
Amplification of inflammatory response
Stimulation of inflammatory cell recruitment
Increase expression and production of chemoattractant molecules, adhesion molecules, chemokines, and growth factors
Alteration of nitric oxide synthesis
Stimulation of oxidative stress response
Induction of hemodynamic abnormalities
Alteration of endothelial cell permeability

immunity in T2D.[11] In this context, inflammation represents a common link participating in the pathogenesis of T1D and T2D,[12] and antiinflammatory therapies have shown efficacy in both types of diabetes, supporting that inflammation may stand as a common background in the pathogenesis of diabetes.[13,14]

In this chapter, we review the knowledge on the inflammatory and immune mechanisms involved in diabetes mellitus and diabetic complications, focused on the role of cytokines and different signaling pathways (Table 7.1).

INSULIN RESISTANCE

IR is one of the primary pathophysiological determinants of diabetes, and considerable effort has been expended to define the mechanisms and origins of this process. Both genetic and environmental factors will influence the development of complex inflammatory mechanisms associated with the performance of the immune system in different organ effectors of insulin, which can be classified into cell intrinsic and extrinsic effects. Cell-intrinsic effects comprise endoplasmic reticulum (ER) stress, intracellular lipid deposition/imbalance, mitochondrial dysfunction, oxidative stress, and anabolic demand, whereas circulating cytokines (tumor necrosis factor-α (TNFα), interleukin (IL)-1β, IL-6, and IL-8) and chemokines (monocyte chemoattractant protein (MCP)-1), adipokines (adiponectin, leptin, and resistin), serum fatty acids, and hypoxia are the dominant extrinsic pathways that modulate peripheral insulin signaling.[15,16]

Despite their biological diversity, a striking majority of these determinants converge on common final inflammatory pathways, with TNFα, IL-1β, and IL-6 being the most common cytokines involved in the development of IR. TNF-α can increase glucose uptake in both visceral and subcutaneous adipocytes by activating the adenosine monophosphate activated protein kinase pathway, whereas it triggers IR in visceral adipocytes by activating Janus Kinase 1/2 (JNK 1/2), enhancing adipocyte lipolysis and increasing the serine/threonine phosphorylation of insulin receptor substrate-1 (IRS-1). IL-1β contributes to IR by impairing insulin signaling in peripheral tissues and macrophages, which leads to the reduced insulin sensitivity of β-cells and possible impaired insulin secretion. The levels of IL-1β in diverse cells are increased during hyperglycemia, playing a critical role in initiating and maintaining inflammation-induced organ dysfunction. In addition, IL-1β might increase systemic inflammation and inhibit insulin action in the major insulin-target cells. Finally, IL-6 is recognized as an inflammatory mediator that causes IR by reducing the expression of glucose transporter-4 and IRS-1. These effects are exerted by the activation of the Janus kinase-signal transducer and activator of transcription signaling pathway and increased the expression of suppressor of cytokine signaling 3. In addition, IL-6 also induces IR by blocking the phosphoinositide 3-kinase pathway and impairing glycogen synthesis by downregulating the expression of microRNA-200s and upregulating that of friend of GATA (GATA is referred to a family of transcription factors characterized by their ability to bind to the DNA sequence "GATA" G: Guanine; A: Adenine; T: Thymine) 2. It was suggested that IR in human skeletal muscle is related to IL-6 stimulation, which induces toll-like receptor-4 gene expression by activating STAT3 (Signal transducer and activator of transcription 3).[15–18]

ISLET INFLAMMATION

Although the presence of islet inflammation is classically acknowledged for autoimmune destruction of β-cells in T1D, new data implicates overlapping pathogenesis between T1D and T2D.[19] Thus, pancreatic islet inflammation has emerged as a key contributor to the loss of functional β-cell mass in both T1D and T2D. Evidence supports the role of proinflammatory cytokines in this context, leading to the development of oxidative stress, ER stress, and mitochondrial dysfunction, which compromises β-cell function and survival, and in addition, recruit macrophages into islets, thus maintaining and enhancing inflammation.[20,21] The traffic of M1 type macrophage (macrophage polarized to a proinflammatory phenotype) and auto-reactive T-cells into the islets results in local cytokine release and cell-mediated immunity that directly triggers β-cell inflammation. This inflammatory response leads to the loss of β-cell function through dysfunction, death, and/or dedifferentiation.[22,23]

DIABETIC COMPLICATIONS

Activation of inflammatory processes contributes not only to the development diabetes, but also to the pathogenesis of diabetic complications. Inflammatory cytokines are now recognized as key factors in ocular, renal, and neural injury associated with diabetes.[24]

Diabetic Retinopathy

Inflammatory processes are involved in ocular structural and molecular alterations related to diabetes, particularly from the early stages of diabetic retinopathy (DR). Dysregulation of diverse biochemical pathways (increased polyol pathway, enhanced advanced glycation end products formation, activation of protein kinase C, and augmented hexosamine pathway) result in different alterations, such as increased production of free radicals, increased oxidative stress, and activation of the renin-angiotensin system that contribute to the upregulation of inflammatory cytokines.[25] In addition, one of the earliest abnormalities observed in DR is the decrease of retinal perfusion. DR is characterized by alterations resulting in altered vascular wall integrity, increased vascular permeability, and retinal ischemia, and hypoxia, which leads to the synthesis and release of many soluble factors in the vitreous, including inflammatory molecules such as cytokines. Ischemia induces the activation of signaling pathway that attracts macrophages into hypoxic areas. Hypoxia-activated macrophages and microglia (the immune cells of the retina) release inflammatory molecules, which in turn stimulate the production of other cytokines and growth factors.[26]

Both vitreous TNF-α concentrations and the TNF-α vitreous/serum ratios of diabetic patients have been reported to be higher than those of the nondiabetic subjects. In addition, a direct correlationship between TNF-α levels and the severity of DR has been found. This cytokine has important effects regarding ocular involvement in diabetes, such as the increase of retinal endothelial permeability by downregulating the expression of tight junction proteins, the induction of oxidation, production of reactive oxygen species, and the stimulation of leukocyte adhesion. Concerning IL-6, the vitreous level of this cytokine is also higher in diabetic patients than in nondiabetic individuals, and again, vitreous

IL-6 levels are correlated with the severity of DR, neovascularization, and macular thickness. IL-6 has a prominent role in neovascularization, a key clinical feature of DR. Thus, not only promote angiogenesis directly but also support angiogenesis by inducing expression of vascular endothelial growth factor. Therefore, IL-6 may be major mediators of retinal inflammation and neovascularization.[27,28]

Diabetic Neuropathy and Diabetic Foot

Diabetic neuropathy (DNP) affects up to 50% of people with diabetes, and the diffuse peripheral neuropathies (distal sensori-motor polyneuropathy and autonomic neuropathy) are major risk factors for the development of diabetic foot syndrome (DFS) and amputation.

Although there are multiple factors involved in the pathogenesis of DNP, a persistent proinflammatory response through activation of the nuclear factor kappa-B and p38 mitogen activated protein kinase signaling, and the important participation of inflammatory cytokines, have been consistently shown to contribute to this complication.[29]

Experimental studies have demonstrated that TNF-α is a key factor in the pathogenesis of DPN, as well as in central and peripheral sensitization of neuropathic pain.[30,31] Moreover, clinical studies have shown that patients with DPN have increased serum TNF-α levels as compared with diabetic subjects without this complication and healthy controls, and that elevated serum concentrations of this cytokine may be independently associated with a higher risk of DPN.[32] In addition, increased levels of proinflammatory cytokines (IL-6, IL-1, and TNF-α) have been correlated with the progression of nerve degeneration in DPN.[33]

DFS is defined, according to the World Health Organization, as "ulceration of the foot (*distally from the ankle and including the ankle*) associated with neuropathy and different grades of ischemia and infection."[34] Complications of foot ulcers are the major cause of hospitalization and amputation in diabetic patients and lead to significant health care costs as evidenced by the fact that 20%–40% of health care resources are spent on diabetes-related diabetic foot.[35]

Diabetic Kidney Disease

Diabetic kidney disease (DKD) is a paradigmatic example of diabetic complications. Nowadays, it is the largest single cause of end-stage renal disease, which has been described as a medical catastrophe of worldwide dimensions, and in addition, renal involvement is a major cause of morbidity and mortality in the diabetic population. Our understanding on the processes leading to kidney involvement in diabetes has evolved from a classical view where renal injury was explained by metabolic and hemodynamic alterations, and by the modification of molecules under hyperglycaemic conditions, to a scenario where genetic and environmental factors trigger a complex series of pathophysiological events with a prominent role of inflammation.

Inflammatory cytokines play key roles in the development and progression of DKD. Renal cells, including endothelial, epithelial, mesangial, and tubular cells, have been demonstrated to synthesize TNF-α, IL-1, and IL-6, and therefore, these molecules, acting in a paracrine or autocrine manner, are able to induce a variety of effects on different renal structures. Renal expression of IL-1 has been demonstrated to be increased in experimental models of diabetes, which has been related to enhanced expression of chemotactic factors and adhesion molecules, as well as with intraglomerular hemodynamic abnormalities related to alteration of prostaglandin synthesis by mesangial cells. In addition, IL-1 directly increases vascular endothelial cell permeability, and is able to increase hyaluronan generation by renal proximal tubular epithelial cells, being enhanced glomerular hyaluronan production involved in the initiation of glomerular hypercellularity in experimental model of diabetes.[36–39]

Diverse studies have demonstrated that IL-6, through different effects including alteration of endothelial permeability, induction of mesangial cell proliferation, and increase of fibronectin expression, participates in development of renal injury in diabetes. Analysis of kidney biopsies from patients with DKD showed that mesangium, interstitium, tubules, and infiltrating cells were positive for IL-6 mRNA. Furthermore, there was a relationship between the severity of diabetic glomerulopathy and the expression of IL-6 mRNA in glomerular cells. More recently, a significant association between IL-6 and glomerular basement membrane thickening, a crucial lesion of DKD and a strong predictor of progression of renal disease, has been demonstrated. In addition, overexpression of IL-6 in the diabetic rat kidney is directly and significantly associated with wet kidney weight, an accurate index of renal hypertrophy, as well as with increased urinary albumin excretion, the hallmark of diabetic nephropathy. Regarding IL-18, this potent inflammatory cytokine induces interferon-γ and leads to production of other inflammatory cytokines, as well as apoptosis of endothelial cells. Increased serum and urinary levels of IL-18 have been observed in patients with DKD, showing an independent relationship with albuminuria and urinary excretion of β-2 microglobulin, a marker of tubulo-interstitial injury.[36–39]

TNF-α posses a variety of biological activities related to the induction of renal damage in diabetes, including cytotoxic effects on renal cells, induction of apoptosis and necrotic cell death, alterations of intraglomerular blood flow and glomerular filtration rate, alteration of endothelial permeability, and induction of reactive oxygen species production. Experimental studies have consistently reported that TNF-α mRNA and protein levels are increased in glomerular and proximal tubule cells from diabetic rats, with a significant role of this cytokine in the development of renal hypertrophy and hyperfunction, two main alterations during the initial stage of DKD. Furthermore, renal cortical mRNA levels and urinary TNFα concentrations have been direct and independently correlated to urinary albumin excretion, with a significant increase of TNFα levels in renal interstitial fluid and urine preceding the development of albuminuria. Finally, clinical studies have found that diabetic patients with nephropathy have higher serum and urinary concentrations of TNF-α than nondiabetic subjects as well as diabetic patients without renal involvement. These works showed a direct and independent association between the levels of this cytokine and clinical markers of glomerular as well as tubule interstitial damage, with a significant rise of serum and urinary TNF-α as DKD progressed.[36–39]

CONCLUSIONS

Inflammatory cytokines exert an important diversity of actions related to diabetes mellitus, since the initial stages of diabetes, with a key participation in development and progression of the disease, to a critical role in more advanced stages and pathogenesis of diabetic complications. The need to identify new therapeutic targets and additional strategies for treating diabetes and its complications is clearly evident. Providing diabetic patient's protection from the development and progression of the disease remains a great challenge. The modulation of inflammatory processes in general, and inflammatory cytokines in particular, might be useful in the prevention or therapy of diabetes, as well as provide new therapeutic targets for prevention and treatment of diabetic complications.

References

1. Danaei G, Finucane MM, Lu Y, Singh GM, Cowan MJ, Paciorek CJ, et al. National, regional, and global trends in fasting plasma glucose and diabetes prevalence since 1980: systematic analysis of health examination surveys and epidemiological studies with 370 country-years and 2.7 million participants. *Lancet* 2011;**378**:31–40.
2. Hameed I, Masoodi SR, Mir SA, Nabi M, Ghazanfar K, Ganai BA. Type 2 diabetes mellitus: from a metabolic disorder to an inflammatory condition. *World J Diabetes* 2015;**6**:598–612.
3. Cabrera SM, Henschel AM, Hessner MJ. Innate inflammation in type 1 diabetes. *Transl Res* 2016;**167**:214–27.

4. Pozzilli P, Guglielmi C. Double diabetes: a mixture of type 1 and type 2 diabetes in youth. *Endocr Dev* 2009;**14**:151–66.
5. Wilkin TJ. The accelerator hypothesis: a review of the evidence for insulin resistance as the basis for type I as well as type II diabetes. *Int J Obes (Lond)* 2009;**33**:716–26.
6. Navarro JF, Mora C. Role of inflammation in diabetic complications. *Nephrol Dial Transplant* 2005;**20**:2061–4.
7. Paneni F, Beckman JA, Creager MA, Cosentino F. Diabetes and vascular disease: pathophysiology, clinical consequences, and medical therapy: part I. *Eur Heart J* 2013;**34**:2436–46.
8. Beckman JA, Paneni F, Cosentino F, Creager MA. Diabetes and vascular disease: pathophysiology, clinical consequences, and medical therapy: part II. *Eur Heart J* 2013;**34**:2444–56.
9. Zaccardi F, Webb DR, Yates T, Davies MJ. Pathophysiology of type 1 and type 2 diabetes mellitus: a 90-year perspective. *Postgrad Med J* 2016;**92**:63–9.
10. Verbeeten KC, Elks CE, Daneman D, Ong KK. Association between childhood obesity and subsequent type 1 diabetes: a systematic review and meta-analysis. *Diabet Med* 2011;**28**:10–18.
11. Odegaard JI, Chawla A. Connecting type 1 and type 2 diabetes through innate immunity. *Cold Spring Harb Perspect Med* 2012;**2**:a007724.
12. Donath MY, Storling J, Maedler K, Mandrup-Poulsen T. Inflammatory mediators and islet beta-cell failure: a link between type 1 and type 2 diabetes. *J Mol Med (Berl)* 2003;**81**:455–70.
13. Baumann B, Salem HH, Boehm BO. Anti-inflammatory therapy in type 1 diabetes. *Curr Diab Rep* 2012;**12**:499–509.
14. Donath MY. Targeting inflammation in the treatment of type 2 diabetes: time to start. *Nat Rev Drug Discov* 2014;**13**:465–76.
15. Khodabandehloo H, Gorgani-Firuzjaee S, Panahi G, Meshkani R. Molecular and cellular mechanisms linking inflammation to insulin resistance and β-cell dysfunction. *Transl Res* 2016;**167**:228–56.
16. Samuel VT, Shulman GI. The pathogenesis of insulin resistance: integrating signaling pathways and substrate flux. *J Clin Invest* 2016;**126**:12–22.
17. Shoelson SE, Lee J, Goldfine AB. Inflammation and insulin resistance. *J Clin Invest* 2006;**116**:1793–801.
18. Chen L, Chen R, Wang H, Liang F. Mechanisms linking inflammation to insulin resistance. *Int J Endocrinol* 2015;**2015**:508409.
19. Imai Y, Dobrian AD, Morris MA, Nadler JL. Islet inflammation: a unifying target for diabetes treatment? *Trends Endocrinol Metab* 2013;**24**:351–60.
20. Imay Y, Dobrian AD, Weaver JR, Butcher MUJ, Cole BK, Galkina EV, et al. Interaction between cytokines and inflammatory cells in islet dysfunction, insulin resistance and vascular disease. *Diabetes Obes Metab* 2013;**15**(Suppl. 3):S117–19.
21. Hasnain SZ, Prins JB, McGuckin MA. Oxidative and endoplasmic reticulum stress in β-cell dysfunction in diabetes. *J Mol Endocrinol* 2016;**56**:R33–54.
22. Weir GC, Bonner-Weir S. Five stages of evolving β-cell dysfunction during progression to diabetes. *Diabetes* 2004;**53**(Suppl. 3):S16–21.
23. Talchai C, Xuan S, Lin HV, Sussel L, Accili D. Pancreatic cell dedifferentiation as a mechanism of diabetic cell failure. *Cell* 2012;**150**:1223–34.
24. Williams MD, Nadler JL. Inflammatory mechanisms of diabetic complications. *Curr Diabetes Rep* 2007;**7**:242–8.
25. Kern TS. Contributions of inflammatory processes to the development of the early stages of diabetic retinopathy. *Exp Diabetes Res* 2007;**2007**:95103.
26. Semeraro F, Cancarini A, dell'Olmo R, Rezzola S, Romano MR, Costagliola C. Diabetic retinopathy: Vascular and inflammatory disease. *J Diabetes Res* 2015;**2015**:582060.

27. Mysliwiec M, Balcerska A, Zorena K, Mysliwska J, Lipowski P, Raczynska K. The role of vascular endothelial growth factor, tumor necrosis factor alpha and interleukin-6 in pathogenesis of diabetic retinopathy. *Diabetes Res Clin Pract* 2008;**79**:141–6.
28. Gustavsson C, Agardh CD, Agardh E. Profile of intraocular tumour necrosis factor-α and interleukin-6 in diabetic subjects with different degrees of diabetic retinopathy. *Acta Ophthalmol* 2013;**91**:445–52.
29. Cameron NE, Cotter MA. Pro-inflammatory mechanisms in diabetic neuropathy: focus on the nuclear factor kappa B pathway. *Curr Drug Targets* 2008;**9**:60–7.
30. Yamakawa I, Kojima H, Terashima T, Katagi M, Oi J, Urabe H, et al. Inactivation of TNF-α ameliorates diabetic neuropathy in mice. *Am J Physiol Endocrinol Metab* 2011;**301**: E844–52.
31. Leung L, Cahill CM. TNF-alpha and neuropathic pain—a review. *J Neuroinflammation* 2010;**7**:27.
32. Mu ZP, Wang YG, Li CQ, Lv WS, Wang B, Jing ZH, et al. Association between tumor necrosis factor-α and diabetic peripheral neuropathy in patients with type 2 diabetes: a meta-analysis. *Mol Neurobiol* 2017;**54**:983–96.
33. Skundric DS, Lisak RP. Role of neuropoietic cytokines in development and progression of diabetic polyneuropathy: from glucose metabolism to neurodegeneration. *Exp Diabesity Res* 2003;**4**:303–12.
34. Jeffcoate WF, Macfarlane RM, Fletcher EM. The description and classification of diabetic foot lesions. *Diabetic Med* 1993;**10**:676–9.
35. Boulton AJ, Vileikyte L, Ragnarson-Tennvall G, Apelqvist J. The global burden of diabetic foot disease. *Lancet* 2005;**366**:1719–24.
36. Martini S, Eichinger F, Nair V, Kretzler M. Defining human diabetic nephropathy on the molecular level: integration of transcriptomic profiles with biological knowledge. *Rev Endocr Metab Disord* 2008;**9**:267–74.
37. Navarro-González JF, Mora-Fernández C. The role of inflammatory cytokines in diabetic nephropathy. *J Am Soc Nephrol* 2008;**19**:433–42.
38. Navarro-González JF, Mora-Fernández C, Muros de Fuentes M, García-Pérez J. Inflammatory molecules and pathways in the pathogenesis of diabetic nephropathy. *Nat Rev Nephrol* 2011;**7**:327–40.
39. Donate-Correa J, Martín-Núñez E, Muros de Fuentes M, Mora-Fernández C, Navarro-González JF. Inflammatory cytokines in diabetic nephropathy. *J Diabetes Res* 2015;**2015**:948417.

CHAPTER

8

Proinflammatory and Regulatory Cytokines in Sarcoidosis

Cristan Herbert[1], *Hasib Ahmadzai*[1], *and Paul S. Thomas*[1,2,3]

[1]UNSW Australia, Sydney, NSW, Australia [2]POWH Clinical School, UNSW, Randwick, NSW, Australia [3]Prince of Wales Hospital, Randwick, NSW, Australia

OUTLINE

Cytokine Effector Functions in Tissues.
DOI: http://dx.doi.org/10.1016/B978-0-12-804214-4.00007-5

Sarcoidosis is a systemic granulomatous disease of uncertain etiology and heterogeneous clinical presentation. The disease is often self-limiting but may be fatal due to pulmonary fibrosis and cardiac or neurological complications. Although the clinical course is generally unpredictable, there is evidence that the inflammatory milieu may influence outcomes; therefore a better understanding of the inflammatory response and the cytokines involved may be useful for informing treatment decisions.

In this chapter, we discuss the cells and cytokine networks involved in the pathogenesis of sarcoidosis. We will address both proinflammatory and regulatory cytokines which have been implicated in the disease.

SARCOIDOSIS: AN ENIGMATIC DISEASE

Sarcoidosis is characterized by the development of noncaseating granulomas which typically affect the lungs and thoracic lymph nodes, but which can also affect the skin, eyes, heart, and nerves.[1] Although self-limiting in the majority of cases, the disease is persistent in one-third of patients and fatal in approximately 5% of cases due to the development of pulmonary fibrosis or complications associated with cardiac or neurological involvement.[2]

Sarcoidosis can develop at any age, although it is more common during the third and fourth decade of life, and women are affected more than men. The incidence varies markedly throughout the world with as many as 64 per 100,000 in some Scandinavian counties,[3] and less than 1 per 100,000 in the United Kingdom.[4] In the United States, the prevalence is approximately three times higher in African Americans (35 per 100,000) than in Caucasian Americans (10.9 per 100,000).[5]

No specific cause of sarcoidosis has been identified. Due to the heterogeneous nature of the disease, it has been suggested that sarcoidosis may in fact represent a reaction pattern common to multiple independent causative agents and be dependent on host factors.[6] Polymorphisms in genes for human leukocyte antigens (HLA) have been associated with sarcoidosis. Specifically, HLA-DRB1*07, *14, and *15 are associated with chronic disease, while HLA-DRB*01 and *03 appear to protect against chronicity.[7] Putative environmental causes include exposure to wood smoke, insecticides, dusts, and certain metals can cause a similar disease (e.g., beryllium).[6] Other potential causes include exposure to antigens of *Mycobacterium tuberculosis*[8] and *Propionibacterium acnes*.[9] Ultimately, it is believed that the etiological agents responsible induce a chronic immune response leading to granuloma formation.

Typically, granulomas function to contain pathogens and protect the surrounding tissue, but as no etiological agent has been identified in sarcoidosis, there is no clear benefit to the host.[10] Sarcoid granulomas consist of a collection of macrophages interspaced with CD4 + T-cells of the T-helper (Th) 1 subtype, epithelioid cells, and multinucleated giant cells. CD8 + T-cells, regulatory T-cells, and fibroblasts may also be involved.[1,11] Four phases of sarcoid granuloma have been described including initiation, accumulation, effector, and resolution,[10] each regulated by the cells involved and the complex network of cytokines they elaborate.

T-CELLS CYTOKINES

During the formation of granulomas, it is hypothesized that the as yet unidentified antigen is internalized by dendritic cells or other antigen presenting cells (APC). APCs present the antigen to naïve T-cells and secrete interleukin (IL)-12, which drives the differentiation of Th1 cells. Th1 cells in turn produce IFN-γ and IL-2. There is evidence that a strong Th1 response is associated with acute onset of disease and spontaneous resolution.[12]

IFN-γ is spontaneously released by Th1 CD4 + T-cells, potently activates alveolar macrophages[13] and promotes expression of the IL-12 receptor which enhances Th1 responses, while suppressing Th2 responses.[14] Levels of IFN-γ are elevated in bronchoalveolar lavage (BAL) fluid from patients with sarcoidosis.[15] Along with IFN-γ, IL-2, a key growth, survival and differentiation factor for T-cells, is also spontaneously released by T-lymphocytes in active sarcoidosis.[16]

Although a primary role for Th1 cells has been established in sarcoidosis, Th17 cells have also been implicated. The differentiation of Th17 cells is promoted by TGF-β and IL-6 and their survival is enhanced by IL-23. Th17 cells produce IL-17A, IL-17F, and IL-22 and their number is increased in blood, lavage fluid, and lung biopsies from patients with sarcoidosis.[17,18] Recently, Ramstein et al. demonstrated elevated numbers of a subset of IFN-γ-secreting Th17 cells, termed Th17.1 cells, in BAL fluid from patients with sarcoidosis.[19] Their data indicate that Th17.1 cells, not Th1 cells, may be the predominant source of IFN-γ in sarcoidosis but further studies are required to fully define the role of Th17, and Th17.1 cells in sarcoidosis.

PROINFLAMMATORY CYTOKINES AND CHEMOKINES

Once a granuloma is established, cellular recruitment, proliferation, and survival are regulated by numerous cytokines and chemokines.

IL-12 is the key cytokine which promotes the differentiation of naïve CD4 + T-cells into Th1 cells.[20] In BAL fluid from patients with sarcoidosis, levels of IL-12 are elevated, along with IL-18.[21] IL-12 upregulates expression of the IL-18 receptor on CD4+ T-cells,[22] and IL-18 in turn upregulates the IL-12 receptor. These cytokines act synergistically to enhance production of IFN-γ and promote granuloma formation.[23] IL-12 also promotes the production of IFN-γ by Th17 cells.[24]

Tumor necrosis factor-α (TNF-α) is critical for the recruitment of monocytes to granulomas. TNF-α is elevated in exhaled breath condensate of patients with sarcoidosis[25] mainly from activated macrophages.[26] TNF-α levels correlate with disease activity[27] and some polymorphisms in the TNF gene are associated with favorable outcomes.[28]

Other mediators are also implicated in sarcoidosis such as the chemokines, and macrophages, epithelioid histocytes, and multinucleated giant cells are the major source of both CC and CXC chemokines.[29,30] Levels of CCL2 and CCL5 are elevated in BAL during all stages of the disease,[29] and expression of the chemokine receptor, CCR5, is increased on BAL cells from patients with sarcoidosis.[31] In addition, CCL2 enhances fibroblast survival[32] and may be involved in the transition from a Th1- to Th2-dominated immune response.[33] Other chemokines have a less clear role as there are conflicting reports on the involvement of CCL3 and CCL4 in sarcoidosis. Elevated levels of CCL3 and CCL4 were reported,[34] but contradicted in a more recent study.[29] Increased levels of CCL15 have been detected in BAL fluid of Stage III patients and associated with disease progression after 2 years.[35] CCL16, which promotes production of IL-12, IL-1β, and TNF by alveolar macrophages, is increased in BAL fluid from patients at all stages.[36]

CXC chemokines also contribute to the recruitment and activation of T-cells and a potential role in sarcoid has been identified for several family members including CXCL9, -10, and -11.[37] The receptor for these cytokines, CXCR3, has been identified on CD4+ and CD8+ T-cells from patients with sarcoidosis.[38] CXCL-11 is increased in patients with Stage I sarcoidosis, while CXCL9 is elevated in all stages.[30] Patients with high initial levels of CXCL9 demonstrated a greater decline in lung function compared to patients with a low initial concentration of CXCL9 suggesting this cytokine may be a useful marker of disease prognosis.[39]

Similarly, CXCL10 is elevated in serum and BAL fluid from patients relative to healthy controls,[40] and may be a marker of active disease.[41] CXCL10 is selectively elevated in sarcoid granulomas and may also help distinguish sarcoidosis from other granulomatous disease.[42]

REGULATORY CYTOKINES

Resolution of sarcoidosis may occur either spontaneously or following corticosteroid therapy, but the mechanisms are incompletely defined. It is likely that resolution is promoted by regulatory T-cells (Tregs) and their cytokines.

Tregs can suppress activation, proliferation, and effector functions of a variety of immune cells.[43] The number of pulmonary Tregs is reduced in patients with chronic sarcoidosis compared to healthy controls, but not in patients in which there was spontaneous resolution.[44] Recently, Broos et al. found no deficit in Tregs either locally or systemically however, Treg displayed impaired survival due to increased susceptibility for apoptosis.[45] Tregs are a major source of regulatory cytokines in sarcoidosis including IL-10, TGF-β, and IL-35.

IL-10 is a potent anti-inflammatory cytokine produced by a variety of cells including Tregs, monocytes, and activated macrophages[46] and can suppress the function of many cells, particularly monocytes and macrophages. IL-10 suppresses production of proinflammatory mediators including IFN-γ from Th1 cells,[47] and IL-12 and nitric oxide by macrophages.[48] In patients with sarcoidosis, the concentration of IL-10 in BAL fluid is elevated.[49]

TGF-β is produced by epithelial cells, T-cells, and macrophages and has the potential to be both proinflammatory and antiinflammatory[50] as it is involved in the differentiation of Th17 cells, can activate fibroblasts, and has been implicated in various fibrotic conditions.[51] However, TGF-β can also inhibit T-cell activation,[52] and earlier studies identified an association between high concentrations of TGF-β in BAL and spontaneous remission of sarcoidosis.[53]

IL-35 is a recently characterized cytokine produced mainly by dendritic cells and regulatory T-cells.[54] Although a member of the IL-12 family, IL-35 appears to be strictly immunosuppressive. IL-35 can inhibit T-cell proliferation[55] and may also promote the development of regulatory T-cells.[56] The contribution of IL-35 to the pathogenesis of sarcoidosis is unclear but due to its potential to modulate Th17 cells and regulatory T-cells, IL-35 is an attractive target for future sarcoid research.[57]

IL-27 is another member of the IL-12 family, with a potential regulatory role.[57] IL-27 stimulates IL-10 secretion and proliferation of regulatory T-cells, can impair production of TNF and IL-1β,[58] and suppress Th17 cells.[59] Expression of mRNA for IL-27 is elevated in peripheral blood mononuclear cells (PBMC) from patients with sarcoidosis suggesting that IL-27 may contribute to peripheral anergy.[60]

TH2 CYTOKINES

While a robust Th1 response may be associated with acute onset followed by remission, it has been suggested that a Th2 immune response may be associated with insidious onset and increased risk of chronicity.[11] Th2 cytokines can influence TGF-β production and may be important in the development of pulmonary fibrosis in sarcoidosis.[61] It is conceivable that the development of pulmonary fibrosis in sarcoidosis may be the result of a switch to a Th2-biased immune response.[62] Th2 cells have been identified in the BAL of patients,[63] and elevated levels of the Th2 cytokine, IL-13, have been demonstrated in the serum of patients with sarcoidosis[64] and the Th2-promoting cytokine, IL-33, in BAL fluid and tissue.[42]

Further research is required to clarify the role of Th2 cytokines in sarcoidosis and whether evidence of a Th2 response might be useful for predicting persistent disease or risk of pulmonary fibrosis.

DISEASE BIOMARKERS

Many cytokines have been investigated as potential biomarkers for sarcoidosis,[12,25,39,41] but no single cytokine has proved reliable, or more useful than the level of serum angiotensin converting enzyme (ACE). Recently, there is increasing interest in microRNAs (miRNA) as biomarkers for sarcoidosis. microRNAs are 20–24 nucleotide noncoding RNA which can cause the translational repression or degradation of target mRNA sequences.[65] Importantly, mRNA regulation can occur even when hybridization is imperfect so that each miRNA can regulate entire families of mRNA, including cytokines.

In patients with sarcoidosis, elevated levels of miRNA have been measured in PBMC[66] and BAL fluid,[67] and miRNA has also been detected in exhaled breath condensate.[68] In a recent study, elevated miR-155 and decreased let7c in BAL fluid was associated with disease progression after 2 years.[69] Research into the potential of miRNAs as biomarkers for sarcoidosis is currently ongoing.

ANTICYTOKINE THERAPY

Enhanced understanding of the immunopathological basis of sarcoidosis has resulted in the development of targeted therapies for sarcoidosis, including anti-TNF-α agents with improvements in patients treated with the monoclonal anti-TNF-α antibody, infliximab.[70,71] More recently another anti-TNF-α antibody, adalimumab, was shown to be an effective alternative in patients who developed intolerance to infliximab.[72] Despite demonstrated efficacy the high cost of these anti-TNF-α agents limits their use in the treatment of sarcoidosis.

SUMMARY

Sarcoidosis is a complex disease in which cytokines produced by Th1/Th17 cells, regulatory T-cells, monocytes, and activated macrophages all play a role. The balance between proinflammatory and regulatory cytokines appears to be crucial in determining the outcome of this disease.

References

1. Iannuzzi MC, Fontana JR. Sarcoidosis: clinical presentation, immunopathogenesis, and therapeutics. *JAMA* 2011;**305**(4):391–9.
2. Judson MA, Boan AD, Lackland DT. The clinical course of sarcoidosis: presentation, diagnosis, and treatment in a large white and black cohort in the United States. *Sarcoidosis Vasc Diffuse Lung Dis* 2012;**29**(2):119–27.
3. Hillerdal G, Nou E, Osterman K, Schmekel B. Sarcoidosis: epidemiology and prognosis. A 15-year European study. *Am Rev Respir Dis* 1984;**130**(1):29–32.
4. James DG. Epidemiology of sarcoidosis. *Sarcoidosis* 1992;**9**(2):79–87.
5. Rybicki BA, Major M, Popovich Jr. J, Maliarik MJ, Iannuzzi MC. Racial differences in sarcoidosis incidence: a 5-year study in a health maintenance organization. *Am J Epidemiol* 1997;**145**(3):234–41.
6. Dubrey S, Shah S, Hardman T, Sharma R. Sarcoidosis: the links between epidemiology and aetiology. *Postgrad Med J* 2014;**90**(1068):582–9.
7. Grunewald J, Brynedal B, Darlington P, Nisell M, Cederlund K, Hillert J, et al. Different HLA-DRB1 allele distributions in distinct clinical subgroups of sarcoidosis patients. *Respir Res* 2010;**11**:25.
8. Oswald-Richter KA, Beachboard DC, Seeley EH, Abraham S, Shepherd BE, Jenkins CA, et al. Dual analysis for mycobacteria and propionibacteria in sarcoidosis BAL. *J Clin Immunol* 2012;**32**(5):1129–40.
9. Eishi Y, Suga M, Ishige I, Kobayashi D, Yamada T, Takemura T, et al. Quantitative analysis of mycobacterial and propionibacterial DNA in lymph nodes of Japanese and European patients with sarcoidosis. *J Clin Microbiol* 2002;**40**(1):198–204.
10. Co DO, Hogan LH, Il-Kim S, Sandor M. T cell contributions to the different phases of granuloma formation. *Immunol Lett* 2004;**92**(1–2):135–42.
11. Iannuzzi MC, Rybicki BA, Teirstein AS. Sarcoidosis. *N Engl J Med* 2007;**357**(21):2153–65.
12. Oswald-Richter KA, Richmond BW, Braun NA, Isom J, Abraham S, Taylor TR, et al. Reversal of global CD4+ subset dysfunction is associated with spontaneous clinical resolution of pulmonary sarcoidosis. *J Immunol* 2013;**190**(11):5446–53.
13. Mollers M, Aries SP, Dromann D, Mascher B, Braun J, Dalhoff K. Intracellular cytokine repertoire in different T cell subsets from patients with sarcoidosis. *Thorax* 2001;**56**(6):487–93.
14. Amsen D, Spilianakis CG, Flavell RA. How are T(H)1 and T(H)2 effector cells made? *Curr Opin Immunol* 2009;**21**(2):153–60.
15. Robinson BW, McLemore TL, Crystal RG. Gamma interferon is spontaneously released by alveolar macrophages and lung T lymphocytes in patients with pulmonary sarcoidosis. *J Clin Invest* 1985;**75**(5):1488–95.
16. Pinkston P, Bitterman PB, Crystal RG. Spontaneous release of interleukin-2 by lung T lymphocytes in active pulmonary sarcoidosis. *N Engl J Med* 1983;**308**(14):793–800.

17. Richmond BW, Ploetze K, Isom J, Chambers-Harris I, Braun NA, Taylor T, et al. Sarcoidosis Th17 cells are ESAT-6 antigen specific but demonstrate reduced IFN-gamma expression. *J Clin Immunol* 2013;**33**(2):446–55.
18. Ten Berge B, Paats MS, Bergen IM, van den Blink B, Hoogsteden HC, Lambrecht BN, et al. Increased IL-17A expression in granulomas and in circulating memory T cells in sarcoidosis. *Rheumatology (Oxford)* 2012;**51**(1):37–46.
19. Ramstein J, Broos CE, Simpson LJ, Ansel KM, Sun SA, Ho ME, et al. IFN-gamma-producing T-helper 17.1 cells are increased in sarcoidosis and are more prevalent than T-helper type 1 cells. *Am J Respir Crit Care Med* 2016;**193**(11):1281–91.
20. Manetti R, Parronchi P, Giudizi MG, Piccinni MP, Maggi E, Trinchieri G, et al. Natural killer cell stimulatory factor (interleukin 12 [IL-12]) induces T helper type 1 (Th1)-specific immune responses and inhibits the development of IL-4-producing Th cells. *J Exp Med* 1993;**177**(4):1199–204.
21. Mroz RM, Korniluk M, Stasiak-Barmuta A, Chyczewska E. Increased levels of interleukin-12 and interleukin-18 in bronchoalveolar lavage fluid of patients with pulmonary sarcoidosis. *J Physiol Pharmacol* 2008;**59**(Suppl. 6):507–13.
22. Yoshimoto T, Takeda K, Tanaka T, Ohkusu K, Kashiwamura S, Okamura H, et al. IL-12 up-regulates IL-18 receptor expression on T cells, Th1 cells, and B cells: synergism with IL-18 for IFN-gamma production. *J Immunol* 1998;**161**(7):3400–7.
23. Shigehara K, Shijubo N, Ohmichi M, Takahashi R, Kon S, Okamura H, et al. IL-12 and IL-18 are increased and stimulate IFN-gamma production in sarcoid lungs. *J Immunol* 2001;**166**(1):642–9.
24. Boniface K, Blumenschein WM, Brovont-Porth K, McGeachy MJ, Basham B, Desai B, et al. Human Th17 cells comprise heterogeneous subsets including IFN-gamma-producing cells with distinct properties from the Th1 lineage. *J Immunol* 2010;**185**(1):679–87.
25. Mohan N, Akter R, Bryant K, Herbert C, Chow S, Thomas PS. Exhaled breath markers of alveolar macrophage activity in sarcoidosis. *Inflamm Res* 2016;**65**(6):471–8.
26. Fehrenbach H, Zissel G, Goldmann T, Tschernig T, Vollmer E, Pabst R, et al. Alveolar macrophages are the main source for tumour necrosis factor-alpha in patients with sarcoidosis. *Eur Respir J* 2003;**21**(3):421–8.
27. Ziegenhagen MW, Benner UK, Zissel G, Zabel P, Schlaak M, Muller-Quernheim J. Sarcoidosis: TNF-alpha release from alveolar macrophages and serum level of sIL-2R are prognostic markers. *Am J Respir Crit Care Med* 1997;**156**(5):1586–92.
28. Wijnen PA, Nelemans PJ, Verschakelen JA, Bekers O, Voorter CE, Drent M. The role of tumor necrosis factor alpha G-308A polymorphisms in the course of pulmonary sarcoidosis. *Tissue Antigens* 2010;**75**(3):262–8.
29. Palchevskiy V, Hashemi N, Weigt SS, Xue YY, Derhovanessian A, Keane MP, et al. Immune response CC chemokines CCL2 and CCL5 are associated with pulmonary sarcoidosis. *Fibrog Tissue Repair* 2011;**4**:10.
30. Busuttil A, Weigt SS, Keane MP, Xue YY, Palchevskiy V, Burdick MD, et al. CXCR3 ligands are augmented during the pathogenesis of pulmonary sarcoidosis. *Eur Respir J* 2009;**34**(3):676–86.
31. Petrek M, Gibejova A, Drabek J, Mrazek F, Kolek V, Weigl E, et al. CC chemokine receptor 5 (CCR5) mRNA expression in pulmonary sarcoidosis. *Immunol Lett* 2002;**80**(3):189–93.
32. Gerke AK, Hunninghake G. The immunology of sarcoidosis. *Clin Chest Med* 2008;**29**(3):379–90, vii.
33. Mantovani A, Sica A, Sozzani S, Allavena P, Vecchi A, Locati M. The chemokine system in diverse forms of macrophage activation and polarization. *Trends Immunol* 2004;**25**(12):677–86.
34. Capelli A, Di Stefano A, Lusuardi M, Gnemmi I, Donner CF. Increased macrophage inflammatory protein-1alpha and macrophage inflammatory protein-1beta levels in

bronchoalveolar lavage fluid of patients affected by different stages of pulmonary sarcoidosis. *Am J Respir Crit Care Med* 2002;**165**(2):236–41.

35. Arakelyan A, Kriegova E, Kubistova Z, Mrazek F, Kverka M, du Bois RM, et al. Protein levels of CC chemokine ligand (CCL)15, CCL16 and macrophage stimulating protein in patients with sarcoidosis. *Clin Exp Immunol* 2009;**155**(3):457–65.
36. Cappello P, Caorsi C, Bosticardo M, De Angelis S, Novelli F, Forni G, et al. CCL16/LEC powerfully triggers effector and antigen-presenting functions of macrophages and enhances T cell cytotoxicity. *J Leukoc Biol* 2004;**75**(1):135–42.
37. Nishioka Y, Manabe K, Kishi J, Wang W, Inayama M, Azuma M, et al. CXCL9 and 11 in patients with pulmonary sarcoidosis: a role of alveolar macrophages. *Clin Exp Immunol* 2007;**149**(2):317–26.
38. Katchar K, Eklund A, Grunewald J. Expression of Th1 markers by lung accumulated T cells in pulmonary sarcoidosis. *J Intern Med* 2003;**254**(6):564–71.
39. Su R, Li MM, Bhakta NR, Solberg OD, Darnell EP, Ramstein J, et al. Longitudinal analysis of sarcoidosis blood transcriptomic signatures and disease outcomes. *Eur Respir J* 2014;**44**(4):985–93.
40. Nureki S, Miyazaki E, Ando M, Ueno T, Fukami T, Kumamoto T, et al. Circulating levels of both Th1 and Th2 chemokines are elevated in patients with sarcoidosis. *Respir Med* 2008;**102**(2):239–47.
41. Geyer AI, Kraus T, Roberts M, Wisnivesky J, Eber CD, Hiensch R, et al. Plasma level of interferon gamma induced protein 10 is a marker of sarcoidosis disease activity. *Cytokine* 2013;**64**(1):152–7.
42. Christophi GP, Caza T, Curtiss C, Gumber D, Massa PT, Landas SK. Gene expression profiles in granuloma tissue reveal novel diagnostic markers in sarcoidosis. *Exp Mol Pathol* 2014;**96**(3):393–9.
43. Sakaguchi S, Miyara M, Costantino CM, Hafler DA. FOXP3 + regulatory T cells in the human immune system. *Nat Rev Immunol* 2010;**10**(7):490–500.
44. Prasse A, Zissel G, Lutzen N, Schupp J, Schmiedlin R, Gonzalez-Rey E, et al. Inhaled vasoactive intestinal peptide exerts immunoregulatory effects in sarcoidosis. *Am J Respir Crit Care Med* 2010;**182**(4):540–8.
45. Broos CE, van Nimwegen M, Kleinjan A, ten Berge B, Muskens F, in 't Veen JC, et al. Impaired survival of regulatory T cells in pulmonary sarcoidosis. *Respir Res* 2015;**16**:108.
46. Sabat R. IL-10 family of cytokines. *Cytokine Growth Factor Rev* 2010;**21**(5):315–24.
47. Fiorentino DF, Zlotnik A, Vieira P, Mosmann TR, Howard M, Moore KW, et al. IL-10 acts on the antigen-presenting cell to inhibit cytokine production by Th1 cells. *J Immunol* 1991;**146**(10):3444–51.
48. Bogdan C, Vodovotz Y, Nathan C. Macrophage deactivation by interleukin 10. *J Exp Med* 1991;**174**(6):1549–55.
49. Barbarin V, Petrek M, Kolek V, Van Snick J, Huaux F, Lison D. Characterization of p40 and IL-10 in the BALF of patients with pulmonary sarcoidosis. *J Interferon Cytokine Res* 2003;**23**(8):449–56.
50. Wahl SM, McCartney-Francis N, Mergenhagen SE. Inflammatory and immunomodulatory roles of TGF-beta. *Immunol Today* 1989;**10**(8):258–61.
51. Goodwin A, Jenkins G. Role of integrin-mediated TGFbeta activation in the pathogenesis of pulmonary fibrosis. *Biochem Soc Trans* 2009;**37**(Pt 4):849–54.
52. Roth WK. TGF beta and FGF-like growth factors involved in the pathogenesis of AIDS-associated Kaposi's sarcoma. *Res Virol* 1993;**144**(1):105–9.
53. Zissel G, Homolka J, Schlaak J, Schlaak M, Muller-Quernheim J. Anti-inflammatory cytokine release by alveolar macrophages in pulmonary sarcoidosis. *Am J Respir Crit Care Med* 1996;**154**(3 Pt 1):713–19.
54. Li X, Mai J, Virtue A, Yin Y, Gong R, Sha X, et al. IL-35 is a novel responsive anti-inflammatory cytokine—a new system of categorizing anti-inflammatory cytokines. *PLoS One* 2012;**7**(3):e33628.

55. Niedbala W, Wei XQ, Cai B, Hueber AJ, Leung BP, McInnes IB, et al. IL-35 is a novel cytokine with therapeutic effects against collagen-induced arthritis through the expansion of regulatory T cells and suppression of Th17 cells. *Eur J Immunol* 2007;**37**(11):3021–9.
56. Collison LW, Chaturvedi V, Henderson AL, Giacomin PR, Guy C, Bankoti J, et al. IL-35-mediated induction of a potent regulatory T cell population. *Nat Immunol* 2010;**11**(12):1093–101.
57. Ringkowski S, Thomas PS, Herbert C. Interleukin-12 family cytokines and sarcoidosis. *Front Pharmacol* 2014;**5**:233.
58. Kalliolias GD, Gordon RA, Ivashkiv LB. Suppression of TNF-alpha and IL-1 signaling identifies a mechanism of homeostatic regulation of macrophages by IL-27. *J Immunol* 2010;**185**(11):7047–56.
59. Fitzgerald DC, Fonseca-Kelly Z, Cullimore ML, Safabakhsh P, Saris CJ, Zhang GX, et al. Independent and interdependent immunoregulatory effects of IL-27, IFN-beta, and IL-10 in the suppression of human Th17 cells and murine experimental autoimmune encephalomyelitis. *J Immunol* 2013;**190**(7):3225–34.
60. Loke WS, Freeman A, Garthwaite L, Prazakova S, Park M, Hsu K, et al. T-bet and interleukin-27: possible TH1 immunomodulators of sarcoidosis. *Inflammopharmacology* 2015;**23**(5):283–90.
61. Patterson KC, Hogarth K, Husain AN, Sperling AI, Niewold TB. The clinical and immunologic features of pulmonary fibrosis in sarcoidosis. *Transl Res* 2012;**160**(5):321–31.
62. Moller DR. Pulmonary fibrosis of sarcoidosis. New approaches, old ideas. *Am J Respir Cell Mol Biol* 2003;**29**(Suppl. 3):S37–41.
63. Baumer I, Zissel G, Schlaak M, Muller-Quernheim J. Th1/Th2 cell distribution in pulmonary sarcoidosis. *Am J Respir Cell Mol Biol* 1997;**16**(2):171–7.
64. Hauber HP, Gholami D, Meyer A, Pforte A. Increased interleukin-13 expression in patients with sarcoidosis. *Thorax* 2003;**58**(6):519–24.
65. Bartel DP. MicroRNAs: target recognition and regulatory functions. *Cell* 2009;**136**(2):215–33.
66. Jazwa A, Kasper L, Bak M, Sobczak M, Szade K, Jozkowicz A, et al. Differential inflammatory microRNA and cytokine expression in pulmonary sarcoidosis. *Arch Immunol Ther Exp (Warsz)* 2015;**63**(2):139–46.
67. Kiszalkiewicz J, Piotrowski WJ, Pastuszak-Lewandoska D, Gorski P, Antczak A, Gorski W, et al. Altered miRNA expression in pulmonary sarcoidosis. *BMC Med Genet* 2016;**17**:2.
68. Sng JJ, Prazakova S, Thomas PS, Herbert C. MMP-8, MMP-9 and neutrophil elastase in peripheral blood and exhaled breath condensate in COPD. *Int J Chron Obstruct Pulmon Dis* 2016;**14**(2):238–44.
69. Dyskova T, Fillerova R, Novosad T, Kudelka M, Zurkova M, Gajdos P, et al. Correlation network analysis reveals relationships between MicroRNAs, transcription factor t-bet, and deregulated cytokine/chemokine-receptor network in pulmonary sarcoidosis. *Mediators Inflamm* 2015;**2015**:121378.
70. Doty JD, Mazur JE, Judson MA. Treatment of sarcoidosis with infliximab. *Chest* 2005;**127**(3):1064–71.
71. Saleh S, Ghodsian S, Yakimova V, Henderson J, Sharma OP. Effectiveness of infliximab in treating selected patients with sarcoidosis. *Respir Med* 2006;**100**(11):2053–9.
72. Crommelin HA, van der Burg LM, Vorselaars AD, Drent M, van Moorsel CH, Rijkers GT, et al. Efficacy of adalimumab in sarcoidosis patients who developed intolerance to infliximab. *Respir Med* 2016;**115**:72–7.

CHAPTER

9

Cytokines and Their Implication in Axon Degeneration and Regeneration Following Peripheral Nerve Injury

Petr Dubový

Masaryk University, Brno, Czech Republic

OUTLINE

Cytokine Effector Functions in Tissues.
DOI: http://dx.doi.org/10.1016/B978-0-12-804214-4.00008-7

INTRODUCTION

Generally, cytokines, and their subfamily the chemokines, are inflammatory mediators induced in injured tissue and are implicated in both physiological and nonphysiological wound healing. Peripheral nerve degeneration is induced by traumatic or iatrogenic injury as well as by various diseases including HIV, diabetic, or chemotherapy-induced neuropathy. Just like in other types of tissue, inflammatory mediators like cytokines have both detrimental and beneficial effects in a nerve lesion. The regulation of cytokines following peripheral nerve injury is a "double-edged sword" with negative roles as in neuropathic pain (NPP) induction and positive effects on axon regeneration. For many years, the upregulation of cytokines after nerve injury was understood in the context of their negative effects like the development of NPP.[1] However, upregulation of some cytokines was detected in the homologous contralateral nerve[2] in that increased axon regeneration was also conditioned.[3] It is now recognized that some cytokines may have beneficial effects on damaged neurons by helping to restart the internal regenerative program and promote axon regeneration in those neurons. Identifying a checkpoint of the discrimination between the detrimental and the beneficial effects of cytokines after nerve injury has been a challenge for basic research in the last few years.

The goal of the present chapter is to summarize published results (both our own and from other groups) that suggest a role for cytokines not only in NPP development but also in axon regeneration after nerve injury.

UPREGULATION OF CYTOKINES DISTAL TO NERVE INJURY

Wallerian degeneration (WD) is a cascade of stereotypical cellular and molecular events distal to injury of nerve fibers considered to be a sort of innate immune reaction or neuroinflammation.[4] The cellular events comprise inflammatory activation of Schwann cells and invasion of various types of immune cells including macrophages that produce cytokines and chemokines. Cytokines play a key role during the early phase of WD to facilitate invasion of macrophages and cleaning of the axon and myelin debris[5] that is an important precondition for axon regeneration. On the other hand, increased expression of cytokines distal to nerve injury is associated with degenerative neuropathology and painful neuropathy, including primary demyelination and axonal degeneration.[6,7]

Local Inflammatory Profiling of Schwann Cells, NPP Induction, and Promotion of Axon Regeneration as a Dual Effect of Cytokines

Regulation of most cytokines in trauma-injured nerve usually occurs in two or three waves. A fine-tuned spatiotemporal expression of cytokines by activated Schwann cells during the early period of WD is predominantly associated with the recruitment of hematogenous macrophages and other immune cells. This cytokine network comprises tumor necrosis factor-α (TNFα), interleukin-1β (IL-1β), monocyte chemoattractant protein-1 (MCP-1), and neuropoietic cytokines like interleukin-6 (IL-6) and leukemia inhibitory factor (LIF) (reviewed in Ref. [8]). The recruited macrophages are responsible mainly for clearing the myelin debris—an essential prerequisite for axon regeneration because myelin contains molecules that inhibit axon growth.[9]

The next wave of cytokines distal to nerve injury is associated with dedifferentiated Schwann cells that simultaneously restore the synthesis of axon-promoting molecules including neurotrophins.[10] Along with the immune cells, the activated Schwann cells distal to nerve injury function as robust sources of both pro (IL-1β, IL-6, TNFα) and antiinflammatory (IL-4, IL-10) cytokines. These inflammatory Schwann cells were detected in close proximity to growing axons suggesting that the cytokines derived from them do not prevent growth of axons following nerve injury.[11] Proinflammatory cytokines (e.g., TNFα and IL-6) have been shown to stimulate neurotrophin-dependent neurite outgrowth. This cytokine–neurotrophin synergic effect is considered to be one of the mechanisms by which local aseptic inflammation can modulate axonal regeneration.[12]

On the other hand, proinflammatory cytokines released by Schwann cells and macrophages sensitize injured as well as undamaged axons to contribute to ectopic discharges associated with induction and maintenance of NPP.[13] The NPP induced by a nerve lesion is a combination of the axon injury itself and the release of algesic mediators produced by cells of WD. A partial nerve injury particularly results in WD alongside some of the axons while the undamaged axons are exposed to molecules including cytokines produced by WD of injured fibers. Nevertheless, NPP can develop without any injury of afferent axons as evidenced by experimental lesions of the ventral roots of spinal nerves containing motor axons.[14] This adds to the evidence that molecules released during WD are involved in the sensitization of undamaged afferent fibers to the development of NPP.

Whether the effect of cytokines after traumatic nerve injury results in NPP or in axon regeneration may depend on the balance of pro and antiinflammatory cytokines.[15] Inflammation-induced nerve injury with overproduction of proinflammatory cytokines always leads to NPP manifestation.[16] However, further experiments are necessary in order to better understand the dual effects (NPP induction and axon regeneration) of cytokines following nerve injury.

The question arises as to what are the endogenous ligands triggering the inflammatory activation of Schwann cells and the upregulation of cytokines. It is known that tissue damage results in the release of damage-associated molecular patterns (DAMPs) that through Toll-like receptors (TLRs) initiate subsequent inflammatory signaling. Degenerating axons produce mitochondria and myelin sheath fragments and together with degradation of endoneurial extracellular matrix molecules constitute potential endogenous ligands of TLRs. Primary in-vitro cultures of rat Schwann cells have been observed to express high levels of TLR2–7 mRNA and protein.[17] Activated Schwann cells distal to a nerve injury displayed increased immunostaining for TLR2, TLR3, and TLR9 when compared with Schwann cells from an intact nerve. However, TLR4 is constitutively expressed in Schwann cells and its immunofluorescence intensity is not significantly affected seven days after injury (Fig. 9.1). These and other experimental results indicate that in-vivo inflammatory induction of Schwann cells and the subsequent release of cytokines is mediated by the local activation of TLRs by DAMPs.[18] TLRs are thus implicated indirectly in NPP induction and axon regeneration after nerve lesion.[19]

EXPRESSION OF CYTOKINES AROUND AND IN AXOTOMIZED PRIMARY SENSORY NEURONS

Increased expression of cytokines is induced around and in axotomized bodies of both motor and primary sensory neurons. The expression of cytokines around and in the primary sensory neurons will be mentioned here because these neurons are critical for the induction and maintenance of NPP. Moreover, these neurons are frequently used as in-vivo and in-vitro models for studying cytokine-mediated promotion of axon growth.

Dorsal root ganglia (DRG) consist of the primary sensory neurons and their bodies are surrounded by satellite glial cells (SGCs) that react immediately upon neuron injury.[20] SGCs surrounding axotomized neuronal bodies upregulate the expression of various cytokines and their receptors and this is an example of inflammatory profiling of these glial cells triggered by nerve injury.[21–25] Cytokines upregulated by activated SGCs may influence the activity of primary sensory neurons leading to disturbances in afferentation and thus inducing NPP.[26] On the other

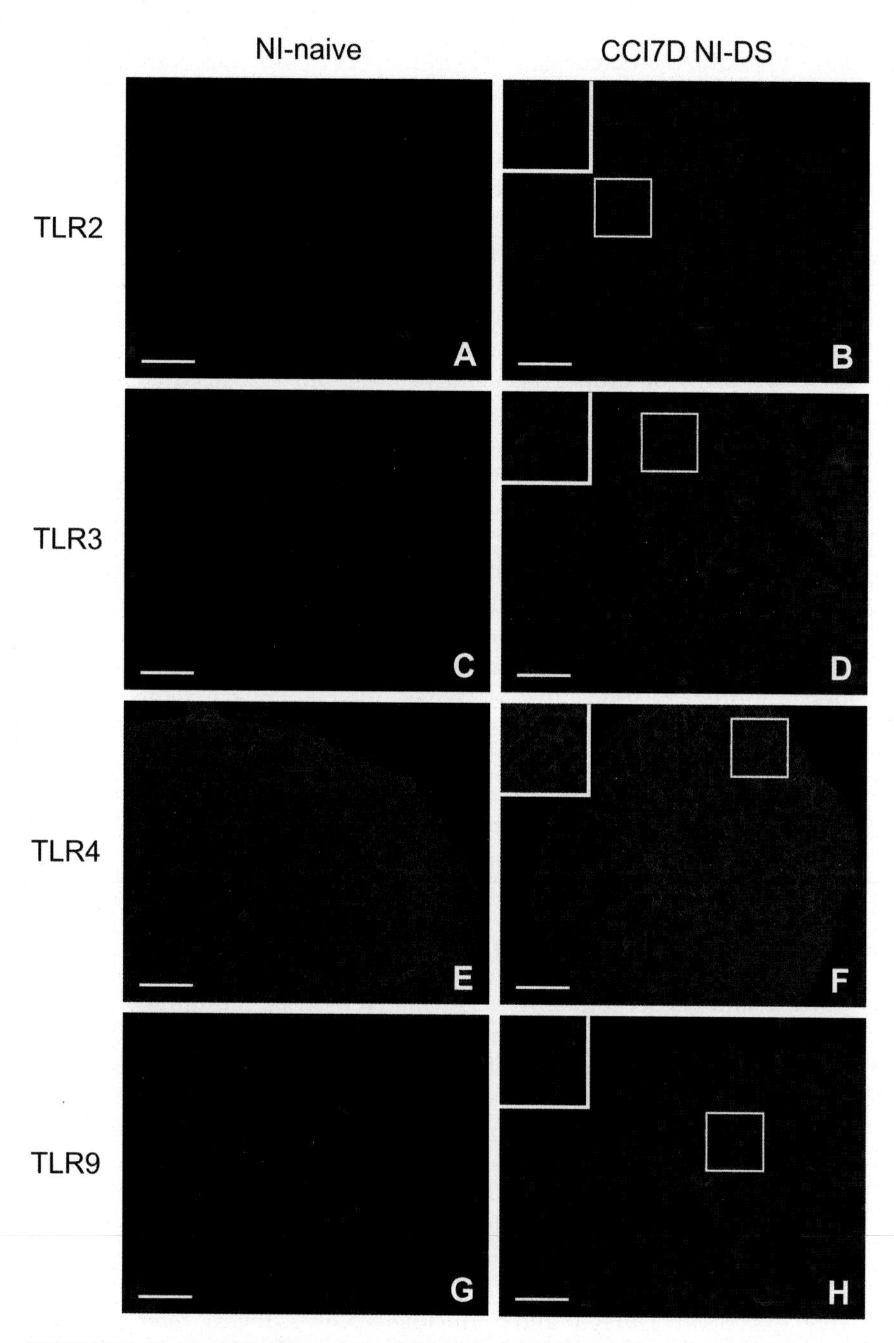

FIGURE 9.1 Immunofluorescence staining for TLR2, TLR3, TLR4, and TLR9 in cross-sections of rat naïve sciatic nerve (NI-naïve) and distal nerve stump 7 days after chronic constriction injury (CCI7D NI-DS). Schwann cells of distal nerve stumps display an increased immunofluorescence. Scale bars = 60 μm (A–D, G, H); 150 μm (E, F).

hand, cytokine-mediated communication between axotomized DRG neurons and their activated SGCs could be involved in neuron maintenance and axon growth promotion.[25,27] Thus, inflammatory profiling of SGCs like Schwann cells has a dual effect on the bodies of injured primary sensory neurons, and further investigation is necessary to identify a checkpoint for the discrimination of the detrimental and beneficial effects of cytokines released by these glial cells following nerve injury.

Peripheral axotomy also induces the recruitment of blood-derived macrophages and their invasion into the DRG neuron envelope. The macrophages of the neuronal envelope are able to influence induction of either NPP or the neuronal regenerative program.[28,29] This dual ability is dependent upon the activation state of the macrophages. Macrophages stimulated to an antiinflammatory state (M2) are axon growth-promoting, but proinflammatory macrophages (M1) are involved in growth-inhibition. Modulation of neuronal activities by macrophages is mediated by mobile factors like cytokines.[30,31] A role for macrophages in the promotion of neurons to axon regeneration has been discussed in comprehensive reviews[29,32] and is not the subject of the current chapter.

Cytokines released by neurons and the activated glial and immune cells surrounding them following nerve injury create an intricate network of mutual relationships. Moreover, the cytokines may act in combination or at a given time in the nerve injury-induced inflammatory process, resulting in conditions that lead to both NPP development as well as axon regeneration. These dual effects of cytokines on axotomized neurons are illustrated by the contrasting effects of TNFα and IL-6 and their signal transduction.

Role of Cytokines in the DRG and NPP Induction

TNFα and its receptors: A nerve injury induces the upregulation of TNFα in both neurons and their SGCs, while tumor necrosis factor receptor-1 (TNFR-1) is upregulated only in neuronal bodies of all sizes. In contrast to TNFR-1, tumor necrosis factor receptor-2 (TNFR-2) is increased mainly in the SGCs and small-sized neurons.[11,21] This pattern of molecular distribution suggests both autocrine and paracrine actions of the endogenous TNFα—via TNFR-1 on neuronal bodies and via TNFR-2 on SGCs and nociceptive neurons.[11] Many observations indicate a correlation between upregulated TNFα and its receptors with the induction of NPP (for a review, see Ref. [33]). Differences in expression of constitutively expressed TNFR-1 and inducible TNFR-2 reveal their different roles in nerve injury-induced hypersensitivity.[34] Electrophysiological results provide evidence that TNFR-2 may have a more important role than TNFR-1 in the excitability of neurons undamaged by axon injury.[35]

In this context, the effects of TNFα on large-sized DRG neurons can be mediated by TNFR-2 produced in SGCs. Moreover, bilateral increase of TNFR-2 in SGCs after unilateral nerve injury[11] may contribute to the expression of mirror-image pain.[36]

IL-6 and its signal-transducing molecules: Injury to a peripheral nerve results in the upregulation of IL-6 and corresponding signal-transducing molecules in DRG neurons and their SGCs. This process is associated with the development of NPP signs.[25,37] The neuronal effects of IL-6 depend on the presence of the soluble IL-6 receptor (sIL-6R). The presence of IL-6/sIL-6R increase susceptibility of the sensory neurons to noxious heat via the glycoprotein-130/Janus kinase/protein kinase Cδ (gp130/Jak/PKCδ) signaling pathway.[38]

Role of Cytokines in the Reactivation of the Intrinsic Regenerative Program of Neurons

TNFα and its receptors: Despite results suggesting a role for TNFα in NPP induction, endogenous TNFα in axotomized DRG neurons may be involved in the regulation of DRG neuron maintenance via activation of nuclear factor kappa-light-chain-enhancer of activated B cells (NFκB).[39] Activation of TNFR-1 may also elicit upregulation of IL-6,[40] which is a critical factor for reactivation of the neuronal regenerative program (see below). This beneficial effect is supported by results indicating a low association of TNFα/IL-6 crosstalk in DRG neurons with NPP induction.[40]

IL-6 and its signal-transducing molecules: Axonal damage due to traumatic nerve injury triggers regeneration programs in the neural bodies. These programs include the Janus kinase/signal transducers and activators of transcription-3 (JAK/STAT3) and phosphoinositide-3-kinase/protein kinase B/mammalian target of rapamycin (PI3K/AKT/mTOR) signaling cascades. It is known that the biological effects of neuropoietic cytokines like IL-6, LIF, and ciliary neurotrophic factor (CNTF) are mediated through the gp130/JAK/STAT3 signaling pathway.[41] Therefore, their involvement in the stimulation of axon regeneration is not surprising. Invivo studies have revealed increased expression of IL-6 and LIF in axotomized primary sensory neurons which are in the growth mode.[25,42] A role for these cytokines in axon regeneration was confirmed in IL-6 and LIF knock-out mice that display impaired peripheral nerve regeneration.[42,43] Moreover, the experimental model of conditioning injury revealed that IL-6 and LIF are involved in the reactivation of the intrinsic regenerative program of DRG neurons.[42,44]

In conclusion, chronic overproduction of cytokines following a nerve injury is implicated in NPP induction. On the other hand, the beneficial effect of cytokines upon axon regeneration depends upon the timing of the production of both pro and antiinflammatory cytokines and upon the control that maintains them at physiological levels. Identification of a checkpoint for the discrimination between the detrimental and the beneficial effects of cytokines is a challenge for basic research which will significantly impact clinical practice.

Acknowledgments

We thank Ms. Dana Kutějová, Ms. Marta Lněníčková, Ms. Jitka Mikulášková, and Ms. Jana Váchová for their skillful technical assistance. This work was supported by project no. 16-08508S of the Grant Agency of the Czech Republic.

References

1. Ramesh G, MacLean AG, Philipp MT. Cytokines and chemokines at the crossroads of neuroinflammation, neurodegeneration, and neuropathic pain. *Mediat Inflamm* 2013;**2013**:1–20.
2. Ruohonen S, Jagodi M, Khademi M, Taskinen HS, Ojala P, Olsson T, et al. Contralateral non-operated nerve to transected rat sciatic nerve shows increased expression of IL-1 beta, TGF-beta 1, TNF-alpha, and IL-10. *J Neuroimmun* 2002;**132**:11–17.
3. Ryoke K, Ochi M, Iwata A, Uchio Y, Yamamoto S, Yamaguchi H. A conditioning lesion promotes in vivo nerve regeneration in the contralateral sciatic nerve of rats. *Biochem Biophys Res Comm* 2000;**267**:715–18.
4. Stoll G, Jander S, Myers RR. Degeneration and regeneration of the peripheral nervous system: From Augustus Waller's observations to neuroinflammation. *J Peripher Nerv Syst* 2002;**7**:13–27.
5. Rotshenker S. Wallerian degeneration: The innate-immune response to traumatic nerve injury. *J Neuroinflamm* 2011;**30**:8.
6. Myers RR, Shubayev VI. The ology of neuropathy: An integrative review of the role of neuroinflammation and TNF-a axonal transport in neuropathic pain. *J Peripher Nerv Syst* 2011;**16**:277–86.
7. Perrin FE, Lacroix S, Aviles-Trigueros M, David S. Involvement of monocyte chemoattractant protein-1, macrophage inflammatory protein-1 alpha and interleukin-1 beta in Wallerian degeneration. *Brain* 2005;**128**:854–66.
8. Shamash S, Reichert F, Rotshenker S. The cytokine network of Wallerian degeneration: tumor necrosis factor-alpha, interleukin-1 alpha, and interleukin-1 beta. *J Neurosci* 2002;**22**:3052–60.
9. Shen YJ, DeBellard ME, Salzer JL, Roder J, Filbin MT. Myelin-associated glycoprotein in myelin and expressed by Schwann cells inhibits axonal regeneration and branching. *Mol Cell Neurosci* 1998;**12**:79–91.
10. Gordon T. The physiology of neural injury and regeneration: The role of neurotrophic factors. *J Commun Disord* 2010;**43**:265–73.
11. Dubovy P, Jancalek R, Kubek T. Role of inflammation and cytokines in peripheral nerve regeneration. *Int Rev Neurobiol* 2013;**108**:173–206.
12. Golz G, Uhlmann L, Ludecke D, Markgraf N, Nitsch R, Hendrix S. The cytokine/neurotrophin axis in peripheral axon outgrowth. *Eur J Neurosci* 2006;**24**:2721–30.

13. Campana WM. Schwann cells: Activated peripheral glia and their role in neuropathic pain. *Brain Behav Immun* 2007;**21**:522–7.
14. Sheth RN, Dorsi MJ, Li YB, Murinson BB, Belzberg AJ, Griffin JW, et al. Mechanical hyperalgesia after an L5 ventral rhizotomy or an L5 ganglionectomy in the rat. *Pain* 2002;**96**:63–72.
15. Dubovy P, Klusakova I, Svizenska IH. Inflammatory profiling of Schwann cells in contact with growing axons distal to nerve injury. *Biomed Res Intern* 2014;**2014**:1–7.
16. Chacur M, Milligan ED, Gazda LS, Armstrong C, Wang HC, Tracey KJ, et al. A new model of sciatic inflammatory neuritis (SIN): induction of unilateral and bilateral mechanical allodynia following acute unilateral peri-sciatic immune activation in rats. *Pain* 2001;**94**:231–44.
17. Goethals S, Ydens E, Timmerman V, Janssens S. Toll-like receptor expression in the peripheral nerve. *Glia* 2010;**58**:1701–9.
18. Boivin A, Pineau I, Barrette B, Filali M, Vallieres N, Rivest S, et al. Toll-like receptor signaling is critical for Wallerian degeneration and functional recovery after peripheral nerve injury. *J Neurosci* 2007;**27**:12565–76.
19. Kim D, Lee S, Lee SJ. Toll-like receptors in peripheral nerve injury and neuropathic pain. Toll-like receptors: Roles in infection and neuropathology. In: Kielian T, editor. Current Topics in Microbiology and Immunology *336*. Berlin: Springer-Verlag Berlin; 2009. p. 169–86.
20. Hanani M. Satellite glial cells in sensory ganglia: from form to function. *Brain Res Rev* 2005;**48**:457–76.
21. Dubovy P, Jancalek R, Klusakova I, Svizenska I, Pejchalova K. Intra- and extraneuronal changes of immunofluorescence staining for TNF-alpha and TNFR1 in the dorsal root ganglia of rat peripheral neuropathic pain models. *Cell Mol Neurobiol* 2006;**26**:1205–17.
22. Dubovy P, Klusakova I, Svizenska I, Brazda V. Spatio-temporal changes of SDF1 and its CXCR4 receptor in the dorsal root ganglia following unilateral sciatic nerve injury as a model of neuropathic pain. *Histochem Cell Biol* 2010;**133**:323–37.
23. Dubovy P, Klusakova I, Svizenska I, Brazda V. Satellite glial cells express IL-6 and corresponding signal-transducing receptors in the dorsal root ganglia of rat neuropathic pain model. *Neuron Glia Biol* 2010;**6**:73–83.
24. Jancalek R, Dubovy P, Svizenska I, Klusakova I. Bilateral changes of TNF-alpha and IL-10 protein in the lumbar and cervical dorsal root ganglia following a unilateral chronic constriction injury of the sciatic nerve. *J Neuroinflamm* 2010;**7**:11.
25. Dubovy P, Brazda V, Klusakova I, Hradilova-Svizenska I. Bilateral elevation of interleukin-6 protein and mRNA in both lumbar and cervical dorsal root ganglia following unilateral chronic compression injury of the sciatic nerve. *J Neuroinflamm* 2013;**10**:55.
26. Takeda M, Takahashi M, Matsumoto S. Contribution of the activation of satellite glia in sensory ganglia to pathological pain. *Neurosci Biobehav Rev* 2009;**33**:784–92.
27. Thippeswamy T, McKay JS, Morris R, Quinn J, Wong LF, Murphy D. Glial-mediated neuroprotection: Evidence for the protective role of the NO-cGMP pathway via neuron-glial communication in the peripheral nervous system. *Glia* 2005;**49**:197–210.
28. Kwon MJ, Kim J, Shin H, Jeong SR, Kang YM, Choi JY, et al. Contribution of macrophages to enhanced regenerative capacity of dorsal root ganglia sensory neurons by conditioning injury. *J Neurosci* 2013;**33**:15095–108.
29. Niemi JP, DeFrancesco-Lisowitz A, Roldan-Hernandez L, Lindborg JA, Mandell D, Zigmond RE. A critical role for macrophages near axotomized neuronal cell bodies in stimulating nerve regeneration. *J Neurosci* 2013;**33**:16236–48.
30. Kigerl KA, Gensel JC, Ankeny DP, Alexander JK, Donnelly DJ, Popovich PG. Identification of two distinct macrophage subsets with divergent effects causing either

neurotoxicity or regeneration in the injured mouse spinal cord. *J Neurosci* 2009;**29**:13435–44.
31. Gordon S, Taylor PR. Monocyte and macrophage heterogeneity. *Nat Rev Immunol* 2005;**5**:953–64.
32. Defrancesco-Lisowitz A, Lindborg JA, Niemi JP, Zigmond RE. The neuroimmunology of degeneration and regeneration in the peripheral nervous system. *Neuroscience* 2015;**302**:174–203.
33. Leung L, Cahill CM. TNF-alpha and neuropathic pain - a review. *J Neuroinflamm* 2010;**7**:27.
34. Schafers M, Sommer C, Geis C, Hagenacker T, Vandenabeele P, Sorkin LS. Selective stimulation of either tumor necrosis factor receptor differentially induces pain behavior in vivo and ectopic activity in sensory neurons in vitro. *Neuroscience* 2008;**157**:414–23.
35. Leo M, Argalski S, Schafers M, Hagenacker T. Modulation of voltage-gated sodium channels by activation of tumor necrosis factor receptor-1 and receptor-2 in small drg neurons of rats. *Mediat Inflamm* 2015;**2015**:1–8. <http://dx.doi.org/10.1155/2015/124942>.
36. Cheng CF, Cheng JK, Chen CY, Lien CC, Chu D, Wang SY, et al. Mirror-image pain is mediated by nerve growth factor produced from tumor necrosis factor alpha-activated satellite glia after peripheral nerve injury. *Pain* 2014;**155**:906–20.
37. Murphy PG, Ramer MS, Borthwick L, Gauldie J, Richardson PM, Bisby MA. Endogenous interleukin-6 contributes to hypersensitivity to cutaneous stimuli and changes in neuropeptides associated with chronic nerve constriction in mice. *Eur J Neurosci* 1999;**11**:2243–53.
38. Obreja O, Biasio W, Andratsch M, Lips KS, Rathee PK, Ludwig A, et al. Fast modulation of heat-activated ionic current by proinflammatory interleukin 6 in rat sensory neurons. *Brain* 2005;**128**:1634–41.
39. Fernyhough P, Smith DR, Schapansky J, Van Der Ploeg R, Gardiner NJ, Tweed CW, et al. Activation of nuclear factor-kappa B via endogenous tumor necrosis factor alpha regulates survival of axotomized adult sensory neurons. *J Neurosci* 2005;**25**:1682–90.
40. Lee KM, Jeon SM, Cho HJ. Tumor necrosis factor receptor 1 induces interleukin-6 upregulation through NF-kappaB in a rat neuropathic pain model. *Eur J Pain* 2009;**13**:794–806.
41. Bauer S, Kerr BJ, Patterson PH. The neuropoietic cytokine family in development, plasticity, disease and injury. *Nat Rev Neurosci* 2007;**8**:221–32.
42. Cafferty WBJ, Gardiner NJ, Gavazzi I, Powell J, McMahon SB, Heath JK, et al. Leukemia inhibitory factor determines the growth status of injured adult sensory neurons. *J Neurosci* 2001;**21**:7161–70.
43. Zhong J, Dietzel ID, Wahle P, Kopf M, Heumann R. Sensory impairments and delayed regeneration of sensory axons in interleukin-6-deficient mice. *J Neurosci* 1999;**19**:4305–13.
44. Cafferty WBJ, Gardiner NJ, Das P, Qiu J, McMahon SB, Thompson SWN. Conditioning injury-induced spinal axon regeneration fails in interleukin-6 knock-out mice. *J Neurosci* 2004;**24**:4432–43.

CHAPTER

10

Cytokines as Orchestrators of Skeletal Muscle Tissue Maintenance and the Inflammation Associated With Acquired Autoimmune and Hereditary Muscle Diseases

Boel De Paepe

Ghent University Hospital, Ghent, Belgium

OUTLINE

Cytokine Effector Functions in Tissues.
DOI: http://dx.doi.org/10.1016/B978-0-12-804214-4.00009-9

SKELETAL MUSCLE TISSUE MAINTENANCE

The striated skeletal muscle tissue, functioning to perform conscious movement under the control of motor neurons, is composed of arranged elongated multinucleate muscle cells anchored to the bones by tendons. For maintaining tissue homeostasis, healthy skeletal muscle can make do with a limited repertory of cytokines. Yet, upon proper stimulation, the muscle tissue has the capacity to substantially expand its arsenal. Under physiological conditions, the cytokines that are produced by muscle fibers, termed myokines, include the interleukin-6 (IL-6), [1] and IL-15.[2] Of the chemotactic cytokines termed chemokines, high quantities of CXCL12 can be detected in normal muscle fibers.[3] Myoblasts in culture have been shown to secrete constitutive low levels of IL-6, IL-8, transforming growth factor β (TGFβ), interferon α (IFNα), and transcripts of the chemokine CCL3.[4] In comparison to the muscle fibers, the intramuscular vasculature displays a broader constitutive chemokine expression pattern with detectable levels of CXCL1-2-3-8-10-11-12 and CCL2-4.[5] In addition, tumor necrosis factor (TNF) -related apoptosis inducing ligand (TRAIL) is expressed in capillaries of healthy muscle.[6] Muscle-produced cytokines not only exert local effects but, in addition, may be secreted and redistributed via the bloodstream to trigger additional systemic effects.

Cytokine induction is associated with skeletal muscle injury which activates an acute inflammatory response that repairs tissue damage. A cytokine signaling cascade is activated by the transcription factor nuclear factor κB, i.e., the master controller of proinflammatory pathways which is also active within skeletal muscle tissues.[7] At sites of injury, myoblast proliferation and differentiation is promoted to ensure muscle regeneration and repair and vascular dilation and remodeling. Satellite cells, the resident myogenic stem cells, are activated, differentiating and fusing to form new or repaired myofibers.[8] Severely injured myofibers undergo necrosis and are invaded by macrophages and T cells. These immune cells produce cytokines, including IL-1, IL-6, IL-8, and TNFα, that influence blood flow and vascular permeability, that way facilitating blood flow to the affected areas and accelerating the inflammatory response. The tissue's response is a complex equilibrium between breakdown and build-up in response to damage. IL-15 is upregulated upon myoblast differentiation[9] and has been shown to promote muscle hypertrophy and counteract muscle wasting in cancer cachexia.[10] The TGFβ superfamily cytokine myostatin, also called growth and differentiating factor 8 (GDF-8), inhibits satellite cell proliferation and differentiation,[11] regulating muscle mass by preventing unnecessary build-up.

It is clear that the skeletal muscle tissue is dynamically regulated by physical activity. The mechanism of an exertion-induced acute inflammatory response in skeletal muscle is well known. Muscle contractions provoke cytokine expression which mediates the metabolic changes associated with training adaptation. Acute exercise elicits a stress-like acute inflammatory response, while exercise training leads to very different chronic adaptive gene expression patterns.[12] The considerable changes in the immune system induced by exercise influence the distribution and concentration of lymphocyte subpopulations. Exercise undoubtedly has systemic effects with changes in gene expression equally present in an individual's nonexercising muscles.[13] Several interleukins are involved in exercise adaptation of skeletal muscle and IL-6 comes forward as one of the main factors. Transcription of IL-6, a gene silent in resting muscles, is rapidly and spectacularly expressed in response to muscle contractions[1] and is released into the bloodstream.[14] Impressive 100-fold increases have been reported for IL-6 in serum,[15] an increase that is transient with a magnitude related to the duration and intensity of physical activity.[16] Interestingly, IL-6 has both pro and anti-inflammatory properties and the latter may carry the most weight in exercise-induced expression. The increase of IL-6 is followed by elevated levels of antiinflammatory IL-10,[17] a promoter of myogenesis and muscle hypertrophy. Another interleukin that strongly responds to exercise is the chemokine IL-8. IL-8 mRNA markedly increases in muscle biopsies taken after exercise. Most often this does not translate into increased plasma concentrations,[18] pointing to localized activities as an angiogenesis stimulator. The involvement of TNFα in muscle is complex and exercise has been observed to either suppress,[19] or induce,[15,20] plasma TNFα levels. The expression of IFNγ, a regulator of muscle regeneration, is increased in the plasma of elite cyclists after multistage competition.[20]

A subset of chemokines is also induced in exercised muscle and expression can often be shown beyond the muscle tissue itself. After an exercise sessions, CCL2 levels in muscle[21,22] and plasma[12] strongly increase. In addition, elevated levels of fractalkine (CX3CL1) are detected in the plasma of individuals subjected to acute exercise[12] and CCL3 and CCL4 plasma levels are increased after a marathon.[23] Variants in the genes of CCL2 and its β-chemokine receptor (CCR) CCR2 have been shown to determine the effect generated by a strength-training program.[24]

AGING

Muscle wasting and loss of skeletal muscle strength are associated with normal aging and result from loss or atrophy of muscle fibers. By the age of 70 years, an average 25%–30% reduction of the cross-sectional area of skeletal muscle[25] and 30%–40% reduction of muscle strength[26] is observed. Satellite cell density decreases with age as does their proliferative capacity.[27] The muscle mass lost with age is replaced by adipose and connective tissue.[28] This way, from the age of 50, 1%–2% of skeletal muscle mass is lost each year,[29] a process referred to as sarcopenia. Exercise cannot fully counteract the age-related changes in skeletal muscle function and it is apparent that the environment of older muscle causes its limited ability to regenerate.[30] Changes in the inflammation at rest and in response to exercise-induced muscle injury are believed to be key factors contributing to sarcopenia. Efficient repair of muscle after injury requires a coordinated and dynamic local inflammatory response. As the tissue regenerates, infiltrating macrophages switch phenotype from proinflammatory phagocytic to antiinflammatory myogenic cells. The lower numbers of macrophages present in the skeletal muscle of elderly people[31] may thus inhibit muscle regeneration following injury. The age-related increase in the expression of TNFα and IL-6 especially is thought to substantially contribute to sarcopenia. The increased IL-6 levels in aging inversely correlate with muscle mass[32] and muscle performance.[33] Increased TNFα levels also relate to age-associated muscle loss,[34] and high IL-6 and TNFα levels correlate with lower muscle strength in elderly individuals.[35] In addition, in aged individuals, increased expression of myostatin,[36] and IL-1β[37] expression has been observed. Thus, aging can be perceived as a systemic and chronic low-grade inflammatory state characterized by prolonged nonprovoked cytokine induction.

IDIOPATHIC INFLAMMATORY MYOPATHIES

The idiopathic inflammatory myopathies (IIM) are a heterogeneous group of autoimmune muscle disorders that include dermatomyositis (DM), of which juvenile and adult forms exist, sporadic inclusion body myositis (IBM), polymyositis (PM), and necrotizing autoimmune myopathy (NAM).[38] In DM, membrane attack complexes form on blood vessel endothelia, causing capillary loss and muscle ischemia. Inflammatory infiltrates locate mostly at perimysial sites and consist of helper T cells (Th), B cells, and dendritic cells (DCs). In PM and IBM, nonnecrotic muscle fibers are invaded by autoaggressive CD8+ T cells and macrophages. The perforins and granzymes they release result in cytotoxic necrosis of the fibers.[39] In IBM, initially the invasion of nonnecrotic muscle fibers can be observed with later on appearance of the typical degenerative pathological features such as rimmed vacuoles and inclusions[40] and accumulation of abnormal protein aggregates.[41] PM is a rarer subgroup of IIM.[42] In NAM, muscle tissue inflammation is sparser and muscle fiber necrosis is the most prominent feature.[43] The chronic inflammation associated with the IIM is complexly regulated by cytokines and their expression differs between subgroups.

Interleukins

Upregulation of interleukins is observed in IIM patients and systemic expression levels of IL-6,[44] IL-17,[45] and IL-18[46] positively correlate with disease severity in DM. Many of the differentially expressed interleukins in the IIM influence the activity of T cells. IL-4 is overexpressed in IIM muscle[47] but muscle infiltrates rich in IL-4-producing Th2 cells are associated with low severity of myositis in DM.[48] Possibly, these Th2 cells contribute to muscle regeneration. For IL-6 on the other hand, an aggravating effect upon disease progression is presumed. Blood cell IL-6 expression is increased in DM compared with control subjects and serum levels correlate strongly with disease activity.[44] IL-12, a positive regulator of myogenic differentiation[49] and stimulator of Th1-driven immune reactions, is increased in IIM muscle[47] and in IBM serum.[50] Also, the consistently observed IL-15 elevation in IIM muscle[51] is able to promote CD8+ cytotoxic T-cell function[52] which probably forms a feed-forward loop to perpetuate the inflammatory response in the IIM. IL-17 in particular seems to have a key role in T cell-driven immunity in the IIM by promoting chronic and destructive inflammation. IL-17 cannot be detected in control muscle but is present in muscle from DM[53–56] and other IIM patients.[47] The IL-17-producing Th subset (Th17) is expanded in peripheral blood from IIM patients.[57] The Th17 inducer IL-23 is also overexpressed in IIM muscle.[47]

Tumor Necrosis Factor Superfamily

The important role played by TNF cytokines in the IIM has long been recognized.[58] Expression of TNFα, also referred to as tumor necrosis factor TNFSF2, is firmly associated with muscle damage in the IIM. TNFα is prominently expressed by immune cells and blood vessel endothelial cells in IIM muscle tissues.[59–61] The soluble forms of its receptors TNF-R55 and TNF-R75 are increased in DM and PM sera.[62] TNF-R75 expression is notably increased near inflammatory infiltrates in all IIM and in DM on the perimysial and perifascicular blood vessel endothelium in DM even remote from inflammation.[59] Polymorphisms in the gene-encoding TNFα have been linked to either an increased risk of, or protection against, the development of DM.[63,64]

It appears that lymphotoxins (LT) are also important factors orchestrating sustained inflammation in the IIM. LTα (TNFSF1) has been implicated in the cytotoxic response of CD8+ T cells toward nonnecrotic muscle fibers in PM.[65] LTβ (TNFSF3) is increased in muscle tissues of DM patients, where it localizes to blood vessels and intramuscular follicle-like structures. The latter contain large numbers of T cells, B cells, and plasmacytoid DCs organized in functional compartments,[66] allowing local maturation of autoaggressive immune cells. LTβ is an early marker for inflammatory muscle disease and its expression precedes inflammatory cell infiltration.[67]

TNFSF13B or B-cell activating factor (BAFF), also termed B-lymphocyte stimulator (BLyS), is crucial for B-cell maturation and survival and autoantibody production by plasma cells. Plenty of B-cells and plasma cells can be found in DM, PM and IBM muscle. In addition to its effects on B cells, BAFF is an important regulator of T-cell activation and differentiation by enhancing Th1-driven immune reactions.[68] BAFF transcripts are markedly upregulated in muscle extracts from DM, PM, and IBM patients.[69] In DM muscle, BAFF localizes to the muscle fibers in perifascicular areas.[70] Interestingly, mononuclear cells infiltrating IIM muscle express IFNα,[71] a potent BAFF inducer. Earlier studies had shown increased BAFF serum levels in DM patients only[72] and not in PM and IBM;[73] however, a recent study documents significantly increased serum BAFF levels both in DM and in PM patients with levels correlating with disease activity.[74] A proliferation-inducing ligand (APRIL) or TNFSF13 is highly homologous to BAFF but serum APRIL levels are unaltered in DM, PM, and IBM patients.[75]

Not all TNF superfamily members are proinflammatory and TNFSF10 or TRAIL is thought to inhibit autoreactive T-cell populations, thus is potentially beneficial to IIM muscle recovery. TRAIL blood levels are increased in most but not all DM patients[76] and in PM, many inflammatory cells are TRAIL positive.[77]

Chemokines

The role of chemokines in the accumulation of inflammatory cells within IIM muscle has been recognized for almost two decades now and cannot be overestimated. Of special interest are the IFN type 1-indicible chemokines. IFNs play a crucial role in the sustained inflammatory response, contributing to leukocyte infiltration and upregulation of major histocompabibility complex (MHC) class I expression. Dysregulation of type 1 IFNs has been implicated in the IIM and seems especially prominent in DM. CXCL9, CXCL10, and CXCL11 are absent from normal muscle but are expressed in DM at variable levels.

Interestingly, expression strongly correlates with the intensity of inflammatory cell infiltration in the muscle specimens[78] and the severity of muscle damage.[48] In part of DM patients, increased blood cell CXCL10 mRNA levels can be present.[76] High plasma levels of CXCL10 distinguish IIM from hereditary muscle diseases, giving a sensitivity of 90%.[79] Elevation of CXCL10 blood levels correlates with DM disease activity.[46,76]

The presence of substantial amounts of DCs, with selective preference to the myeloid or plasmacytoid lineage depending on the IIM subtype,[80,81] has been shown in the IIM and correlates with the expression of CXCL12, also termed stromal cell-derived factor (SDF-1), an important attractant for DCs. The chemokine is produced locally by muscle fibers and blood vessel endothelial cells.[3] This possibly leads to the accumulation of α-chemokine receptor (CXCR) CXCR4+ T cells in IIM muscle[48,66] with the chemokine receptor functioning as an important costimulator of the T-cell receptor.

In most IIM, large numbers of muscle-infiltrating macrophages can be found and chemokines involved in macrophage function are strongly activated in these diseases. Elevation of monocyte chemoattractant protein CCL2 is increased in the blood of DM patients,[44] where CCL2 and CCL7 blood levels correlate with disease activity.[76] Macrophage inflammatory protein CCL19 mRNA levels are significantly higher in mononuclear cells within DM muscle tissue than their counterparts found scattered in control muscle.[66]

DUCHENNE MUSCULAR DYSTROPHY

Muscular dystrophies are inherited diseases characterized by progressive weakness and loss of muscle mass. The most common dystrophies are Duchenne muscular dystrophy (DMD) and Becker muscular dystrophy both of which are caused by defects in the dystrophin gene. DMD is an X-linked muscle disease, with a prevalence of one in 3500 boys worldwide, caused by a severe deficiency of dystrophin protein. When the sarcolemmal dystrophin-associated glycoprotein complex is disturbed, the linkage between the muscle cell's cytoskeleton and the extracellular matrix is compromised, which leads to sarcolemmal instability and increased vulnerability to mechanical stress.[82] Contraction of dystrophin deficient myofibers produces severe damage and generates cycles of muscle fiber necrosis and regeneration. Necrotizing myofibers are attacked by macrophages; few T cells, B cells, and DCs are also found within the inflammatory areas. This inflammatory response leads to fibrosis and muscle contractile dysfunction. Disease intensity is aggravated by a complexly regulated interplay of soluble factors and adhesion molecules.[83] Studies in the murine disease model mdx have contributed substantially to our understanding of the cytokine-induced effects. However, caution is needed when extrapolating results as the clinical picture of the murine model is less severe and follows a different time course than the human disease.

Transforming Growth Factor β

TGFβ is highly expressed in muscle samples from mdx mice[84] as well as in human DMD muscle.[85] DMD is characterized by intense tissue fibrosis and TGFβ seems to function as a major fibrogenic factor. TGFβ levels are higher in DMD muscle than in DM,[54] where levels correlate with the degree of connective tissue proliferation and not with severity of inflammation.[86] The TGFβ family member myostatin also displays antimyogenic properties and has been put forward as a target for DMD therapy.[87] Its inhibition may stimulate muscle cell proliferation and recovery in patients.

Interleukins

The interleukin profile of DMD is not deviant from that of the IIM, yet the extent to which individual cytokines are expressed differs. TNFα and IL-6 are also increased in DMD patients[88] but IL-6 levels are substantially lower than those found in DM.[54] Contrary, compared to DM, higher levels of IL-17 are present in DMD muscle and patients with highest IL-17 levels in their muscle biopsy have poorer functional outcome at six years of age.[54]

Chemokines

As muscular dystrophy is characterized by often very pronounced immune cell infiltration inside the muscle tissue, it is not surprising that many chemokines are upregulated in these diseases. In mdx mice, increased levels of CCL2, CCL5, CCL7, and the receptors CXCR4, CCR1, and CCR2 are present in muscle.[89,90] A study of DMD quadriceps also shows an increase of CXCL12 messenger expression.[91] Expression of CXCL1, CXCL2, CXCL3, CXCL8, and CXCL11 absent from normal muscle fibers is induced in DMD myofibers. CXCL11, CXCL12, and the ligand-receptor couple CCL2–CCR2 is upregulated on blood vessel endothelium of DMD patients.[92] Of special interest is that the chemokine profile of DMD muscle is substantially different from IIM, reflecting the secondary versus primary inflammatory nature of

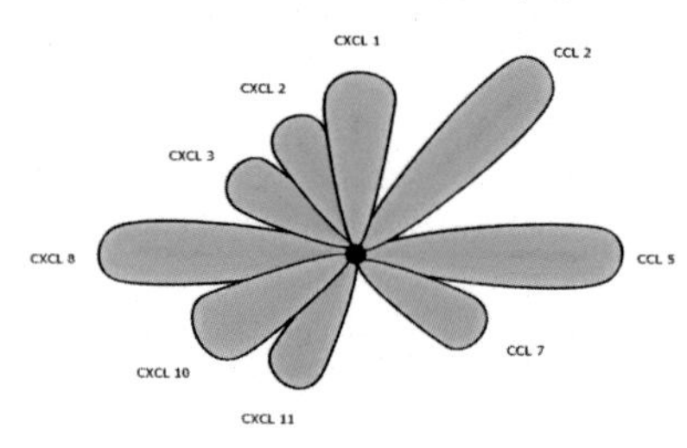

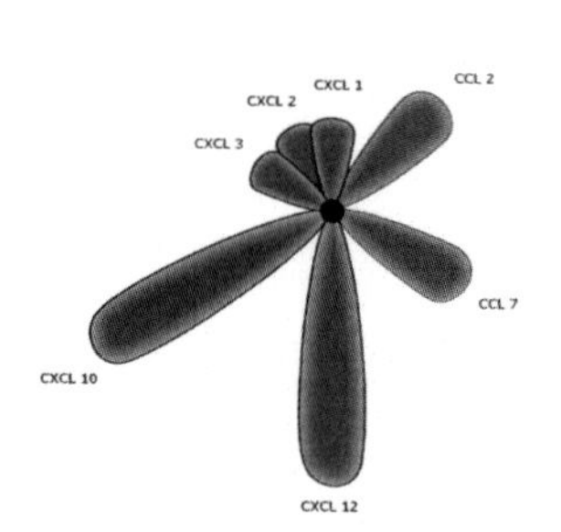

FIGURE 10.1 Chemokine profile of muscle-infiltrating CD68 + macrophages. Width of chemokine arms represents the level of expression observed in the majority of macrophages.

these different diseases. CCL2 is expressed in much lower levels in DMD muscle compared to DM.[54] The chemokine expression pattern of muscle-infiltrating CD68+ macrophages, with expression of high levels of CXCL8, CCL2, and CCL5, is strikingly different from their expression pattern observed in the IIM (Fig. 10.1).

CONCLUSION

Cytokines have key roles in myositis, both in diseases with primary autoimmune origin as well as in inflammation secondary to hereditary muscle dysfunction. Their prominence and specific expression patterns make them amenable therapeutic targets for innovative more selective immunosuppression.[93]

References

1. Hiscock N, Chan MH, Bisucci T, Darby IA, Febbraio MA. Skeletal myocytes are a source of interleukin-6 mRNA expression and protein release during contraction: Evidence of fiber type specificity. *FASEB J* 2004;**18**:992–4.
2. Grabstein KH, Eisenman J, Shaneback K, Rauch C, Srinivasan S, Fung V, et al. Cloning of a T cell growth factor that interacts with the beta chain of the interleukin-2 receptor. *Science* 1994;**264**:965–8.
3. De Paepe B, Schroder JM, Martin JJ, Racz GZ, De Bleecker JL. Localization of the α-chemokine SDF-1 and its receptor CXCR4 in idiopathic inflammatory myopathies. *Neuromuscular Disord* 2004;**14**:265–73.
4. Nagaraju K, Raben N, Merritt G, Loeffler L, Krik K, Plotz P. A variety of cytokines and immunologically relevant surface molecules are expressed by normal human skeletal muscle cells under proinflammatory stimuli. *Clin Exp Immunol* 1998;**113**:407–14.
5. De Paepe B, Creus KK, De Bleecker JL. Chemokines in idiopathic inflammatory myopathies. *Front Biosci* 2008;**13**:2548–77.

6. Danielsson O, Nilsson C, Lindvall B, Emerudh J. Expression of apoptosis related proteins in normal and diseased muscle: A possible role for Bcl-2 in protection of striated muscle. *Neuromuscular Disord* 2009;**19**:412–17.
7. Creus KK, De Paepe B, Werbrouck BF, Vervaet V, Weis J, De Bleecker JL. Idiopathic inflammatory myopathies and the classical NF-κB complex: Current insights and implications for therapy. *Autoimm Rev* 2009;**8**:627–31.
8. Schultz E, Jaryszak DL, Valliere CR. Response of satellite cells to focal skeletal muscle injury. *Muscle Nerve* 1985;**8**:217–22.
9. Quinn LS, Strait-Bodey L, Anderson BG, Argiles JM, Havel PJ. Interleukin-15 stimulates adiponectin secretion by 3 T3-L1 adipocytes: evidence for a skeletal muscle-to-fat signaling pathway. *Cell Biol Int* 2005;**29**:449–57.
10. Carbo N, Lopez-Soriano J, Costelli P, Busquets S, Alvarez B, Baccino IM, et al. Interleukin-15 antagonizes muscle protein waste in tumour-bearing rats. *Br J Cancer* 2000;**83**:526–31.
11. Langley B, Thomas M, Bishop A, Sharma M, Gilmour S, Kambadur R. Myostatin inhibits myoblast differentiation by down-regulating MyoD expression. *J Biol Chem* 2002;**277**:49831–40.
12. Catoire M, Mensink M, Boekschoten MV, Hangelbroek R, Muller M, Schrauwen P, et al. Identification of human exercise-induced myokines using secretome analysis. *Physiol Genomics* 2014;**46**:256–67.
13. Catoire M, Mensink M, Boekshoten MV, Hangelbroek R, Muller M, Schrauwen P, et al. Pronounced effects of acute endurance exercise on gene expression in resting and exercising human skeletal muscle. *PLoS One* 2012;**7**:e51066.
14. Steensberg A, Keller C, Starkie RL, Osada T, Febbraio MA, Pedersen BK. IL-6 and TNF-α expression in, and release from, contracting human skeletal muscle. *Am J Physiol Endocrinol Metab* 2002;**283**:E1272–8.
15. Ostowski K, Rohde T, Zacho M, Asp S, Pedersen BK. Evidence that IL-6 is produced in skeletal muscle during intense long-term muscle activity. *J Physiol* 1998;**508**:949–53.
16. Febbraio MA, Pedersen BK. Muscle-derived interleukin-6: Mechanisms for activation and possible biological roles. *FASEB J* 2002;**16**:1335–47.
17. Ostrowski K, Rohde T, Asp S, Schjerling P, Pedersen BK. Pro- and anti-inflammatory cytokine balance in strenuous exercise in humans. *J Physiol* 1999;**515**:287–91.
18. Akerstrom TC, Steensberg A, Keller P, Keller C, Penkowa M, Pedersen BK. Exercise induces interleukin-8 expression in human skeletal muscle. *J Physiol* 2005;**563**:507–16.
19. Starkie R, Ostrowski SR, Jauffred S, Febbraio M, Pedersen BK. Exercise and IL-6 infusion inhibit endotoxin-induced TNF-alpha production in humans. *FASEB J* 2003;**17**:884–6.
20. Cordova Martinez A, Martorell Pons M, Sureda Gomila A, Tur Mari JA, Pons Biescas A. Changes in circulating cytokines and markers of muscle damage in elite cyclists during a multi-stage competition. *Clin Physiol Funct Imaging* 2015;**35**:351–8.
21. Hubal MJ, Chen TC, Thompson PD, Clarkson PM. Inflammatory gene changes associated with the repeated-bout effect. *Am J Physiol Regul Integr Comp Physiol* 2008;**294**: R1628–37.
22. Della Gatta PA, Granham AP, Peake JM, Cameron-Smith D. Effect of exercise training on skeletal muscle cytokine expression in the elderly. *Brain Behave Immun* 2014;**34**:80–6.
23. Klarlund Pedersen B. Exercise and cytokines. *Immunol Cell Biol* 2000;**78**:532–5.
24. Harmon BT, Orkunoglu-Suer EF, Adham K, Larkin JS, Gordish-Dressman H, Clarkson PM, et al. CCL2 and CCR2 variants are associated with skeletal muscle strength and change in strength with resistance training. *J Appl Physiol* 2010;**109**:1779–85.

25. Porter MM, Vandervoort AA, Lexell J. Aging of human muscle: Structure, function and adaptability. *Scand J Med Sci Sports* 1995;**5**:129–42.
26. Skelton DA, Greig CA, Davies JM, Young A. Strenght, power and related functional ability of healthy people aged 65–89 years. *Age Ageing* 1994;**23**:371–7.
27. Renault V, Thornell LE, Butler-Browne G, Mouly V. Human skeletal muscle satellite cells: Aging, oxidative stress and the mitotic clock. *Exp Gerontol* 2002;**37**:1229–36.
28. Lexell J. Human aging, muscle mass, and fiber type composition. *J Gerontol A Biol Sci Med Sci* 1995;**50**:11–16.
29. Hughes VA, Frontera WR, Roubenoff R, Evans WJ, Singh MA. Longitudinal changes in body composition in older men and women: Role of body weight change and physical activity. *Am J Clin Nutr* 2002;**76**:473–81.
30. Conboy IM, Rando TA. Aging, stem cells and tissue regeneration: Lessons from muscle. *Cell Cycle* 2005;**4**:407–10.
31. Hamada K, Vannier E, Sacheck JM, Witsell AL, Roubenoff R. Senescence of human skeletal muscle impairs the local inflammatory cytokine response to acute eccentric exercise. *FASEB J* 2005;**19**:264–6.
32. Visser M, Pahor M, Taaffe DR, Goodpaster BH, Simonsick FM, Newman AB, et al. Relationship of interleukin-6 and tumor necrosis factor-alpha with muscle mass and muscle strength in elderly men and women: The health ABC study. *J Gerontol A Biol Sci Med Sci* 2002;**57**:M326–32.
33. Bautmans I, Njemini R, Lambert M, Demanet C, Mets T. Circulating acute phase mediatros and skeletal muscle performance in hospitalized geriatric patients. *J Gerontol A Biol Sci Med Sci* 2005;**60**:361–7.
34. Pedersen M, Bruunsgaard H, Weis N, Hendel HW, Andreassen BU, Eldrup E, et al. Circulating levels of of TNF-alpha and IL-6-relation to truncal fat mass and muscle mass in healthy elderly individuals and in patients with type-2 diabetes. *Mech Ageing Dev* 2003;**124**:495–502.
35. Yende S, Waterer GW, Tolley EA, Newman AB, Bauer DC, Taaffe DR, et al. Inflammatory markers are associated with ventilatory limitation and muscle dysfunction in obstructive lung disease in well functioning elderly subjects. *Thorax* 2006;**61**:10–16.
36. Leger B, Derave W, De Bock K, Hespel P, Russell AP. Human sarcopenia reveals an increase in SOCS-3 and myostatin and a reduced efficiency of Akt phosphorylation. *Rejuvenation Res* 2008;**11**:163–75.
37. Przybyla B, Gurley C, Harvey JF, Bearden E, Kortebein P, Evans WJ, et al. Aging alters macrophage properties in human skeletal muscle both at rest and in response to acute resistance exercise. *Exp Gerontol* 2006;**41**:320–7.
38. Dalakas MC. Pathogenesis and therapies of immune-mediated myopathies. *Autoimm Rev* 2012;**11**:203–6.
39. Goebels N, Michaelis D, Engelhardt M, Hubers S, Bender A, Pongratz D, et al. Differential expression of perforin in muscle-infiltrating T cells in polymyositis and dermatomyositis. *J Clin Invest* 1996;**97**:2905–10.
40. van de Vlekkert J, Hoogendijk JE, de Visser M. Myositis with endomysial cell invasion indicates inclusion body myositis even if other criteria are not fulfilled. *Neuromuscular Disord* 2015;**25**:451–6.
41. Askanas V, Engel WK, Nogalska A. Pathogenic considerations in sporadic inclusion-body myositis, a degenerative muscle disease associated with aging and abnomalities of myoproteostasis. *J Neuropathol Exp Neurol* 2012;**71**:680–93.
42. Chahin N, Engel AG. Correlation of muscle biopsy, clinical course, and outcome in PM and sporadic IBM. *Neurology* 2008;**70**:418–24.

43. Vattemi G, Mirabella M, Guglielmi V, Lucchini M, Tomelleri G, Ghirardello A, et al. Muscle biopsy features of idiopathic inflammatory myopathies and differential diagnosis. *Auto-Immun Highlights* 2014;**5**:77–85.
44. Bilgic H, Ytterberg SR, Amin S, McNallan KT, Wilson JC, Koeuth T, et al. Interleukin-6 and type 1 interferon-regulated genes and chemokines mark disease activity in dermatomyositis. *Arthritis Rheum* 2009;**60**:3436–46.
45. Notarnicola A, Lapadula G, Natuzzi D, Lundberg IE, Iannone F. Correlation between serum levels of IL-15 and IL-17 in patients with idiopathic inflammatory myopathies. *Scand J Rheumatol* 2015;**44**:224–8.
46. Bellutti Enders F, van Wijk F, Scholman R, Hofer M, Prakken BJ, van Royen-Kerkhof A, et al. Correlation of CXCL10, tumor necrosis factor receptor type II, and galectin 9 with disease activity in juvenile dermatomyositis. *Arthritis Rheum* 2014;**66**:2281–9.
47. Kondo M, Murakawa Y, Harashima N, Kobayashi S, Yamaguchi S, Harada M. Roles of proinflammatory cytokines and the Fas/Fas ligand interaction in the pathogenesis of inflammatory myopathies. *Immunology* 2009;**128**:e589–99.
48. Fujiyama T, Ito T, Ogawa N, Suda T, Tokura Y, Hashizume H. Preferential infiltration of interleukin-4-producing CXCR4+ T cells in the lesional muscle but not skin of patients with dermatomyositis. *Clin Exp Immunol* 2014;**177**:110–20.
49. Romanazzo S, Forte G, Miroshima K, Taniguchi A. IL-12 involvement in myogenic differentiation of C2C12 in vitro. *Biomater Sci* 2015;**3**:469–79.
50. Allenbach Y, Chaara W, Rosenzwajg M, Six A, Prevel N, Mingozzi F, et al. Th1 response and systemic Treg deficiency in inclusion body myositis. *PLoS One* 2014;**9**: e88788.
51. Sugiura T, Kawaguchi Y, Takagi K, Ohta S, Fukasawa C, Ohsako-Higami S, et al. Increased CD40 expression on muscle cells of polymyositis and dermatomyositis: role of CD40-CD40 ligand interaction in IL-6, IL-8, IL-15, and monocyte chemoattractant protein-1 production. *J Immunol* 2002;**14**:917–24.
52. Huang PL, Hou MS, Wang SW, Chang CL, Liou YH, Liao NS. Skeletal muscle interleukin 15 promotes CD8+ T-cell function and autoimmune myositis. *Skeletal Muscle* 2015;**5**:e33.
53. Khanna S, Reed AM. Immunopathogenesis of juvenile dermatomyositis. *Muscle Nerve* 2010;**41**:581–92.
54. De Pasquale L, D'Amico A, Verardo M, Petrini S, Bertini E, De Benedetti F. Increased muscle expression of interleukin-17 in Duchenne muscular dystrophy. *Neurology* 2012;**78**:1309–14.
55. Chevrel G, Page G, Granet C, Streichenberger N, Varennes A, Miossec P. Interleukin-17 increases the effects of IL-1 beta on muscle cells: arguments for the role of T cells in the pathogenesis of myositis. *J Neuroimmunol* 2003;**137**:125–33.
56. Tournadre A, Porcherot M, Chérin P, Marie I, Hachulla E, Miossec P. Th1 and Th17 balance in inflammatory myopathies: interaction with dendritic cells and possible link with response to high-dose immunoglobulins. *Cytokine* 2009;**46**:297–301.
57. Espinosa-Ortega F, Gomez-Martin D, Santana-De Anda K, Romo-Tena J, Villasenor-Ovies P, Alcocer-Varela J. Quantitative T cells subsets profile in peripheral blood from patients with idiopathic inflammatory myopathies: Tilting the balance towards proinflammatory and pro-apoptotic subsets. *Clin Exp Immunol* 2014;**179**:520–8.
58. De Paepe B, Creus KK, De Bleecker JL. The tumor necrosis factor superfamily of cytokines in the inflammatory myopathies: Potential targets for therapy. *Clin Dev Immunol* 2012;e369432.
59. De Bleecker JL, Meire VI, Declercq W, Van Aken EH. Immunolocalization of tumor necrosis factor-alpha and its receptors in inflammatory myopathies. *Neuromuscular Disord* 1999;**9**:239–46.

60. Kuru S, Inukai A, Liang Y, Doyu M, Takano A, Sobue G. Tumor necrosis factor-α expression in muscles of polymyositis and dermatomyositis. *Acta Neuropathol* 2000;**99**:585–8.
61. Tews DS, Goebel HH. Cytokine expression profile in idiopathic inflammatory myopathies. *J Neuropathol Exp Neurol* 1996;**55**:342–7.
62. Gabay C, Gay-Croisier F, Roux-Lombard P, Meyer O, Maineti C, Guerne PA, et al. Elevated serum levels of interleukin-1 receptor antagonist in polymyositis/dermatomyositis. A biological marker of disease activity with a possible role in the lack of acute-phase protein response. *Arthritis Rheum* 1994;**37**:1744–51.
63. Fedczyna TO, Lutz J, Pachman LM. Expression of TNFalpha by muscle fibers in biopsies from children with untreated juvenile dermatomyositis: association with the TNFalpha-308A allele. *Clin Immunol* 2001;**100**:236–9.
64. Mamyrova G, O'Hanlon TP, Sillers L, Malley K, James-Newton L, Parks CG, et al. Cytokine gene polymorphisms as risk and severity factors for juvenile dermatomyositis. *Arthritis Rheum* 2008;**58**:3941–50.
65. Liang Y, Inukai A, Kuru S, Kato T, Doyu M, Sobue G. The role of lymphotoxin in pathogenesis of polymyositis. *Acta Neuropathol* 2000;**100**:521–7.
66. Lopez de Padilla CM, Vallejo AN, Lacomis D, McNallan K, Reed AM. Extranodal lymphoid microstructures in inflamed muscle and disease severity of new-onset juvenile dermatomyositis. *Arthritis Rheum* 2009;**60**:1160–72.
67. Creus KK, De Paepe B, Weis J, De Bleecker JL. The multifaceted character of lymphotoxin β in inflammatory myopathies and muscular dystrophies. *Neuromuscular Disord* 2012;**22**:712–19.
68. Mackay F, Leung H. The role of the BAFF/APRIL system on T cell function. *Semin Immunol* 2006;**18**:284–9.
69. Salajegheh M, Pinkus JL, Amato AA, Morehouse C, Jallal B, Yao Y, et al. Permissive environment of B-cell maturation in myositis muscle in the absence of B-cell follicles. *Muscle Nerve* 2010;**42**:576–83.
70. Baek AH, Suh GI, Hong JM, Suh BC, Shim Choi YS. The increased expression of B cell activating factor (BAFF) in patients with dermatomyositis. *Neuromuscular Disord* 2010;**20**:634.
71. Lundberg IE, Helmers SB. The type I interferon system in idiopathic inflammatory myopathies. *Autoimm* 2010;**43**:239–43.
72. Matsushita T, Hasegawa M, Yanaba K, Kodera M, Takehara K, Sato S. Elevated serum BAFF levels in patients with systemic sclerosis: enhanced BAFF signaling in systemic sclerosis B lymphocytes. *Arthritis Rheum* 2006;**54**:192–201.
73. Krystufkova O, Vallerskog T, Barbasso Helmers S, Mann H, Putova I, Belacek J, et al. Increased serum levels of B cell activating factor (BAFF) in subsets of patients with idiopathic inflammatory myopathies. *Ann Rheum Dis* 2009;**68**:836–43.
74. Peng QL, Shu XM, Wang DX, Wang Y, Lu X, Wang GC. B-cell activating factor as a serological biomarker for polymyositis and dermatomysoitis. *Biomark Med* 2014;**8**:395–403.
75. Szodoray P, Alex P, Knowlton N, Centola M, Dozmorov I, Csipo I, et al. Idiopathic inflammatory myopathies, signified by distinctive peripheral cytokines, chemokines and the TNF family members B-cell activating factor and a proliferation inducing ligand. *Rheumatol* 2010;**49**:1876–7.
76. Baechler EC, Bauer JW, Slattery CA, Ortmann WA, Espe KJ, Novitzke J, et al. An interferon signature in the peripheral blood of dermatomyositis patients is associated with disease activity. *Mol Med* 2007;**13**:59–68.
77. Danielsson O, Nilsson C, Lindvall B, Ernerudh J. Expression of apoptosis related proteins in normal and diseased muscle: a possible role for Bcl-2 in protection of striated muscle. *Neuromuscular Disord* 2009;**19**:412–17.

78. Fall N, Bove KE, Stringer K, Lovell DJ, Brunner HI, Weiss J, et al. Association between lack of angiogenic response in muscle tissue and high expression of angiostatic ELR-negative CXC chemokines in patients with juvenile dermatomyositis. *Arthritis Rheum* 2005;**52**:3175–80.
79. Uruha A, Noguchi S, Sato W, Nishimura H, Mitsuhashi S, Yamamura T, et al. Plasma IP-10 level distinguishes inflammatory myopathy. *Neurology* 2015;**85**:1–2.
80. Greenberg SA, Pinkus GS, Amato AA, Pinkus JL. Myeloid dendritic cells in inclusion-body myositis and polymyositis. *Muscle Nerve* 2007;**35**:17–23.
81. Greenberg SA, Pinkus JL, Pinkus GS, Burlesou T, Sanoudou D, Tawil R, et al. Interferon alpha/beta-mediated innate immune mechanisms in dermatomyositis. *Ann Neurol* 2005;**57**:664–78.
82. Deconinck N, Dan B. Pathophysiology of Duchenne muscular dystrophy: Current hypotheses. *Pediatr Neurol* 2007;**36**:1–7.
83. Villalta SA, Rosenberg AS, Bluestone JA. The immune system in Duchenne muscular dystrophy: Friend or foe. *Rare Diseases* 2015;**3**:e1010966.
84. Gosslin LE, Martinez DA. Impact of TNF-alpha blockade on TGF-beta1 and type 1 collagen mRNA expression in dystrophic muscle. *Muscle Nerve* 2009;**30**:244–6.
85. Bernasconi P, Di Blasi C, Mora M, Morandi L, Galbiati S, Confalonieri P, et al. Transforming growth factor-beta1 and fibrosis in congenital muscular dystrophies. *Neuromuscular Disord* 1999;**9**:28–33.
86. Confalonieri P, Bernasconi P, Cornelio F, Mantegazza R. Transforming growth factor-beta 1 in polymyositis and dermatomyositis correlates with fibrosis but not with mononuclear cell infiltrates. *J Neuropathol Exp Neurol* 1997;**56**:479–84.
87. Amthor H, Hoogaars WM. Interference with myostatin/ActRIIB signaling as a therapeutic strategy for Duchenne muscular dystrophy. *Curr Gene Ther* 2012;**12**:245–59.
88. Messina S, Vita GL, Aguennouz M, Sframeli M, Romeo S, Rodolico C, et al. Activation of NF-kappaB pathway in Duchenne muscular dystrophy: relation to age. *Acta Myol* 2011;**30**:16–23.
89. Porter JD, Guo W, Merriam AP, Khanna S, Cheng G, Zhou X, et al. Persistent overexpression of specific CC class chemokines correlates with macrophage and T-cell recruitment in mdx skeletal muscle. *Neuromuscular Disord* 2003;**13**:223–5.
90. Demoule A, Divangahi M, Danialou G, Gvozdic D, Larkin G, Bao W, et al. Expression and regulation of CC class chemokines in the dystrophic (mdx) diaphragm. *Am J Respir Cell Mol Biol* 2005;**33**:178–85.
91. Pescatori M, Broccolini A, Minetti C, Bertini E, Bruno C, D'amico A, et al. Gene expression profiling in the early phases of DMD: a constant molecular signature characterizes DMD muscle from early postnatal life throughout disease progression. *FASEB J* 2007;**21**:1210–26.
92. De Paepe B, De Bleecker J. Upregulation of chemokines and their receptors in Duchenne muscular dystrophy: Potential for attenuation of myofiber necrosis. *Med Inflamm* 2013;e540370.
93. De Paepe, Zschuntzsch J. Scanning for therapeutic targets within the cytokine network of idiopathic inflammatory myopathies. *Int J Mol Sci* 2015;**16**:18683–713.

CHAPTER

11

Airway Epithelial Cytokines in Asthma and Chronic Obstructive Pulmonary Disease

Rakesh K. Kumar and Cristan Herbert

UNSW Australia, Sydney, NSW, Australia

OUTLINE

The principal functions of epithelial tissues are traditionally described as (i) covering, lining, and protecting surfaces (ii) absorption and (iii) secretion.[1] However, as is increasingly being recognized, surface epithelial cells are active participants in cell-to-cell communication

Cytokine Effector Functions in Tissues.
DOI: http://dx.doi.org/10.1016/B978-0-12-804214-4.00010-5

in response to environmental stimuli. This intercellular messaging is primarily via the release of cytokines.

In this chapter, we will focus on the epithelial cells of the conducting airways (i.e., the tracheobronchial tree, excluding the alveolar epithelium) as a source of cytokines. We will examine the contribution of these airway epithelial cells (AEC) to both innate host defences and the adaptive immunological response.[2] In particular, our emphasis will be on how cytokines produced by AEC are relevant to the pathogenesis of asthma and chronic obstructive pulmonary disease (COPD) as well as on mechanisms regulating the production of these cytokines.

AEC: THE FRONT LINE

The airway epithelium is continuously exposed to diverse environmental stimuli, primarily delivered via inhalation. Notably, these include infectious agents (viruses, bacteria, and fungi), noninfectious particulate matter (irritant particles, allergen-loaded particles), and noxious gases.

Most of these environmental agents elicit a host response with initial recognition of the stimulus being primarily dependent on surface receptors on AEC, especially pattern recognition receptors.[3] These include the toll-like receptors,[4] the nucleotide-binding oligomerization domain (NOD)-like receptors,[5] and the cytosolic retinoic acid-inducible gene I (RIG-I)-like receptors.[6] In addition, AEC express protease-activated receptors on their surface.[7] Oxidant stress injury by many environmental irritants also contributes to activation of AEC.[8]

The innate host defence response to such stimuli includes the synthesis and secretion of a variety of cytokines by AEC which can recruit and activate a range of cells involved in inflammation and the adaptive immunological response. The different groups of cytokines known to be produced by AEC (Table 11.1) are discussed further below.

Proinflammatory Cytokines

The cytokines interleukin (IL)-1β, IL-6, and tumor necrosis factor α (TNF-α) play key roles in amplifying the inflammatory response. They have multiple overlapping effects that contribute to microvascular changes, altered expression of endothelial adhesion molecules for leukocytes as well as activation of a range of cells participating in inflammation and healing.[9] Most environmental stimuli cause upregulated expression of mRNA for these cytokines by AEC with secretion of the active proteins. Notably, they are produced following infection by respiratory viruses such as rhinovirus

TABLE 11.1 Major Classes of Cytokines Produced by Airway Epithelial Cells

Proinflammatory cytokines	IL-1β, IL-6, TNF-α
	IL-11, IL-32
Chemokines	CCL1, CCL2, CCL3, CCL5, CCL11, CCL17, CCL19, CCL20
	CX3CL1
	CXCL1, CXCL5, CXCL8, CXCL9, CXCL10, CXCL11
Interferons	IFNβ
	IFNλ1, IFNλ2/3
Other cytokines able to influence the adaptive immune response	IL-33, IL-25, TSLP
	G-CSF, GM-CSF, IL-15, BAFF
Growth factors	TGF-β, EGF, PDGF, VEGF

Abbreviations: CCL, CC chemokine ligand; CXCL, CXC chemokine ligand; G-CSF, granulocyte colony stimulating factor; GM-CSF, granulocyte-macrophage colony stimulating factor; BAFF, B cell activating factor; TSLP, thymic stromal lymphopoietin; EGF, epidermal growth factor; PDGF, platelet-derived growth factor; VEGF, vascular endothelial growth factor

and influenza virus[10] as well as following exposure to various forms of ambient particulate matter.[11,12] Expression of the proinflammatory cytokine IL-32 by AEC is similarly increased in response to viral infection and in the setting of oxidative stress.[13,14]

Chemokines

A key feature of the response of AEC to injury is the production of a variety of chemokines. The chemokine response to respiratory viral infections has been elucidated in experimental studies using poly I:poly C double-stranded RNA (dsRNA) as a surrogate stimulus.[15,16] There is marked upregulation of expression of a range of CCL chemokines (notably CCL2, 3, 5, 17, and 20) and CXCL chemokines (notably CXCL1, 5, 8, 9, 10, and 11) as well as of CX3CL1. These contribute to recruitment of natural killer (NK) cells, T lymphocytes, monocytes, dendritic cells (DC), neutrophils, and eosinophils (Fig. 11.1).

Interferons

As part of the innate host defence response to respiratory viral infections, AEC upregulate expression of interferons (IFN), both type I (especially IFNβ1) and type III (IFNλ1 and IFNλ2/3).[17] In turn, this leads to

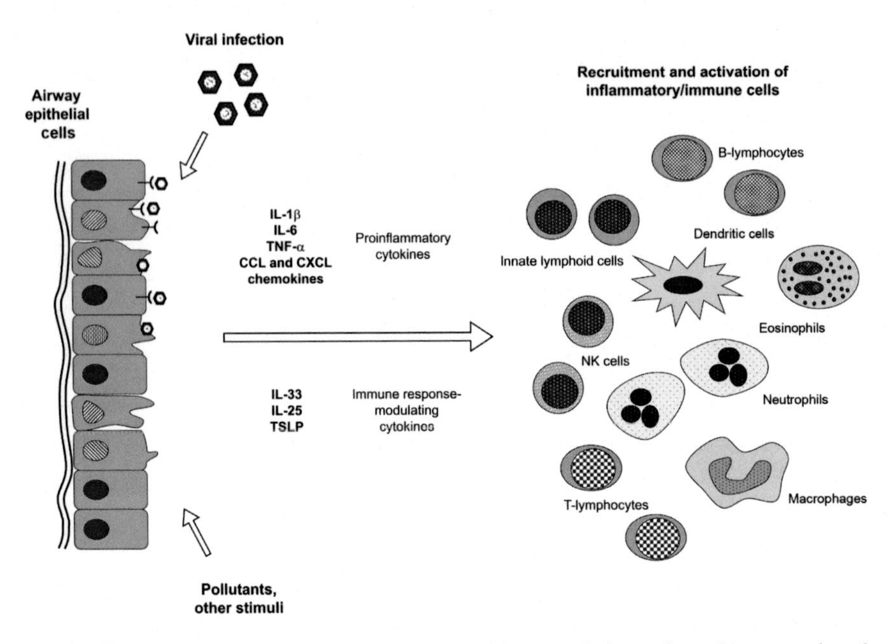

FIGURE 11.1 Diagrammatic representation of how AEC-derived cytokines, produced in response to viral infection and exposure to pollutants or other environmental stimuli, can drive inflammatory and immunological responses.

upregulation of a large family of IFN-stimulated genes, including many of the chemokines referred to above.[18]

Other Important Immune Response-Modulating Cytokines

Recently, there has been a great deal of interest in AEC-derived cytokines that contribute to initiating or enhancing a Th2-biased immunological response. Notably, these include thymic stromal lymphopoietin (TSLP), IL-25, and IL-33[19] which are discussed further below.

Other noteworthy immune response-modulating cytokines produced by AEC include granulocyte-macrophage colony-stimulating factor (GM-CSF), which is important in polarization of DC,[20] and IL-15 which contributes to DC maturation.[21] AEC may also help to drive the antiviral antibody response via production of B-cell activating factor of the TNF family (BAFF).[22]

Growth Factors

AEC are active participants in healing and repair responses following injury. In this context, they secrete a variety of growth factors, notably of the epidermal growth factor (EGF), transforming growth factor (TGF)-β and vascular endothelial growth factor (VEGF) families.[23,24]

AEC-DERIVED CYTOKINES IN THE INDUCTION OF ASTHMA AND COPD

There is now consensus that AEC-derived cytokines make a crucial contribution to the induction phase of childhood allergic asthma. In particular, TSLP has an important role in activating DC and together with cytokines such as GM-CSF and IL-33, may enable DC to induce CD4 + T lymphocytes to differentiate toward a Th2 phenotype.[8,19]

IL-33 and IL-25 also have key roles with respect to recruitment and activation of innate lymphoid cells that produce type 2 cytokines (ILC2).[25,26] These cells which are a major source of both IL-5 and IL-13 appear to be critically important in the induction of an allergic immunological response.[27]

Chemokines produced by AEC may also be significant in the induction phase of allergic responses via their direct chemoattractant effects on T lymphocytes. In this context, it is notable that CCL1, CCL17, and CCL22 promote Th2 cell recruitment.[2]

The role of AEC-derived cytokines in the development of COPD is less clear. However, cigarette smoke and ambient particulates induce AEC to release proinflammatory cytokines (IL-1β, IL-6, and TNF-α) and chemokines (CXCL8 and CCL20), so it is plausible that they contribute to the initiation and regulation of innate and adaptive responses in this condition as well.[28]

AEC-DERIVED CYTOKINES IN EXACERBATIONS OF ASTHMA AND COPD

AEC are the primary target of many respiratory viral infections which are the main trigger for acute exacerbations of both asthma and COPD.[29,30] As mentioned above, viral infection elicits the production of interferons and chemokines. Of special relevance in the antiviral response is that the IFN-stimulated chemokines CXCL9, CXCL10, and CXCL11 promote Th1 cell recruitment.[2]

There is accumulating evidence that IL-33 may be particularly significant in terms of promoting the inflammatory response in asthmatic exacerbations. In experimental models of an acute exacerbation of chronic asthma, we have demonstrated that enhanced allergic inflammation, including that associated with a nonallergic trigger, is inhibited by treatment with a neutralizing antibody to IL-33.[31,32] The importance of IL-33 has been confirmed in studies of human asthmatic exacerbations.[33] Other animal and human studies have also suggested a role for IL-25 in asthmatic exacerbations.[34] However, relatively little is known about the contribution of AEC-derived cytokines to the pathogenesis of exacerbations of COPD.

REGULATION OF THE AEC-DERIVED CYTOKINE RESPONSE

Influence of the Local Cytokine Environment

In diseases such as asthma and COPD, the cytokine environment associated with the underlying chronic inflammation may alter the responses of AEC to infection or environmental stimuli. Enhanced release of proinflammatory cytokines and chemokines by AEC in response to viral infection has been demonstrated in allergic asthmatics as compared with healthy controls.[35,36] Consistent with this, we and others have shown that following infection with human rhinovirus or exposure to dsRNA, AEC pretreated with the Th2 cytokines IL-4 and IL-13 exhibit increased expression of a number of proinflammatory and antiviral genes (e.g., IL-6, IL-32, CXCL8, CXCL10, CXCL11, CCL5, and IFN-λ1; Fig. 11.2).[16,37,38] These results are supported by in-vivo human experiments demonstrating upregulation of cytokines in response to dsRNA in the setting of allergic inflammation.[39] Interaction between cytokines may also be important in vivo e.g., in asthmatics, increased release of CCL11 and CCL22 may be related to enhanced release of TSLP by virus-infected AEC.[40]

Much less is known about the effects of the cytokine environment in COPD in which Th1 and Th17 cytokines predominate.[41] Nevertheless, there is good evidence that in the metaplastic airway epithelium of smokers, expression of proinflammatory cytokines (IL-1β, IL-6, and TNF-α) is markedly enhanced.[42] Furthermore, a recent study has demonstrated that costimulation of human AEC with IL-17A and dsRNA significantly upregulates the expression of several proinflammatory cytokines (e.g., CXCL1, CXCL8, and G-CSF).[43]

Specific Molecular Mechanisms

The biological effects of some AEC-derived cytokines may be influenced by cosecreted regulatory molecules e.g., AEC secrete IL-1β in conjunction with the inhibitory IL-1 receptor antagonist and a soluble form of IL-1 receptor.[44] At the cellular level, concomitant upregulation of

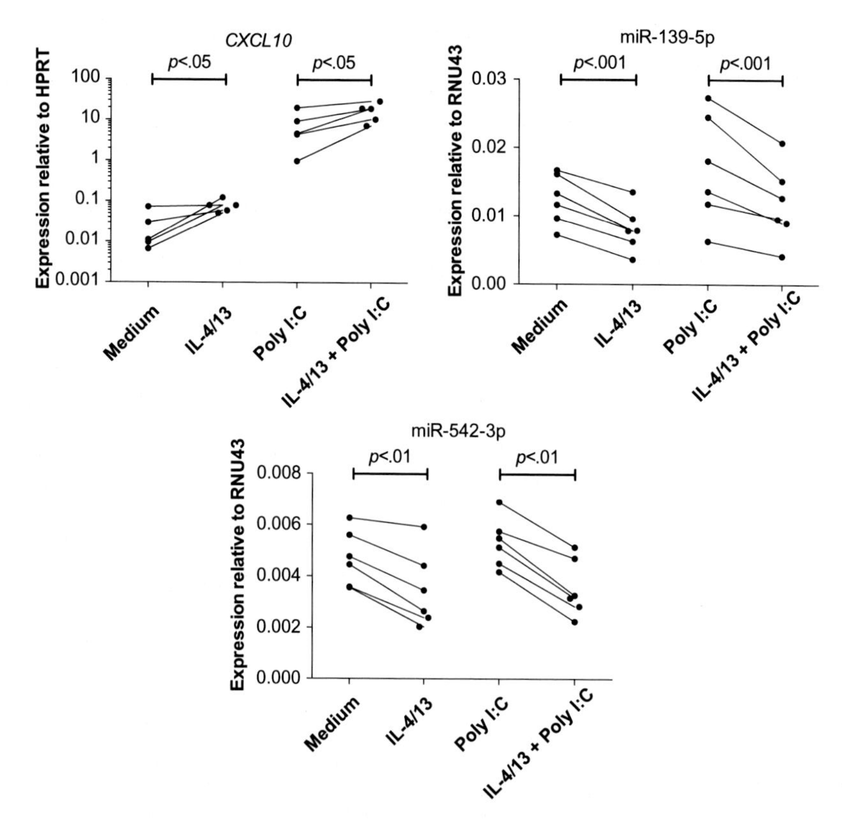

FIGURE 11.2 Following stimulation with double-stranded RNA, AEC that had been exposed to the Th2 cytokines IL-4 and IL-13 exhibited enhanced expression of the chemokine CXCL10. This was associated with decreased expression of the corresponding negative regulatory miRNAs.[16,48]

molecules such as suppressor of cytokine synthesis 1 in activated AEC may also have a regulatory role.[45]

Currently, there is much interest in the epigenetic regulation of cytokine production with a particular focus on the role of microRNA (miRNA) in controlling transcription and/or mRNA degradation. Differential expression of miRNA has been demonstrated in human asthmatic as compared with healthy control airway epithelium, with evidence that proinflammatory cytokine expression is directly regulated by some of these miRNA.[46,47] Furthermore, we recently demonstrated that in an allergic environment, altered regulation of selected miRNA may promote enhanced inflammatory and antiviral responses of human AEC in the setting of exacerbations of asthma (Fig. 11.2).[48] However, much remains to be learned about the role of miRNA and other noncoding RNA in regulating the cytokine response of AEC.

References

1. Mescher AL. *Epithelial tissue. Junqueira's basic histology*. 13 ed. McGraw-Hill; 2013. p. 73–97.
2. Kato A, Schleimer RP. Beyond inflammation: airway epithelial cells are at the interface of innate and adaptive immunity. *Curr Opin Immunol* 2007;**19**:711–20.
3. Hallstrand TS, Hackett TL, Altemeier WA, Matute-Bello G, Hansbro PM, Knight DA. Airway epithelial regulation of pulmonary immune homeostasis and inflammation. *Clin Immunol* 2014;**151**:1–15.
4. Kawai T, Akira S. The role of pattern-recognition receptors in innate immunity: update on Toll-like receptors. *Nat Immunol* 2010;**11**:373–84.
5. Fukata M, Vamadevan AS, Abreu MT. Toll-like receptors (TLRs) and Nod-like receptors (NLRs) in inflammatory disorders. *Semin Immunol* 2009;**21**:242–53.
6. Thompson MR, Kaminski JJ, Kurt-Jones EA, Fitzgerald KA. Pattern recognition receptors and the innate immune response to viral infection. *Viruses* 2011;**3**:920–40.
7. Knight DA, Lim S, Scaffidi AK, Roche N, Chung KF, Stewart GA, et al. Protease-activated receptors in human airways: Upregulation of PAR-2 in respiratory epithelium from patients with asthma. *J Allergy Clin Immunol* 2001;**108**:797–803.
8. Kumar RK, Siegle JS, Kaiko GE, Herbert C, Mattes JE, Foster PS. Responses of airway epithelium to environmental injury: Role in the induction phase of childhood asthma. *J Allergy (Cairo)* 2011;**2011**:257017.
9. Alvine TD, Knopick PL, Nilles ML, Bradley DS. *Inflammatory mediators*. eLS: *Wiley*. 2015p. 1–9. Available from: http://dx.doi.org/10.1002/9780470015902.a0000945.pub2.
10. Yamaya M. Virus infection-induced bronchial asthma exacerbation. *Pulm Med* 2012;**2012**:834826.
11. Sakamoto N, Hayashi S, Gosselink J, Ishii H, Ishimatsu Y, Mukae H, et al. Calcium dependent and independent cytokine synthesis by air pollution particle-exposed human bronchial epithelial cells. *Toxicol Appl Pharmacol* 2007;**225**:134–41.
12. Kumar RK, Shadie AM, Bucknall MP, Rutlidge H, Garthwaite L, Herbert C, et al. Differential injurious effects of ambient and traffic-derived particulate matter on airway epithelial cells. *Respirology* 2015;**20**:73–9.
13. Kudo M, Ogawa E, Kinose D, Haruna A, Takahashi T, Tanabe N, et al. Oxidative stress induced interleukin-32 mRNA expression in human bronchial epithelial cells. *Respir Res* 2012;**13**:19.
14. Ioannidis I, McNally B, Willette M, Peeples ME, Chaussabel D, Durbin JE, et al. Plasticity and virus specificity of the airway epithelial cell immune response during respiratory virus infection. *J Virol* 2012;**86**:5422–36.
15. Matsukura S, Kokubu F, Kurokawa M, Kawaguchi M, Ieki K, Kuga H, et al. Synthetic double-stranded RNA induces multiple genes related to inflammation through Toll-like receptor 3 depending on NF-kappaB and/or IRF-3 in airway epithelial cells. *Clin Exp Allergy* 2006;**36**:1049–62.
16. Herbert C, Zeng QX, Shanmugasundaram R, Garthwaite L, Oliver BG, Kumar RK. Response of airway epithelial cells to double-stranded RNA in an allergic environment. *Transl Respir Med* 2014;**2**:11.
17. Khaitov MR, Laza-Stanca V, Edwards MR, Walton RP, Rohde G, Contoli M, et al. Respiratory virus induction of alpha-, beta- and lambda-interferons in bronchial epithelial cells and peripheral blood mononuclear cells. *Allergy* 2009;**64**:375–86.
18. Liu SY, Sanchez DJ, Aliyari R, Lu S, Cheng G. Systematic identification of type I and type II interferon-induced antiviral factors. *Proc Natl Acad Sci USA* 2012;**109**:4239–44.
19. Papazian D, Hansen S, Wurtzen PA. Airway responses towards allergens—from the airway epithelium to T cells. *Clin Exp Allergy* 2015;**45**:1268–87.
20. Ritz SA, Stampfli MR, Davies DE, Holgate ST, Jordana M. On the generation of allergic airway diseases: from GM-CSF to Kyoto. *Trends Immunol* 2002;**23**:396–402.

21. Regamey N, Obregon C, Ferrari-Lacraz S, van Leer C, Chanson M, Nicod LP, et al. Airway epithelial IL-15 transforms monocytes into dendritic cells. *Am J Respir Cell Mol Biol* 2007;**37**:75–84.
22. McNamara PS, Fonceca AM, Howarth D, Correia JB, Slupsky JR, Trinick RE, et al. Respiratory syncytial virus infection of airway epithelial cells, in vivo and in vitro, supports pulmonary antibody responses by inducing expression of the B cell differentiation factor BAFF. *Thorax* 2013;**68**:76–81.
23. Leigh R, Oyelusi W, Wiehler S, Koetzler R, Zaheer RS, Newton R, et al. Human rhinovirus infection enhances airway epithelial cell production of growth factors involved in airway remodeling. *J Allergy Clin Immunol* 2008;**121**:1238–45.
24. Iwanaga K, Elliott MS, Vedal S, Debley JS. Urban particulate matter induces pro-remodeling factors by airway epithelial cells from healthy and asthmatic children. *Inhal Toxicol* 2013;**25**:653–60.
25. Mjosberg JM, Trifari S, Crellin NK, Peters CP, van Drunen CM, Piet B, et al. Human IL-25- and IL-33-responsive type 2 innate lymphoid cells are defined by expression of CRTH2 and CD161. *Nat Immunol* 2011;**12**:1055–62.
26. Bartemes KR, Kephart GM, Fox SJ, Kita H. Enhanced innate type 2 immune response in peripheral blood from patients with asthma. *J Allergy Clin Immunol* 2014;**134**:671–8.
27. Halim TY, Krauss RH, Sun AC, Takei F. Lung natural helper cells are a critical source of Th2 cell-type cytokines in protease allergen-induced airway inflammation. *Immunity* 2012;**36**:451–63.
28. Gao W, Li L, Wang Y, Zhang S, Adcock IM, Barnes PJ, et al. Bronchial epithelial cells: The key effector cells in the pathogenesis of chronic obstructive pulmonary disease? *Respirology* 2015;**20**:722–9.
29. Jackson DJ, Sykes A, Mallia P, Johnston SL. Asthma exacerbations: Origin, effect, and prevention. *J Allergy Clin Immunol* 2011;**128**:1165–74.
30. Zwaans WA, Mallia P, van Winden ME, Rohde GG. The relevance of respiratory viral infections in the exacerbations of chronic obstructive pulmonary disease—a systematic review. *J Clin Virol* 2014;**61**:181–8.
31. Bunting MM, Shadie AM, Flesher RP, Nikiforova V, Garthwaite L, Tedla N, et al. Interleukin-33 drives activation of alveolar macrophages and airway inflammation in a mouse model of acute exacerbation of chronic asthma. *Biomed Res Int* 2013;**2013**:250938.
32. Shadie AM, Herbert C, Kumar RK. Ambient particulate matter induces an exacerbation of airway inflammation in experimental asthma: role of interleukin-33. *Clin Exp Immunol* 2014;**177**:491–9.
33. Jackson DJ, Makrinioti H, Rana BM, Shamji BW, Trujillo-Torralbo MB, Footitt J, et al. IL-33-dependent type 2 inflammation during rhinovirus-induced asthma exacerbations in vivo. *Am J Respir Crit Care Med* 2014;**190**:1373–82.
34. Beale J, Jayaraman A, Jackson DJ, Macintyre JD, Edwards MR, Walton RP, et al. Rhinovirus-induced IL-25 in asthma exacerbation drives type 2 immunity and allergic pulmonary inflammation. *Sci Transl Med* 2014;**6**:256ra134.
35. Hackett TL, Singhera GK, Shaheen F, Hayden P, Jackson GR, Hegele RG, et al. Intrinsic phenotypic differences of asthmatic epithelium and its inflammatory responses to RSV and air pollution. *Am J Respir Cell Mol Biol* 2011;**45**:1090–100.
36. Rohde G, Message SD, Haas JJ, Kebadze T, Parker H, Laza-Stanca V, et al. CXC-chemokines and antimicrobial peptides in rhinovirus-induced experimental asthma exacerbations. *Clin Exp Allergy* 2014;**44**:930–9.
37. Cakebread JA, Haitchi HM, Xu Y, Holgate ST, Roberts G, Davies DE. Rhinovirus-16 induced release of IP-10 and IL-8 is augmented by Th2 cytokines in a pediatric bronchial epithelial cell model. *PLoS One* 2014;**9**:e94010.
38. Herbert C, Do K, Chiu V, Garthwaite L, Chen Y, Young PM, et al. Allergic environment enhances airway epithelial pro-inflammatory responses to rhinovirus infection. *Clin Sci* 2017;**131**:499–509.

39. Brandelius A, Andersson M, Uller L. Topical dsRNA challenges may induce overexpression of airway antiviral cytokines in symptomatic allergic disease. A pilot in vivo study in nasal airways. *Respir Med* 2014;**108**:1816–19.
40. Nino G, Huseni S, Perez GF, Pancham K, Mubeen H, Abbasi A, et al. Directional secretory response of double stranded RNA-induced thymic stromal lymphopoetin (TSLP) and CCL11/eotaxin-1 in human asthmatic airways. *PLoS One* 2014;**9**:e115398.
41. Brusselle GG, Joos GF, Bracke KR. New insights into the immunology of chronic obstructive pulmonary disease. *Lancet* 2011;**378**:1015–26.
42. Herfs M, Hubert P, Poirrier AL, Vandevenne P, Renoux V, Habraken Y, et al. Proinflammatory cytokines induce bronchial hyperplasia and squamous metaplasia in smokers: Implications for chronic obstructive pulmonary disease therapy. *Am J Respir Cell Mol Biol* 2012;**47**:67–79.
43. Mori K, Fujisawa T, Kusagaya H, Yamanaka K, Hashimoto D, Enomoto N, et al. Synergistic proinflammatory responses by IL-17A and Toll-like receptor 3 in human airway epithelial cells. *PLoS One* 2015;**10**:e0139491.
44. Yang Y, Bin W, Aksoy MO, Kelsen SG. Regulation of interleukin-1beta and interleukin-1beta inhibitor release by human airway epithelial cells. *Eur Respir J* 2004;**24**:360–6.
45. Gielen V, Sykes A, Zhu J, Chan B, Macintyre J, Regamey N, et al. Increased nuclear suppressor of cytokine signaling 1 in asthmatic bronchial epithelium suppresses rhinovirus induction of innate interferons. *J Allergy Clin Immunol* 2015;**136**:177–88.
46. Solberg OD, Ostrin EJ, Love MI, Peng JC, Bhakta NR, Hou L, et al. Airway epithelial miRNA expression is altered in asthma. *Am J Respir Crit Care Med* 2012;**186**:965–74.
47. Martinez-Nunez RT, Bondanese VP, Louafi F, Francisco-Garcia AS, Rupani H, Bedke N, et al. A microRNA network dysregulated in asthma controls IL-6 production in bronchial epithelial cells. *PLoS One* 2014;**9**:e111659.
48. Herbert C, Sebesfi M, Zeng QX, Oliver BG, Foster PS, Kumar RK. Using multiple online databases to help identify microRNAs regulating the airway epithelial cell response to a virus-like stimulus. *Respirology* 2015;**20**:1206–12.

CHAPTER

12

Eosinophil Cytokines in Allergy

Paige Lacy

University of Alberta, Edmonton, AB, Canada

OUTLINE

Cytokine Effector Functions in Tissues.
DOI: http://dx.doi.org/10.1016/B978-0-12-804214-4.00011-7

INTRODUCTION

Eosinophils are granulated white blood cells that form part of the innate immune system, and circulate at low levels in healthy individuals, preferentially recruiting to the gut mucosa after their egress from the bone marrow.[1] Classically, eosinophils were considered to be protective against helminthic parasitic infections as part of their immune function, although recent studies have cast some doubt on this concept.[2] A potential role for eosinophils may exist for other infectious pathogens such as viruses, particularly respiratory viruses.[3–5] Other findings suggest that eosinophils may serve a role in immunosuppression of proinflammatory signals in homeostasis, particularly in the lungs and gut.[6,7] While the specific immunological role of eosinophils is still incompletely understood, their contribution to human diseases such as allergy and asthma is more well defined. In many types of allergic inflammation, eosinophils are strikingly elevated, and they have been found to increase in blood, as well as accumulate in the gut mucosa, skin, and lung tissues.[8–10] In allergic diseases such as atopic asthma, the numbers of eosinophils broadly correlate with severity of disease. Recruitment of eosinophils to tissues is orchestrated by a diversity of cells and mediators involving antigen-presenting cells, mast cells, T cells, B cells, epithelial cells, macrophages, neutrophils, and other cell types that release eosinophil-attracting factors.

Eosinophils are a source of over 35 immunomodulatory cytokines, chemokines, and growth factors (collectively referred to as cytokines for this chapter) that have a marked effect on the progression of immune and inflammatory responses (Fig. 12.1).[11,12] This chapter is focused on the types of cytokines that are released from eosinophils and their role in the propagation of immune responses.

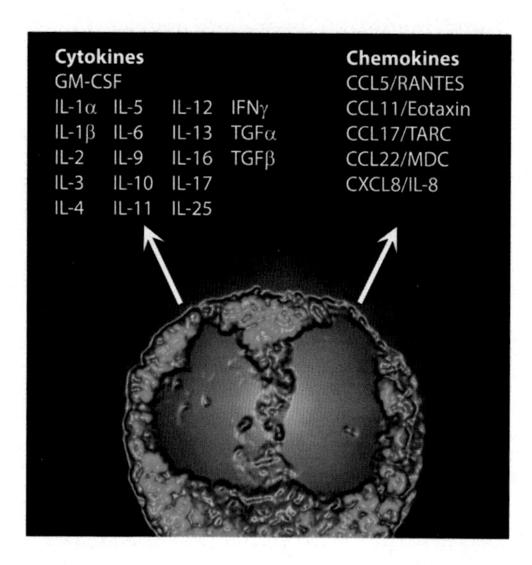

FIGURE 12.1 **Cytokines, chemokines, and growth factors secreted by eosinophils that have a postulated role in allergic inflammation.** Shown is a rendered image from CCL5/RANTES immunofluorescence in a human peripheral blood eosinophil.

ROLE OF EOSINOPHILS IN ALLERGY

Eosinophils, or eosinophil-like cells such as heterophils, have been reported in invertebrates as well as vertebrates, suggesting an evolutionarily conserved role for eosinophils in immune modulation and host protection.[5,13] Their role in immunity is still under intense investigation, with early evidence showing an association of eosinophils with helminthic parasite infections. However, evidence from studies with transgenic mice that lack eosinophils, such as the PHIL and ΔdblGATA strains,[14,15] have indicated that the role of eosinophils in protection of the host against helminths is more complex than previously recognized.[16] More recent studies suggest a role for eosinophils in protection against respiratory virus infections.[4] These recent findings contribute to an evolving area of research on the function of eosinophils in immunity.

In marked contrast, eosinophils are well known for their association with allergic inflammation. Organs, such as the lung and skin, that normally harbor very few eosinophils undergo a dramatic increase in tissue-recruited cells, including the lungs and upper airways, in a subset of patients with allergic asthma.[9] Eosinophils recruit together with other inflammatory cells to sites of allergen exposure, such as the peribronchial regions of the lungs, which results in allergic exacerbations. Upon their activation, eosinophils are believed to contribute to bronchial symptoms including bronchoconstriction, mucus secretion, and coughing.

The inflammatory processes underlying allergic exacerbations are controlled by a complex network of cytokines that regulate bone marrow progenitor differentiation, migration to inflammatory foci, cell adhesion receptor expression, and immunoglobulin E (IgE) responses. The infiltration of eosinophils to inflammatory responses may further exacerbate inflammation by their ability to release a plethora of cytokines. Along with cytokines, eosinophils release a range of cationic proteins that are cytotoxic, including major basic protein (MBP), eosinophil peroxidase (EPX), eosinophil-derived neurotoxin (EDN), and eosinophil cationic protein (ECP).[17] These factors have independent proinflammatory effects on tissues and are implicated in the exacerbation of allergic responses.

A large body of evidence from animal models has suggested that allergic inflammation arises from inappropriate polarization of the innate and adaptive immune systems toward a Th2 response, since a greater expression of Th2 cytokines (interleukin (IL)-4, IL-5, and IL-13) occurs in allergy.[9] The primary roles for these cytokines are to promote IgE switching in B cells, which leads to binding of allergen-specific IgE to mast cells, along with enhancement of T cell differentiation to a Th2 phenotype, and suppression of Th1 responses. Th2 responses result in the maintenance of an abnormally hypersensitive status in reaction to usually innocuous allergens. These cytokines also serve to increase blood and tissue eosinophilia, enhance eosinophil survival in tissues, and initiate eosinophil degranulation. By prolonging eosinophil survival in tissues, Th2 cytokines are thought to prolong the ability of eosinophils to actively release their immunomodulatory mediators into tissues.

However, while the majority of asthma cases in patients fit the Th2 phenotype described in animal models, certainly not all of them may be categorized in this way.[9] While the proportion of asthmatics with eosinophilia is not known, approximately 50% of patients with mild to severe asthma have some form of blood eosinophilia.[9] Thus, eosinophils may be important contributors to allergic inflammation in around half of all cases of asthma.

Taken together, the network of cytokine signaling underlying allergic inflammation and asthma is highly complex. Th2 responses are strongly evoked in animal models of asthma and allergy; however, not all human asthma phenotypes are driven by Th2 cytokines. The contribution of eosinophil-derived cytokines to allergic diseases is not fully understood, although recent developments suggest that they may be implicated in the manifestation of allergic inflammation.

DEGRANULATION RESPONSES IN EOSINOPHILS

Eosinophils are densely packed with a unique type of secretory granules called crystalloid granules. These are so named because of the core of the granule which contains crystallized MBP, a highly charged cationic protein with cytotoxic effects upon its release.[10] The crystalline core of the eosinophil secretory granule may be visualized by its electron-dense properties using transmission electron microscopy.[10] Surrounding the core of the eosinophil granule is the matrix, which contains additional cationic proteins, including EPX, ECP, and EDN. Other components are also stored in the matrix including cytokines, which are released along with eosinophil granule proteins, although via distinct trafficking mechanisms (Fig. 12.2).

There are four major pathways by which eosinophil crystalloid granules are released to the cell exterior by a process known as degranulation. The first is classical exocytosis, in which the membrane of individual granules fuses with the plasma membrane to form contiguous structures.[18,19] The second is compound exocytosis, whereby the crystalloid granules first fuse with each other (via homotypic fusion) and then fuse with the cell membrane through a single fusion pore.[20–22] The third pathway of degranulation is known as piecemeal degranulation, where small, rapidly mobilizable secretory vesicles bud from the surface of crystalloid granules and shuttle materials to the cell membrane as part of the eosinophil's tubulovesicular system.[23–26] Finally,

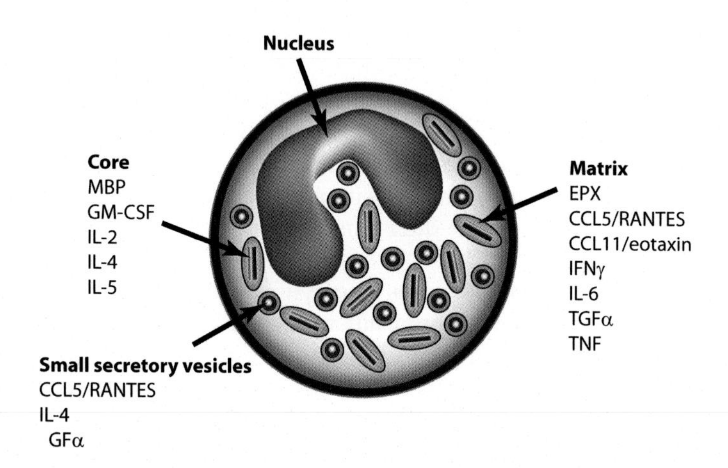

FIGURE 12.2 **Sites of storage and transport for eosinophil-derived cytokines.** The eosinophil crystalloid granule consists of two subcompartments: the core, made up of crystallized MBP, and the matrix, which contains EPX, among other granule components. Cytokines are shown in their specific locations in secretory organelles. Small secretory vesicles also transport cytokines, including CCL5/RANTES, IL-4, and TGFα.

degranulation may occur through necrosis or "cytolysis," in which intact, membrane-bound crystalloid granules are released following the loss of cell membrane integrity.[26–28]

Degranulation from eosinophils is usually associated with marked tissue damage, as eosinophil granule proteins are cytotoxic and can induce epithelial damage, leading to increased inflammation.[8,10] Excessive release of MBP, EPX, and ECP correlates with increased disease severity in allergy and asthma.[10] Along with the release of eosinophil granule proteins, prestored immunomodulatory cytokines are also released, many of which are differentially secreted in distinct pathways from those of eosinophil granule proteins.[29]

Agonists that potently evoke degranulation from eosinophils include platelet-activating factor (PAF),[30,31] opsonized bacteria or other surfaces,[32] complement factors (especially C5a),[33] and immunoglobulin complexes.[34] In addition, a range of cytokines also induce or promote eosinophil degranulation, including granulocyte/macrophage colony-stimulating factor (GM-CSF), interferon-γ (IFNγ), IL-3, IL-5, and CCL11/eotaxin.[24,35–38] Most of these agonists are elevated in allergic inflammation and are predicted to contribute to eosinophil degranulation, leading to widespread tissue injury.

CYTOKINES PRODUCED BY HUMAN EOSINOPHILS

Human eosinophils produce over 35 cytokines, many of which are also expressed in their murine counterparts (Table 12.1). In most cases, both the mRNA transcript and protein expression have been identified for each cytokine in peripheral blood eosinophils obtained from nonatopic as well as atopic subjects. Interestingly, 10 of these cytokines have been found as preformed mediators packaged within eosinophil crystalloid granules, suggesting that eosinophil-derived cytokines may act immediately and directly on their microenvironment within minutes of cell activation.

Eosinophils from tissue sources also synthesize and release numerous cytokines, suggesting that eosinophils retain the ability to generate cytokines after their recruitment into peripheral tissues. Examples of tissue eosinophils that express cytokines have been observed in the respiratory system (nasal polyps, bronchial biopsies, bronchoalveolar lavage (BAL), sputum samples), the gastrointestinal tract (celiac mucosal biopsies), and in the skin.

The following sections elaborate on the expression of individual cytokines in eosinophils derived from blood and tissues. The most frequently used techniques for determining intracellular sites of cytokine

TABLE 12.1 Eosinophil-Derived Cytokines, Chemokines, and Growth Factors in Human and Mouse Studies

Mediator detected in human eosinophils	Molecule detected	Mean quantity of protein stored (per 10^6 cells)	Release factors	Intracellular localization of stored protein	Reference
A. Cytokines					
A proliferation-inducing ligand (APRIL)	mRNA	–	–	–	39
	Protein				
Granulocyte/ Macrophage colony-stimulating factor (GM-CSF)	mRNA	15 pg	Ionomycin	Crystalloid granules (core)	40–46
	Protein		LPS		
Interferon-γ (IFNγ)	mRNA	997 pg	Cytokines	Crystalloid granules, small secretory vesicles	11,47,48
Interleukin-1α	mRNA	–	PMA	–	49,50
	Protein				
Interleukin-1β	mRNA	–	[Constitutively released]	–	51,52
	Protein				
Interleukin-2	mRNA	6 pg	Serum-coated particles	Crystalloid granules (core)	53–56
	Protein		PHA		
			CD28 cross-linking		
Interleukin-3	mRNA	–	Ionomycin	–	41,57
	Protein		IFNγ		
Interleukin-4	mRNA	108 pg	Immune complexes	Crystalloid granules (core)	58–67

(*Continued*)

TABLE 12.1 (Continued)

Mediator detected in human eosinophils	Molecule detected	Mean quantity of protein stored (per 10^6 cells)	Release factors	Intracellular localization of stored protein	Reference
	Protein		Serum-coated particles Cytokines		
Interleukin-5	mRNA	–	Immune complexes	Crystalloid granules (core/ matrix?)	43,62, 68–73
	Protein				
Interleukin-6	mRNA	356 pg	Cytokines	Crystalloid granules (matrix)	48,74–76
	Protein				
Interleukin-9	mRNA	–	–	Crystalloid granules	77–79
	Protein				
Interleukin-10	mRNA	455 pg	Cytokines	Crystalloid granules	48,61,80
	Protein				
Interleukin-11	mRNA	–	–	–	81
	Protein				
Interleukin-12	mRNA	186 pg	Cytokines	Crystalloid granules	48,82
	Protein				
Interleukin-13	mRNA	50 pg	Cytokines	Crystalloid granules	83,84
	Protein				
Interleukin-16	mRNA	–	[Constitutively released]	–	85,86
	Protein				
Interleukin-17	Protein	–	–	–	87

(Continued)

TABLE 12.1 (Continued)

Mediator detected in human eosinophils	Molecule detected	Mean quantity of protein stored (per 10^6 cells)	Release factors	Intracellular localization of stored protein	Reference
Interleukin-25	mRNA	–	Cytokines	–	88–90
	Protein				
Tumor necrosis factor-α (TNF)	mRNA	909 pg	Immune complexes	Crystalloid granules	44,48,61, 91–94
	Protein		TNF		
			LPS		
			Cytokines		
B. Chemokines					
CCL3/ Macrophage inflammatory protein-1α (MIP-1α)	mRNA	–	–	–	92,95
	Protein				
CCL5/Regulated on activation, normal T cell expressed and secreted (RANTES)	mRNA	7000 pg	Serum-coated particles	Crystalloid granules, small secretory vesicles	24,85,96
	Protein		IFNγ		
CCL11/Eotaxin	mRNA	16–22 pg	C5a, immune complexes	Crystalloid granules	97–99
	Protein				
CCL13/Monocyte chemoattractant protein-4 (MCP-4)	mRNA	13 pg	Immune complexes	Crystalloid granules	99
	Protein				
CCL17/Thymus activation regulated chemokine (TARC)	mRNA	–	TNF + IFNγ or IL-4	–	100

(Continued)

TABLE 12.1 (Continued)

Mediator detected in human eosinophils	Molecule detected	Mean quantity of protein stored (per 10^6 cells)	Release factors	Intracellular localization of stored protein	Reference
	Protein				
CCL22/ Macrophage-derived chemokine (MDC)	mRNA	–	TNF + IFNγ or IL-4	–	100
	Protein				
CCL23/Myeloid progenitor inhibitory factor 1 (MPIF-1)	mRNA	–	Cytokines	–	101
	Protein				
CXCL1/Groα	mRNA	95 pg	Cytokines	Crystalloid granules	102
	Protein				
CXCL5/ Epithelial-derived neutrophil-activating peptide 78 (ENA-78)	mRNA	1500 pg	Cytokines	–	103
	Protein				
CXCL8/ Interleukin-8	mRNA	–	C5a	–	44,61, 104–108
	Protein		fMLP		
			GM-CSF—RANTES or PAF		
			Immune complexes		
			TNFα, cytokines		
			LPS		

(*Continued*)

TABLE 12.1 (Continued)

Mediator detected in human eosinophils	Molecule detected	Mean quantity of protein stored (per 10^6 cells)	Release factors	Intracellular localization of stored protein	Reference
CXCL9/ Monokine induced by γ interferon (MIG)	mRNA	–	TNF + IFNγ or IL-4	–	100,109
	Protein				
CXCL10/ Interferon γ induced protein 10 (IP-10)	mRNA	–	TNF + IFNγ or IL-4	–	100,109
	Protein				
CXCL11/ Interferon-inducible T cell α chemoattractant (I-TAC)	mRNA	–	IFNγ	–	109
	Protein				
C. Growth factors					
Heparin-binding epidermal growth factor-like binding protein (HB-EGF-LBP)	mRNA	–	–	–	110
Nerve growth factor (NGF)	mRNA	10 pg	–	–	111
	Protein				
Platelet-derived growth factor, B chain (PDGF-B)	mRNA	–	–	–	112
Stem cell factor (SCF)	mRNA	9 pg	Chymase	Crystalloid granules	113
	Protein				

(*Continued*)

TABLE 12.1 (Continued)

Mediator detected in human eosinophils	Molecule detected	Mean quantity of protein stored (per 10^6 cells)	Release factors	Intracellular localization of stored protein	Reference
Transforming growth factor-α (TGFα)	mRNA	–	Cytokines	–	114–118
	Protein				
Transforming growth factor-β (TGF-β)	mRNA	–	Cytokines	–	116, 118–125
	Protein				
Vascular endothelial growth factor (VEGF)	mRNA	–	Cytokines	Crystalloid granules	126–129
	Protein				

storage in eosinophils include immunocytochemistry, immunofluorescence, subcellular fractionation, immunogold labeling, and immunofluorescence using confocal microscopy analysis. In this chapter, a specific focus will be made on eosinophil-derived cytokines that have a potential role in allergy or that have been identified in tissues from allergic patients.

GM-CSF (Granulocyte/Macrophage Colony-Stimulating Factor)

The principal role of GM-CSF in the immune system is to stimulate the production of granulocytes (neutrophils, eosinophils, and basophils) in the bone marrow. In allergic disease, GM-CSF facilitates allergic airway inflammation and promotes eosinophilic inflammation, Th2 responses, mucus production, and airway hyperresponsiveness together with other cytokines.[130] Further, GM-CSF promotes cutaneous anaphylaxis reactions in mice expressing humanized immune systems.[131] In eosinophils, GM-CSF promotes maturation, differentiation, and survival by delaying apoptosis.[132] GM-CSF also enhances survival of adherent eosinophils in an autocrine manner[133] and degranulation as well as superoxide production from human eosinophils.[38]

Eosinophils isolated from human peripheral blood produce significant levels of GM-CSF.[40,41,46] Early studies identified GM-CSF as a prominent cytokine that was released by eosinophils following stimulation by calcium ionophore (ionomycin) or bacterial lipopolysaccharide.[41,44] It was also demonstrated that GM-CSF prolonged eosinophil survival in an autocrine manner, and that the effect of GM-CSF on delay of apoptosis was inhibited by the immunosuppressive drug cyclosporin A.[41]

In confirmation of findings in blood eosinophils, tissue eosinophils express GM-CSF. For example, GM-CSF$^+$ eosinophils have been detected in nasal biopsies from patients with allergic rhinitis[42] and bronchial biopsies from asthmatic subjects.[43] Following endobronchial challenge in human asthmatics, eosinophils are recruited to the peribronchial regions, and these were shown to express GM-CSF.[43] Moreover, sputum eosinophils from asthmatic subjects express GM-CSF as evaluated by immunocytochemistry.[45] Thus, these findings suggest that GM-CSF is a critical eosinophil-derived cytokine that serves an important role in maintaining viability as well as eosinophil effector functions in allergy.

GM-CSF is one of the 10 cytokines that have so far been identified as preformed mediators that are stored within eosinophil crystalloid granules[46] (Fig. 12.2). The intragranular localization of GM-CSF appears to be associated with the matrix subcompartment that surrounds the crystalline core.[46] This early observation was among the first reports describing the ability of eosinophils to store preformed cytokines, which could be later released in a rapid manner in response to inflammatory signals.

Interferon-γ

IFNγ is the sole member of the type II class of interferons, and serves as a potent proinflammatory Th1 cytokine. Its release is mainly from innate immune NK and NKT cells in response to viral, bacterial and some protozoal infections. IFNγ is classically known for its role in downregulation of Th2-driven allergic inflammation, leading to reduced eosinophilia, in mouse models of allergic inflammation.[134,135] In allergic asthma, IFNγ is predominantly associated with virus-induced asthma exacerbations.[134]

Stimulation of eosinophils by IFNγ has been proposed to promote allergic inflammation through a number of mechanisms in response to virus infections, suggesting a role for IFNγ in autocrine signaling in eosinophils.[134] The effects of IFNγ on eosinophils are to enhance cytotoxicity,[136] increase cytokine expression such as GM-CSF,[40] induce secretion of IL-3, IL-6, and CCL5/RANTES,[24,57,76,137] and promote IL-5- or GM-CSF-induced superoxide release.[138] IFNγ also enhances eosinophil survival, although not as potently as GM-CSF or IL-5,[136] while it suppresses degranulation responses from eosinophils induced by secretory IgA,[36] suggesting a complex effect by this cytokine on eosinophil secretory functions.

Perhaps counterintuitively to the classical model of the Th2 skewed responses in allergy, IFNγ is also expressed in eosinophils, cells which are usually considered to be associated with Th2 responses. IFNγ is among the most abundantly secreted cytokines generated by human eosinophils (Table 12.1).[47,48] Secretion of IFNγ from eosinophils was shown to occur following stimulation with Th1, Th2, and inflammatory signals such as TNFα.[11] Eosinophils also store significant quantities of IFNγ, suggesting that this may be released immediately within minutes of stimulation.

In Balb/c mice, eosinophil-derived IFNγ has bioactive effects on surrounding tissues was shown to contribute to airway hyperresponsiveness in a manner independent of T cells,[139] indicating an important immunomodulatory role for eosinophil-derived cytokines in lung inflammation.

The intracellular sites of storage for IFNγ in eosinophils have not yet been determined, and this remains an area of interest for future studies.

Interleukin-1α

IL-1α, also known as hematopoietin 1, is a member of the IL-1 family of 11 cytokines and is produced mainly by macrophages, neutrophils, epithelial cells, and endothelial cells. This cytokine was originally named IL-1 until it was recognized that this consisted of two distinct cytokines, IL-1α and IL-1β. IL-1α is a potent proinflammatory cytokine associated with inflammation, fever, and sepsis. In allergic inflammation, IL-1α is thought to promote epithelial release of GM-CSF and IL-33, leading to allergic sensitization in mice exposed to house dust mite (HDM).[140] Eosinophils respond to IL-1α by undergoing degranulation[141] and adhesion to endothelial cells.[142,143]

IL-1α was the first cytokine discovered in eosinophils, where it was characterized for its expression in peritoneal eosinophils from mice infected with the larvae of the parasite *Mesocestoides corti*.[50] Human eosinophils were first shown to synthesize IL-1α in association with cytokine-induced human leukocyte antigen DR (HLA-DR) expression.[49] IL-1α mRNA transcripts and protein expression were elevated by stimulation with phorbol ester myristate (PMA), suggesting that human eosinophils can process antigens (such as allergens), and express the costimulatory cytokine IL-1α, for HLA-DR presentation to T cells. Later studies confirmed that eosinophils may function as antigen-presenting cells in mouse models of allergic inflammation.[144–147] The release of IL-1α from eosinophils may serve an important role as antigen-presenting cells in promoting allergen presentation to T cells in subjects with eosinophilia.

No studies have indicated intracellular sites of IL-1α storage or trafficking in eosinophils. As IL-1α is an unusual cytokine that does not contain a signal peptide fragment, it is unlikely to be transported out of the cell via vesicular trafficking. The proposed transport mechanisms for IL-1α are described elsewhere.[148,149]

Interleukin-1β

IL-1β, also called human leukocytic pyrogen, lymphocyte-activating factor and other names, and has similar properties to IL-1α in immune function by promoting inflammatory responses, fever, and sepsis. Like IL-1α, IL-1β is a member of the IL-1 family of 11 cytokines, which regulates and initiates proinflammatory reactions. IL-1β is implicated in allergic disease, including atopic dermatitis and bronchial asthma, and has been shown to support T cell survival, upregulate IL-2 receptor expression on lymphocytes, enhance DC recruitment, increase antibody production from B cells, and promote B cell proliferation as well as elevate Th2 cytokines.[150] For these reasons, IL-1β is considered to be an important therapeutic target for treatment of allergic inflammation.[150] In eosinophils, IL-1β acts in a similar manner to IL-1α by inducing degranulation[36,141] and cell adhesion,[143] although different eosinophil subpopulations responded selectively to each isoform of IL-1.

Interestingly, it was only recently that eosinophils were shown to constitutively release IL-1β as an immunomodulatory cytokine.[51] Eosinophil-derived IL-1β was shown to be involved in promoting the synthesis and secretion of IL-17 from activated $CD4^+$ T cells.[51] Eosinophils also elaborate IL-1β in the gastrointestinal lamina propria where they promote secretory IgA production.[52] Since Th17 cells have been implicated in the pathogenesis of allergic airway inflammation,[151] the ability of eosinophils to generate IL-1β indicates that this may be an important mechanism whereby eosinophils promote differentiation of T cells to Th17 phenotype.

Similarly to IL-1α, IL-1β lacks a signal peptide sequence, and there are no reports indicating the sites of storage or transport of IL-1β in eosinophils.

Interleukin-2

An essential growth factor for T cells, IL-2 is a critical cytokine for key functions of the immune system and has pivotal roles in tolerance and immunity. Its predominant target is T cells, particularly in the thymus where it prevents the development of autoimmunity. In peripheral lymphoid tissues, IL-2 promotes differentiation of T cells into effector T cells and memory T cells. In allergic inflammation, IL-2 contributes to the establishment of the allergic phenotype following early phase allergen exposure leading to IL-4 release,[152] and was recently demonstrated to be important in the generation of Th2 memory cells that promote allergic airway responses in asthma.[153] IL-2 also has a direct effect on eosinophils which express the IL-2 receptor (CD25), by promoting chemotaxis.[154]

Eosinophils constitutively express IL-2 at the mRNA and protein levels, suggesting that eosinophil-derived IL-2 may contribute to T cell proliferation in peripheral tissues.[53–56] The release of IL-2 may be induced by serum-coated particles and phytohemagglutinin (PHA) from human peripheral blood eosinophils. Eosinophil-derived IL-2 was shown to be bioactive, as supernatants from eosinophils treated with anti-CD28 induced T cell proliferation and MHC II expression on a colon carcinoma cell line.[55] Few studies have reported the expression of IL-2 in tissue eosinophils, however, and this remains an area of interest for future investigations to determine the role of eosinophil-derived IL-2 in allergic disease.

Intracellular sites of storage for IL-2 have been characterized in eosinophils. The majority of IL-2 was found stored as a preformed mediator within eosinophil crystalloid granules,[53] with some cytoplasmic staining evident in a minority of freshly isolated peripheral blood eosinophils. Within the granules, IL-2 appeared to be predominantly localized to the crystalline core.[53] These findings suggesting that eosinophils have the capacity to synthesize, store, and release IL-2 from the crystalloid granules, allowing the granules to serve as a reservoir for the rapid release of IL-2 in inflammatory reactions associated with allergy.

Interleukin-3

An essential growth factor for myelocytic cells and granulocytes, IL-3 is required as a pluripotent growth factor along with GM-CSF to promote the proliferation of precursors toward a myeloid phenotype. It is produced by a number of immune cells, primarily T cells, NK cells, and mast cells. In allergic diseases, IL-3 is elevated following allergen exposure in the skin[155] and in bronchial asthma.[155–158] This cytokine is frequently observed to increase in correlation with GM-CSF and IL-5 during allergic responses.[155,157,158] The function of IL-3 in these conditions is suggested to be through augmentation of IgE synthesis from B cells[159] and increased histamine release from basophils.[160] In eosinophils, IL-3 functions together with IL-5 and GM-CSF to promote differentiation from human bone marrow progenitor cells into mature eosinophils.[161] IL-3 alone is insufficient to serve as a chemoattractant for mature peripheral blood eosinophils, although low concentrations are capable of enhancing chemotactic effects of PAF and other agonists.[162] Eosinophils from atopic patients show an increased sensitivity to IL-3 compared to normal individuals.[163] IL-3 also promotes eosinophil survival and activation to generate hypodense eosinophils,[164] and is capable of promoting degranulation and superoxide production from human eosinophils.[38]

IL-3 was among the first cytokines to be characterized in eosinophils, and its release can be evoked by GM-CSF.[41] Eosinophils isolated from human peripheral blood synthesize IL-3 at the mRNA and protein levels, and secrete IL-3 in response to calcium ionophore and IFNγ stimulation.[41,57] Thus, eosinophils marginated into tissue sites that actively secrete IL-3 are likely to prolong their own survival by autocrine signaling. While IL-3 has been shown to increase in allergic inflammation based on biopsies from individuals with atopic rhinitis, the expression of IL-3 in tissue eosinophils was not directly demonstrated but instead showed a correlation between numbers of eosinophils and IL-3 mRNA$^+$ cells.[157,158]

To date, the intracellular sites of storage or trafficking of IL-3 have not yet been elucidated, and this remains an area of interest.

Interleukin-4

Of all the cytokines studied in eosinophils to date, perhaps the best characterized has been IL-4, a cytokine essential for the development of the Th2 response and the production of IgE by B cells, placing IL-4 in a central effector role in allergic inflammation. In eosinophils, IL-4 has been shown to promote chemotaxis in cells obtained from atopic patients.[165] Interestingly, early studies suggested that stimulation of eosinophils with IL-4 may also promote the development of Th1 responses. IL-4-treated eosinophils released the Th1 cytokine IL-12, which in turn induced the expression of IFNγ from Th1 cells during culture of T cells with eosinophil-conditioned media.[82] Conversely, IL-4 also promotes the expansion of Th2 cytokine-producing eosinophils in vivo in a mouse model of asthma, both in the airways and in bone marrow progenitor cells.[166] The finding that eosinophils may release Th1 or Th2 cytokines, depending on the stimulation and microenvironment, suggests that eosinophils possess a more nuanced control of cytokine responses in allergic inflammation than may be previously recognized.

Human peripheral blood eosinophils from atopic donors produce IL-4 and release this cytokine in response to serum-coated surfaces and cytokines.[59] Eosinophils purified from the blood of healthy donors also produce and secrete IL-4 in response to CCL11/eotaxin,[63,64] suggesting that IL-4 is constitutively expressed in blood eosinophils under normal conditions. This is consistent with mouse studies of eosinophil-derived cytokines. In IL-4 reporter 4get mice, eosinophils constitutively express IL-4 transcript at early stages of ontogeny.[167] Moreover, instillation of IL-4 into mice leads to proliferation of IL-4-expressing eosinophils, suggesting that IL-4 promotes differentiation of bone marrow progenitor cells into eosinophils that express Th2 cytokines including IL-4.[166] An earlier study showed that, among the non-T and non-B cell populations in lung and spleen tissues in mice infected with *Nippostrongylus brasiliensis*, eosinophils are among the most abundant IL-4-expressing cells.[168] Using 4get mice, eosinophils were found to be the most prevalent IL-4-expressing cells that infiltrated the lungs of mice infected with *N. brasiliensis*.[169] Eosinophils expressing IL-4 have also been observed in mice infected with fungal *Cryptococcus neoformans*, in which they formed the majority of cells expressing IL-4.[170]

IL-4 production from eosinophils has been shown to occur in the spleen following adjuvant stimulation of B cell responses, based on observations from eosinophil-deficient ΔdblGATA mice injected with alum.[147] Alum injection into ΔdblGATA mice led to attenuation of early B cell priming and IgM production, suggesting that eosinophils may play an important role in promoting the adaptive immune response to vaccines containing adjuvants.

In parallel with findings on blood-derived human eosinophils, tissue eosinophils from human subjects also express IL-4 as shown in airway, lung, and skin biopsy samples. For example, in nasal biopsies obtained from subjects with allergic rhinitis, tissue eosinophils expressed IL-4.[58] As many as 44% of the tissue eosinophils present in nasal polyp tissues were found to be positive for IL-4. Moreover, following allergen-induced cutaneous late-phase reactions, the majority of tissue-infiltrating eosinophils (84%) were IL-4$^+$ after 6 h.[58] This was further demonstrated in lung tissues, where eosinophils from bronchial biopsies taken from atopic asthmatics and normal nonatopic subjects were shown to express IL-4 mRNA.[72] Finally, in skin biopsies of allergic individuals, around 20% of tissue eosinophils were positive for IL-4 mRNA at 24 h following challenge, and this increased to 50%–60% for protein expression of IL-4.[62]

Eosinophil-derived IL-4 is also one of the few eosinophil-derived cytokines that have been shown to have a direct bioactive role in tissues in numerous reports. Several studies have suggested that eosinophil-derived IL-4 is important in priming naïve T cells and activation of mast cell IL-5 release during helminthic parasite infection, for example with *Schistosoma mansoni*.[171] However, recent studies suggest that eosinophil-derived IL-13 instead of IL-4 may play a more prominent role in establishing allergic inflammation in a mouse model.[172] The pathophysiological significance of eosinophil-derived IL-4 in human disease is yet to be determined.

IL-4 is one of at least 10 cytokines that have been identified as a preformed, stored mediator that is located within the crystalloid granules of eosinophils[59] (Fig. 12.2). While this report suggested IL-4 colocalized with the core of the crystalloid granules, later studies using immunogold electron microscopy suggest that IL-4 was located within crystalloid granules as well as small secretory vesicles that are granule-associated vesiculotubular carriers (so-called eosinophil sombrero vesicles).[63,64,66,67] These small secretory vesicles are important in trafficking of IL-4, and their membrane trafficking mechanisms are described in more detail elsewhere.[29]

In summary, IL-4 production from eosinophils may be important in a variety of allergy and immune responses that are only just beginning to be understood. These observations will continue to shape our understanding of the biological role of eosinophils in allergy.

Interleukin-5

Similarly, to IL-4, IL-5 has been extensively investigated for its role in allergic inflammation. While IL-5 does not have a direct role in skewing immune responses toward a Th2 phenotype, it is important in downstream Th2-associated events in response to allergens. IL-5 is an essential cytokine involved in the terminal differentiation and proliferation of precursor cells in the bone marrow into an eosinophilic phenotype. This cytokine acts on $CD34^+$ cells together with IL-3 and GM-CSF to promote eosinophil development.[173,174] IL-5 has an important effector role in mouse and human eosinophils, including prolongation of survival, induction of chemotaxis, priming, and degranulation.[36,38,175–177] Anti-IL-5, as well as anti-IL-5 receptor, are promising new treatments for eosinophilic asthma and may be able to promote a steroid-sparing regime for atopic asthmatics.[178–182] Interestingly, eosinophils also express IL-5, suggesting that IL-5 may have an autocrine or paracrine role in promoting eosinophil differentiation, survival, and activation.

Eosinophil expression of IL-5 was first characterized in tissue biopsies from human subjects. A significant percentage of eosinophils in gut mucosa tissue samples from patients with active celiac disease were found to be positive for IL-5, and following administration of a gluten-free diet, the numbers of $IL\text{-}5^+$ eosinophils declined.[68] However, $IL\text{-}5^+$ eosinophils are not associated with all disorders of the gut, as intestinal mucosal eosinophils in Crohn's disease were not found to be positive for IL-5 transcript.[70]

In allergic airway disease, eosinophils in nasal biopsies from subjects with allergic rhinitis correlated with elevated IL-5$^+$ mRNA in tissue cells.[157] Endobronchial or segmental challenge of atopic asthmatics led to increased IL-5$^+$ eosinophils infiltrating the airways.[43] Eosinophils also express IL-5 mRNA and protein in bronchial biopsies of atopic asthmatics as well as normal nonatopic subjects.[72] In complementary studies, ~20% of tissue eosinophils were positive for IL-5 mRNA in skin biopsies of allergic individuals 24 h following challenge, which increased in correlation with IL-5 protein expression.[62,73]

Human peripheral blood eosinophils also express IL-5, and its release may be evoked by immune complexes.[70,71] Moreover, IL-5 is also found as a preformed, stored cytokine located within the eosinophil crystalloid granule.[70,71] Immunogold electron microscopy analysis of IL-5 protein storage indicated that IL-5 was localized to the core of the crystalloid granule.[71] Thus, eosinophils have the capability of releasing IL-5 into surrounding tissues that has been already synthesized and stored within its secretory granules, suggesting that eosinophils may be able to promote the survival and activation of other newly recruited cells in an autocrine manner.

Interleukin-6

IL-6 is a pleiotropic proinflammatory cytokine that has a diversity of roles in the immune system, with a predominant function in stimulation of immune responses after trauma or infection, particularly in acute phase reactions. This cytokine is also important in allergic asthma, where it is elevated in serum and bronchoalveolar lavage samples both at baseline and following allergen challenge.[183–185] Recent studies indicate that IL-6 has a potential role in determining the adaptive immune response in allergy. Specifically, in vitro studies show that IL-6 promoted differentiation of effector CD4$^+$ T cells to a Th2 phenotype and suppressed Th1 differentiation.[186] Together with this, IL-6 is an essential cofactor together with IL-4 in isotype switching of B cells to produce IgE.[187] IL-6 also has a role in priming of granulocytes, suggesting a direct role in activation of the innate immune system.

Eosinophils express IL-6 and this is one of the 10 cytokines that has been identified as being stored as a preformed mediator in eosinophil crystalloid granules[76] (Fig. 12.2). Stimulation of human peripheral blood eosinophils in vitro by IFNγ and other inflammatory cytokines induces intracellular mobilization of IL-6 prior to its release.[48,74–76] Whether eosinophil-derived IL-6 is important in allergic inflammation is not known.

Interleukin-9

IL-9 was first characterized as a T cell and mast cell growth factor which is predominantly generated by T cells. The production of IL-9 is associated with the Th2 phenotype in allergy, and IL-9 has a prominent role in the establishment of the allergic phenotype in mouse models.[188] IL-9 is also important in human allergy as segmental allergen challenge of atopic asthmatics led to increased IL-9 expression in lymphocytes present in bronchoalveolar lavage fluid.[189] In the mouse model of eosinophilic tissue inflammation, IL-9 acts by promoting an influx of eosinophils and enhancing their differentiation, maturation, and survival.[78,190]

Eosinophils may also augment established Th2 responses by secreting IL-9 in response to proinflammatory cytokines TNFα and IL-1β.[191] The expression of IL-9 mRNA and protein products was shown in both peripheral blood eosinophils obtained from atopic asthmatics and normal healthy individuals. Tissue eosinophils also demonstrated IL-9 mRNA expression in asthmatic airways.[77] These studies indicate that eosinophils have the capacity to express and secrete IL-9 in allergic inflammation.

The intracellular sites of storage and trafficking of IL-9 has not yet been reported for eosinophils, and this is an area of ongoing investigation.

Interleukin-10

The immunosuppressive cytokine IL-10 plays a major role in immune regulation and the promotion of immune tolerance. Its former name is human cytokine synthesis inhibitory factor, and it is primarily produced by monocytes, Th2 cells, and other innate and adaptive immune cells. The main function of IL-10 is to downregulate the production of Th1 cytokines and to prevent antigen presentation in macrophages. IL-10 is also known for its immunosuppressive role in allergic inflammation by suppressing cytokine secretion.[192,193] In parallel with this, IL-10 inhibits the effects of bacterial lipopolysaccharide on survival and cytokine production by human peripheral blood eosinophils.[44]

Human peripheral blood eosinophils have been shown to express and release IL-10.[61] Tissue eosinophils from nasal mucosa samples also demonstrated increased IL-10 expression after nasal allergen challenge.[194] These findings indicate that eosinophil-derived IL-10 may enhance allergic inflammation, as IL-10 acts in concert with IL-4 to mediate the growth, differentiation, and isotype switching of activated B cells.[195] However, because IL-10 is known for its immunosuppressive properties in allergic inflammation,[192,193] this may suggest that eosinophil-derived IL-10 may have a more complex role in the modulating the allergic phenotype than these reports suggest. In helminth infections, eosinophil-derived IL-10 was shown to have a role in proliferation of myeloid dendritic cells and $CD4^+$ T cells, leading to protection of intracellular *Trichinella spiralis* larvae.[80] This striking observation suggests that eosinophils may serve a protective role for helminth larvae, counterintuitive to the classically held notion that eosinophils kill helminths to protect the host. These findings show that a significant functional diversity exists for eosinophils and their cytokines that was not previously appreciated until the development of eosinophil-deficient mouse strains.

The intracellular sites of synthesis, storage, and trafficking of IL-10 have not yet been elucidated in eosinophils, and this remains to be investigated in future studies.

Interleukin-11

A multifunctional cytokine derived from the bone marrow stromal cells, IL-11 has an important role in platelet function and bone development. In allergy, little is known regarding IL-11 function in Th2 responses, although its expression is upregulated in bronchial biopsy specimens from patients with asthma,[81] suggesting that it may have a role in chronic remodeling in asthmatic airways. In this study, eosinophils were shown to express IL-11 mRNA along with airway epithelial cells. Mouse models of allergic airway inflammation have demonstrated a role for IL-11 in inhibiting asthma-associated inflammation while promoting airway fibrosis.[196] There have been no other reports of eosinophil-derived IL-11, so the function of eosinophil-derived IL-11 in allergic diseases has not been determined and its intracellular location not yet been investigated.

Interleukin-12

IL-12 is a Th1 cytokine that is predominantly generated by cells in the innate immune system. Its role is associated with the differentiation of naïve T cells into Th1 cells. In allergy, IL-12 is known to downregulate allergic inflammation following its release, along with IFNγ.[135] Recombinant human IL-12 has been shown to reduce peripheral blood and sputum eosinophils in patients with mild allergic asthma, although no effects on airway hyper-responsiveness or the late asthmatic reaction were observed.[197] In human peripheral blood eosinophils, IL-12 has been demonstrated to induce apoptosis in an antagonistic manner with IL-5.[198]

Eosinophils also produce IL-12 in response to proinflammatory cytokine treatment. Stimulation of human peripheral blood eosinophils with IL-4 led to the secretion of IL-12, which in turn induced expression of

IFNγ from Th1 cells.[82] IL-12 is also one of a number of cytokines secreted by eosinophils stimulated by Th1, Th2, and proinflammatory cytokines.[48] In patients with dermatitis, it has been shown that eosinophil-derived IL-12 induced a switch from Th2 to Th1 responses in late phase allergic skin reactions.[82,199]

The location of IL-12 protein expression in eosinophils has not yet been determined.

Interleukin-13

IL-13 is a prominent Th2 cytokine that is released by many cell types, particularly from Th2 cells. The effects of IL-13 on immune cells are similar to those of IL-4 as these cytokines share a common receptor subunit (the α subunit), although IL-13 plays a predominant role in allergic inflammation by promoting bronchial hyperreactivity and overproduction of mucus.[200] IL-13 also promotes isotype switching of B cells to produce IgE.[201] In the mouse model, IL-13 is an important mediator of allergic asthma,[200] and was shown to induce eosinophil recruitment to the airways in an IL-5 and CCL11/eotaxin-dependent manner.[202] More recently, anti-IL-13 has been promoted as a treatment for severe asthma in human patients, although its effects have not resulted in improvements in asthma control and other clinical parameters.[203] Further studies are in process to determine if targeting of more than one cytokine may result in improved asthma symptoms.

The receptor for IL-13 is expressed in an eosinophilic cell line,[204] although there are few reports describing its direct effects on eosinophils. Conversely, IL-13 is synthesized and secreted by eosinophils.[83,84] IL-13 synthesis and expression has been characterized in human peripheral blood eosinophils from patients with bronchial asthma, atopic dermatitis, and hypereosinophilic syndrome.[84] In addition, tissue eosinophils from nasal polyps were shown to express IL-13, and its release could be induced by activation with cytokines or CD28 ligation.[83,84] Eosinophil-derived IL-13 was also shown to be bioactive by inducing CD23 expression on B cells.[83,84]

A recent study has demonstrated that eosinophil-derived IL-13 may be required for allergic airway responses.[205] This was determined using IL-13$^{-/-}$ eosinophils that were adoptively transferred into ΔdblGATA mice. In the computational model for this study, it was found that IL-13 production by eosinophils was integral to the development of allergic asthma.[205]

The intracellular site of storage of IL-13 in eosinophils is associated with the crystalloid granules, suggesting that IL-13 is a preformed mediator that is released upon stimulation of degranulation.[83] This was determined by subcellular fractionation and immunogold electron microscopy analysis of human peripheral blood eosinophils.

Interleukin-16

IL-16 was formerly known as lymphocyte chemoattractant factor, and serves as a proinflammatory factor that is chemotactic for immune cells, including T cells, monocytes, and eosinophils. It acts through CD4 as its signal-transducing receptor, and is elevated in allergic responses in nasal tissues[206] and bronchoalveolar lavage fluid of histamine-challenged asthma patients.[207] In human eosinophils, IL-16 promotes leukotriene C_4 generation and IL-4 secretion,[208] suggesting that this cytokine activates eosinophils to facilitate an allergic inflammatory response.

Human peripheral blood eosinophils synthesize and secrete IL-16 that was shown to be bioactive for T cells by inducing cell migration.[85,86] Thus, human eosinophils may have the ability to alter $CD4^+$ T cell and memory T cell activities. IL-16 was also demonstrated to be stored as a preformed mediator in eosinophils, with intense granular staining.[86] Further studies on the intracellular storage and trafficking of IL-16 would shed more light on how this is secreted from eosinophils.

Interleukin-17

The predominant role of IL-17 in immune responses is mainly proinflammatory, and it is the major cytokine secreted by Th17 cells. In severe asthma, IL-17 is considered to be a key factor in the pathophysiology of airway disease,[209] and it has been shown to enhance Th2-mediated eosinophilic airway inflammation in mouse models of asthma.[210] A direct effect of IL-17 on eosinophils has not yet been reported.

Eosinophils have been shown to express IL-17 in peripheral blood, sputum, and bronchoalveolar lavage fluid.[87] Levels of IL-17 were significantly higher in peripheral blood eosinophils from subjects with asthma than in control subjects. These findings indicate that eosinophil-derived IL-17 may be an important factor in eosinophilic asthma. However, there are no reports regarding the intracellular sites of synthesis or storage of IL-17 in eosinophils.

Interleukin-25

Also known as IL-17E, IL-25 is produced by epithelial cells along with numerous innate immune cells, and plays a prominent role in enhancing Th2 cytokine production.[211] IL-25 may be important in promoting Th2 cytokine-mediated allergic inflammation along with thymic stromal lymphopoietin (TSLP) by activating the function of adaptive Th2 memory cells through dendritic cells[88] The expression of IL-25 is elevated in tissue biopsies from patients with chronic asthma and atopic dermatitis.[88]

While they do not express significant levels of IL-25 receptor and do not respond to IL-25 treatment, eosinophils purified from the peripheral blood of atopic and nonatopic subjects generate IL-25 constitutively and upon activation with eosinophil-specific cytokines.[88] It was further demonstrated that eosinophil-derived IL-25 induced IL-5 production from Th2 memory cells, along with modest levels of IL-4 and IL-13. Eosinophil-derived IL-25 production was confirmed in independent studies of Churg–Strauss patients, where eosinophils were found to be the main IL-25-producing cells in blood.[89]

No studies have been reported describing the intracellular sites of storage and trafficking of IL-25 in human eosinophils, and this remains an area of interest for future investigations.

CCL5/RANTES (Regulated on Activation, Normal T Cell Expressed and Secreted)

Chemokine (C-C motif) ligand 5 (CCL5), also known as RANTES, is classified as a chemotactic cytokine, or chemokine. It is highly chemotactic for T cells, eosinophils, and basophils, and acts on the G protein-coupled CCR5 receptor for its effector actions. CCL5/RANTES acts as a major regulator of local immune responses, and targets immune cells to sites of inflammation. In allergic diseases, CCL5/RANTES is elevated along with other cytokines in tissue samples from atopic patients, where it is thought to have a role in recruitment of T cells and eosinophils.[212–214] The production of CCL5/RANTES during respiratory virus infections is suggested to exacerbate allergic airway disease.[134] In eosinophils, CCL5/RANTES not only induces chemotaxis but also induces cell activation by elevation of Ca^{2+}, along with upregulation of adhesion molecules,[215,216] respiratory burst,[38,217] and degranulation.[37,38]

Peripheral blood eosinophils from humans express CCL5/RANTES mRNA and protein.[85,96] The release of CCL5/RANTES from eosinophils may be evoked by cytokines such as IFNγ and opsonized particles.[24,96,137] Expression of CCL5/RANTES in tissue eosinophils from human subjects was also confirmed in studies which showed that around 15% of the total CCL5/RANTES^{+} population of cells in nasal mucosal biopsies from seasonal rhinitis patients were eosinophils.[213] In late-phase cutaneous reactions following allergen challenge in atopic subjects, eosinophils recruited to the skin also expressed increased CCL5/RANTES mRNA and protein.[96] These findings suggest that both peripheral and tissue eosinophils express CCL5/RANTES, and that this is released during allergic inflammation.

Eosinophil-derived CCL5/RANTES has been shown to be bioactive, as it has direct chemotactic effects on lymphocytes in culture.[96] In these studies, the bioactivity of eosinophil-derived CCL5/RANTES was inhibited by treatment of eosinophil supernatants with antibody to CCL5/RANTES. Thus, human eosinophils have the capacity to regulate the function of T cells and to elicit the accumulation of eosinophils through an autocrine mechanism.

Eosinophils elaborate CCL5/RANTES from preformed stores within the matrix of their crystalloid granules.[24] The release of CCL5/RANTES was inducible by IFNγ, which led to intracellular mobilization of this chemokine prior to its release.[24] Specifically, IFNγ evoked the redistribution of CCL5/RANTES from the crystalloid granules to a rapidly mobilized pool of small secretory vesicles within 10 minutes of stimulation, leaving MBP^{+} crystalloid granules behind.[24] After 16 h of IFNγ stimulation, CCL5/RANTES was found replenished in crystalloid granules, suggesting that eosinophils have the ability to undergo sustained rounds of cytokine secretion. These findings indicate the eosinophils have the ability to selectively and differentially release cytokines and granule proteins in response to specific inflammatory stimulation.

The mechanism of selective cytokine release from eosinophils has been associated with a tubulovesicular system which consists of the small secretory granules containing CCL5/RANTES along with IL-4.[218] This membrane transport system is hypothesized to form the mechanism associated with piecemeal degranulation, a type of degranulation that is commonly observed in tissues from allergic subjects.[28] The tubulovesicular system allows the trafficking of granule contents to the cell surface through vesicles and tubules that directly bud from the surface of the crystalloid granule.[67,148,218,219] This specialized mechanism of cytokine transport was first characterized with CCL5/RANTES in eosinophils.[24]

Some of the specific membrane fusion machinery has also been investigated in association with CCL5/RANTES secretion. The small secretory vesicles containing CCL5/RANTES were found to colocalize with the SNARE (soluble *N*-ethylmaleimide sensitive factor attachment protein receptor), VAMP-2 (vesicle-associated membrane protein-2), in human peripheral blood eosinophils.[220] SNARE proteins are the hypothetical universal fusion proteins that regulate docking of granules and vesicles to target membranes including the plasma membrane.[221] Fusion of CCL5/RANTES^{+} small secretory vesicles is proposed to require binding to cognate target membrane SNAREs, known as SNAP-23 and syntaxin-4.[222]

CCL11/Eotaxin

CCL11/Eotaxin is an important eosinophil-specific chemokine that is associated with the recruitment of eosinophils into sites of inflammation. It is generated in the lungs of asthmatic patients and has a role in targeting eosinophils at inflammatory foci.[97,223,224] Mouse models of allergic airway inflammation indicate a central role for CCL11/eotaxin in recruitment of eosinophils to the airways.[225,226] Gene knockout of CCL11/eotaxin leads to markedly reduced tissue eosinophil numbers, which is associated with reduced allergic inflammation in the gastrointestinal system, subcutaneous regions, and the airways.[227] CCL11/eotaxin acts on cells through it G protein-coupled receptor, CCR3, and eosinophils respond to CCR3 ligand binding by undergoing chemotaxis, Ca^{2+} mobilization, degranulation, and respiratory burst.[37,217,223,228–230]

Human peripheral blood eosinophils express CCL11/eotaxin and release this chemokine in response to stimulation by complement factors and immune complexes.[98] In human airway tissues, endobronchial or segmental challenge with allergen evokes elevated numbers of eosinophils positive for CCL11/eotaxin labeling.[97]

The intracellular sites of storage of CCL11/eotaxin have not yet been described in detail beyond an apparent association with the crystalloid granules in eosinophils based on conventional immunocytochemical staining.[98] This remains an area of interest for future studies.

CCL17/TARC and CCL22/MDC

The roles of CCL17/TARC (thymus- and activation-regulated chemokine) and CCL22/MDC (monocyte-derived chemokine) in the immune system are associated with induction of chemotaxis in T cells, and particularly Th2 cells, by binding to the chemokine receptor CCR4. These two chemokines are found secreted in parallel from dendritic cells and macrophages. In allergic inflammation, CCL17/TARC has been shown to be elicited upon segmental allergen exposure in asthmatics,[231] which correlated with elevated recruitment of Th2 cells that preferentially express CCR4. Interestingly, eosinophils do not respond significantly to CCL17/TARC stimulation in correlation with an absence of CCR4 expression, at least in mouse cells.[232] Eosinophils were also shown to elicit CCL17/TARC and CCL22/MDC in a mouse model of allergic inflammation, leading to effector T cell recruitment and the establishment of airway hyperresponsiveness.[233]

While there is no detectable CCR4 in eosinophils, human eosinophils express and release CCL17/TARC and CCL22/MDC in response to stimulation by TNF and IFNγ or IL-4.[100] These are among the few cytokines that are actively synthesized and not stored by eosinophils. These intriguing findings suggest that eosinophils may contribute to allergic inflammation by promoting the recruitment of Th2 cells through the release of Th2-specific chemokines. Moreover, the release of CCL17/TARC and CCL22/MDC may occur through a distinct intracellular trafficking pathway from those associated with preformed, stored cytokines. This remains an area of interest for future studies.

CXCL8/Interleukin-8

CXCL8/IL-8 (chemokine C-X-C motif ligand 8, along with its mouse homolog, keratinocyte-derived chemokine, CXCL1/KC) is widely expressed throughout the body, with predominant secretion from macrophages, epithelial cells, airway smooth muscle cells, and endothelial cells. Its role in immunity is to attract neutrophils and other cell types through CXCR1 and CXCR2 to sites of inflammation and infection. In allergic inflammation, CXCL8/IL-8 has been associated with virus-induced exacerbations of asthma and airway hyperresponsiveness.[234] Eosinophils from atopic individuals respond to CXCL8/IL-8 by undergoing chemotaxis in vitro[235]; however, CXCL8/IL-8 challenge of nasal mucosa in allergic patients did not result in significant tissue eosinophil recruitment.[236]

Peripheral blood eosinophils from normal and atopic individuals are very well characterized for their ability to express and release CXCL11/IL-8 in response to stimulation by a variety of factors and cytokines.[44,61,104–108] Further, eosinophils accumulating in the airways following allergen challenge express CXCL8/IL-8.[107] These findings indicate that eosinophils may recruit neutrophils and other cells expressing receptors for CXCL8/IL-8 to sites of allergic inflammation. The role of eosinophil-derived CXCL8/IL-8 in the context of allergic inflammation has not yet been determined, and its intracellular sites of transport to the cell membrane are yet to be investigated.

TGFα (Transforming Growth Factor-α)

A cytokine with growth factor functions, TGFα is a mitogenic factor with a role in cell proliferation, differentiation, and maturation. Its receptor is epidermal growth factor receptor (EGFR) and it has important roles in wound healing and tissue remodeling in response to inflammatory stimuli. The role of TGFα in allergic inflammation has not been described in detail, although it is hypothesized to be important in keratinocyte-mediated wound healing in response to allergen challenge.[237] Further, remodeling in asthmatic airways is postulated to be associated with TGFα production in response to IL-13 in airway epithelial cells.[238]

TGFα was the second cytokine to be described for its expression in eosinophils. It was shown to be expressed at mRNA and protein levels in tissue eosinophils infiltrating interstitial regions adjacent to human colonic or oral carcinomas.[114] This was soon followed by the observation that TGFα$^+$ tissue eosinophils could be detected in a rabbit model of healing cutaneous wounds,[115] and in the nasal mucosa of individuals with allergic rhinitis.[116] The release of TGFα from human eosinophils may be triggered by cytokine stimulation.[118]

Detailed immunogold analysis of human peripheral blood eosinophils showed that TGFα is present as a preformed, stored mediator in eosinophil crystalloid granules as well as small secretory vesicles,[117] suggesting that this cytokine is trafficked through the tubulovesicular system associated with IL-4 and CCL5/RANTES transport. These findings indicate that eosinophils utilize their tubulovesicular system for differential release of several key cytokines in a pathway that may be distinct from classical or compound exocytosis of crystalloid granules.

TGFβ (Transforming Growth Factor-β)

First identified in human platelets, TGFβ is a member of the TGFβ superfamily of cytokines that is secreted by a variety of cell types. TGFβ is produced as three highly homologous isoforms, and is bound as an immobilized, latent precursor form to extracellular matrix proteins in many tissues throughout the body.[239] On its release into an active form, it controls many types of cellular function including proliferation, differentiation, apoptosis, and angiogenesis, as part of its role in wound healing, and most cells possess receptors for TGFβ.[239,240] TGFβ is also an important target for drug treatment in numerous clinical trials specifically focused on cancer therapy, but may also be a promising target for allergic disease.[240] In allergic disease, TGFβ plays a prominent role in the development of asthma, allergic rhinitis, and eczema, along with gastrointestinal complaints.[241,242] Mutations in the TGFβ receptor (TGFβR) were recently discovered in patients with Loeys–Dietz syndrome, resulting in elevated rates of allergic disease and increased eosinophil numbers.[241,242] High levels of TGFβ have been found in patients with asthma, suggesting an important role in airway remodeling.[243] Eosinophils also respond to low concentrations of TGFβ by undergoing chemotaxis,[244] although TGFβ inhibits IL-5 activation of eosinophils.[245]

Perhaps unsurprisingly, human eosinophils from both blood and tissue sources also express TGFβ, and there are numerous reports demonstrating this.[61,116,118–125] In normal individuals, the airway epithelium appears to be the major site of TGFβ expression, while in asthmatics, many reports have shown that eosinophils recruited to the airways are the main source of TGFβ.[124,243,246,247] Eosinophils from bronchial biopsies of patients with severe asthma were shown to exhibit greater expression of TGFβ1 than those of normal control subjects, with up to 75% of tissue eosinophils being positive for TGFβ1.[124]

Moreover, eosinophil-derived TGFβ has bioactive effects in vitro, as TGFβ from eosinophils has been shown to regulate fibroblast proliferation and differentiation, suggesting a role for eosinophils in wound healing.[248] In support of this finding, eosinophils were found to express TGFβ along with IL-13 following intradermal allergen challenge, leading to increased repair and remodeling in human atopic skin.[249]

However, in spite of substantial studies showing that eosinophils are a major source of TGFβ, and another report describing TGFα storage in eosinophil crystalloid granules and secretory vesicles,[117] there are no reports of the intracellular sites of TGFβ synthesis and/or storage in eosinophils.

TNF (Tumor Necrosis Factor-α)

A major proinflammatory cytokine that has pleiotropic roles in local and systemic inflammation, TNF is one of several cytokines that is a target for pharmacological intervention in numerous chronic inflammatory disorders. TNF is also implicated in airway inflammation in asthma, and may be important in refractory asthma.[250,251] TNF is a highly potent activator of monocytes, T cells, neutrophils, and endothelial cells, and also acts by enhancing eosinophil adhesion and cytotoxicity.[252,253] TNF contributes to allergic inflammation by promoting antigen-specific IgE production and induction of Th2 cytokines.[254] Eosinophils respond to TNF activation by undergoing respiratory burst,[255] and releasing matrix metalloprotease-9,[256] but do not undergo full degranulation, based on EDN release.[36] However, TNF acts synergistically with IL-5 to induce degranulation.[38]

Eosinophils synthesize and secrete TNF, and represent a potential source of this proinflammatory cytokine in allergy and chronic inflammation.[48,92,94] Purified peripheral blood eosinophils from atopic individuals spontaneously release TNF in culture, and normal eosinophils stimulated with immobilized immunoglobulins or cytokines express mRNA for this cytokine.[61]

TNF is one of the 10 cytokines that has been detected as a preformed, stored mediator in eosinophil granules based on immunogold electron microscopy[91] (Fig. 12.2). Ultrastructural immunogold analysis demonstrated that TNF was localized to the matrix compartment of eosinophil crystalloid granules in patients with hypereosinophilic syndrome[91] and Crohn's disease.[257] These findings indicate the eosinophil-derived TNF may orchestrate inflammatory processes in an exacerbating or modulatory manner.

SUMMARY

It is clear from many studies that eosinophils have the capacity to release a large array of cytokines and that many of these have been demonstrated to have immunoregulatory roles in allergic responses in both human and mouse models of allergic inflammation. A large proportion of cytokines are found as preformed mediators stored in eosinophil crystalloid granules or similar secretory organelles, and many of these have been demonstrated to have bioactive roles in immunity. Many other cytokines derived from eosinophils that are not described in detail in this chapter may also contribute to allergic inflammation, and these are listed in Table 12.1. Note that this chapter does not include all of the functions of eosinophil-derived cytokines in homeostasis, fat deposition, antiviral effects, and diseases such as cancer. There have been a series of recent striking developments implicating eosinophils in many other immune-related functions, that demonstrate a pivotal role for eosinophils and their cytokines in plasma cell maintenance in the bone marrow,[39] of immunosuppressive effects in the lungs[6] and gastrointestinal tract,[7] as well as implicating a role for these intriguing cells in the formation of beige fat.[258,259] What is evident is that eosinophils are only now being recognized as important immune modulatory cells in maintaining the health of the organism and repressing damaging signals from pathogens and their impact on the immune system.

Taken together, these reports indicate that eosinophils serve an important role in immunity, with a major role for their cytokines, chemokines, and growth factors in augmentation of inflammatory responses in allergic diseases. We look forward to future studies in which the specific functions of each of these eosinophil-derived cytokines is determined for its contribution to allergic inflammation, and to investigate the mechanisms of cytokine trafficking and release for the purpose of development of novel therapeutic targets.

Disclosures: The author has no financial conflicts of interest to disclose.

References

1. Jung Y, Rothenberg ME. Roles and regulation of gastrointestinal eosinophils in immunity and disease. *J Immunol* 2014;**193**:999–1005.
2. Nutman TB. Immune responses in helminth infections. In: Lee JJ, Rosenberg HF, editors. *Eosinophils in health and disease*. New York: Elsevier; 2013. p. 312–20.
3. Lacy P. 28 days later: eosinophils stop viruses. *Blood* 2014;**123**:609–11.
4. Percopo CM, Dyer KD, Ochkur SI, Luo JL, Fischer ER, Lee JJ, et al. Activated mouse eosinophils protect against lethal respiratory virus infection. *Blood* 2014;**123**:743–52.
5. Rosenberg HF, Dyer KD, Domachowske JB. Interactions of eosinophils with respiratory virus pathogens. In: Lee JJ, Rosenberg HF, editors. *Eosinophils in health and disease*. New York: Elsevier; 2013. p. 281–90.
6. Mesnil C, Raulier S, Paulissen G, Xiao X, Birrell MA, Pirottin D, et al. Lung-resident eosinophils represent a distinct regulatory eosinophil subset. *J Clin Invest* 2016; **126**:3279–95.
7. Sugawara R, Lee EJ, Jang MS, Jeun EJ, Hong CP, Kim JH, et al. Small intestinal eosinophils regulate Th17 cells by producing IL-1 receptor antagonist. *J Exp Med* 2016;**213**:555–67.
8. Rothenberg ME, Hogan SP. The eosinophil. *Annu Rev Immunol* 2006;**24**:147–74.
9. Wenzel SE. Asthma phenotypes: the evolution from clinical to molecular approaches. *Nat Med* 2012;**18**:716–25.
10. Hogan SP, Rosenberg HF, Moqbel R, Phipps S, Foster PS, Lacy P, et al. Eosinophils: biological properties and role in health and disease. *Clin Exp Allergy* 2008;**38**:709–50.
11. Spencer LA. Release of cytokines and chemokines from eosinophils. In: Lee JJ, Rosenberg HF, editors. *Eosinophils in health and disease*. Elsevier; 2013. p. Waltham, MA 230–243
12. Davoine F, Lacy P. Eosinophil cytokines, chemokines, and growth factors: emerging roles in immunity. *Front Immunol* 2014;**5**:570.
13. Lee JJ, Jacobsen EA, Mcgarry MP, Schleimer RP, Lee NA. Eosinophils in health and disease: the LIAR hypothesis. *Clin Exp Allergy* 2010;**40**:563–75.
14. Lee JJ, Dimina D, Macias MP, Ochkur SI, Mcgarry MP, O'neill KR, et al. Defining a link with asthma in mice congenitally deficient in eosinophils. *Science* 2004;**305**: 1773–6.
15. Humbles AA, Lloyd CM, Mcmillan SJ, Friend DS, Xanthou G, Mckenna EE, et al. A critical role for eosinophils in allergic airways remodeling. *Science* 2004;**305**:1776–9.
16. Rosenberg HF, Dyer KD, Foster PS. Eosinophils: changing perspectives in health and disease. *Nat Rev Immunol* 2013;**13**:9–22.
17. Lacy P, Adamko DJ, Moqbel R. The human eosinophil. In: Greer JP, Arber DA, Glader B, List AF, Means RT, Paraskevas F, Rogers GM, Foerster J, editors. Wintrobe's *clinical* hematology. Philadelphia, PA: Lippincott Williams & Wilkins; 2013. p. 214–35.
18. Lindau M, Nusse O, Bennett J, Cromwell O. The membrane fusion events in degranulating guinea pig eosinophils. *J Cell Sci* 1993;**104**:203–10.
19. Nusse O, Lindau M, Cromwell O, Kay AB, Gomperts BD. Intracellular application of guanosine-5′-*O*-(3-thiotriphosphate) induces exocytotic granule fusion in guinea pig eosinophils. *J Exp Med* 1990;**171**:775–86.
20. Scepek S, Lindau M. Focal exocytosis by eosinophils—compound exocytosis and cumulative fusion. *EMBO J* 1993;**12**:1811–17.
21. Scepek S, Moqbel R, Lindau M. Compound exocytosis and cumulative degranulation by eosinophils and their role in parasite killing. *Parasitol Today* 1994;**10**:276–8.
22. Scepek S, Coorssen JR, Lindau M. Fusion pore expansion in horse eosinophils is modulated by Ca^{2+} and protein kinase C via distinct mechanisms. *EMBO J* 1998; **17**:4340–5.

23. Karawajczyk M, Seveus L, Garcia R, Bjornsson E, Peterson CG, Roomans GM, et al. Piecemeal degranulation of peripheral blood eosinophils: a study of allergic subjects during and out of the pollen season. *Am J Respir Cell Mol Biol* 2000;**23**:521–9.
24. Lacy P, Mahmudi-Azer S, Bablitz B, Hagen SC, Velazquez JR, Man SF, et al. Rapid mobilization of intracellularly stored RANTES in response to interferon-γ in human eosinophils. *Blood* 1999;**94**:23–32.
25. Dvorak AM, Furitsu T, Letourneau L, Ishizaka T, Ackerman SJ. Mature eosinophils stimulated to develop in human cord blood mononuclear cell cultures supplemented with recombinant human interleukin-5. Part I. Piecemeal degranulation of specific granules and distribution of Charcot-Leyden crystal protein. *Am J Pathol* 1991;**138**: 69–82.
26. Erjefalt JS, Andersson M, Greiff L, Korsgren M, Gizycki M, Jeffery PK, et al. Cytolysis and piecemeal degranulation as distinct modes of activation of airway mucosal eosinophils. *J Allergy Clin Immunol* 1998;**102**:286–94.
27. Saffari H, Hoffman LH, Peterson KA, Fang JC, Leiferman KM, Pease LF, et al. Electron microscopy elucidates eosinophil degranulation patterns in patients with eosinophilic esophagitis. *J Allergy Clin Immunol* 2014;**133**:1728–34.
28. Erjefalt JS, Greiff L, Andersson M, Matsson E, Petersen H, Linden M, et al. Allergen-induced eosinophil cytolysis is a primary mechanism for granule protein release in human upper airways. *Am J Respir Crit Care Med* 1999;**160**:304–12.
29. Spencer LA, Bonjour K, Melo RC, Weller PF. Eosinophil secretion of granule-derived cytokines. *Front Immunol* 2014;**5**:496.
30. Kroegel C, Yukawa T, Dent G, Venge P, Chung KF, Barnes PJ. Stimulation of degranulation from human eosinophils by platelet-activating factor. *J Immunol* 1989;**142**: 3518–26.
31. Dyer KD, Percopo CM, Xie Z, Yang Z, Kim JD, Davoine F, et al. Mouse and human eosinophils degranulate in response to platelet-activating factor (PAF) and lysoPAF via a PAF-receptor-independent mechanism: evidence for a novel receptor. *J Immunol* 2010;**184**:6327–34.
32. Winqvist I, Olofsson T, Olsson I. Mechanisms for eosinophil degranulation; release of the eosinophil cationic protein. *Immunology* 1984;**51**:1–8.
33. Zeck-Kapp G, Kroegel C, Riede UN, Kapp A. Mechanisms of human eosinophil activation by complement protein C5a and platelet-activating factor: similar functional responses are accompanied by different morphologic alterations. *Allergy* 1995;**50**: 34–47.
34. Abu-Ghazaleh RI, Fujisawa T, Mestecky J, Kyle RA, Gleich GJ. IgA-induced eosinophil degranulation. *J Immunol* 1989;**142**:2393–400.
35. Velazquez JR, Lacy P, Mahmudi-Azer S, Moqbel R. Effects of interferon-γ on mobilization and release of eosinophil-derived RANTES. *Int Arch Allergy Immunol* 1999;**118**:447–9.
36. Fujisawa T, Abu-Ghazaleh R, Kita H, Sanderson CJ, Gleich GJ. Regulatory effect of cytokines on eosinophil degranulation. *J Immunol* 1990;**144**:642–6.
37. Fujisawa T, Kato Y, Nagase H, Atsuta J, Terada A, Iguchi K, et al. Chemokines induce eosinophil degranulation through CCR-3. *J Allergy Clin Immunol* 2000;**106**:507–13.
38. Horie S, Gleich GJ, Kita H. Cytokines directly induce degranulation and superoxide production from human eosinophils. *J Allergy Clin Immunol* 1996;**98**:371–81.
39. Chu VT, Frohlich A, Steinhauser G, Scheel T, Roch T, Fillatreau S, et al. Eosinophils are required for the maintenance of plasma cells in the bone marrow. *Nat Immunol* 2011;**12**:151–9.
40. Moqbel R, Hamid Q, Ying S, Barkans J, Hartnell A, Tsicopoulos A, et al. Expression of mRNA and immunoreactivity for the granulocyte/macrophage colony-stimulating factor in activated human eosinophils. *J Exp Med* 1991;**174**:749–52.

41. Kita H, Ohnishi T, Okubo Y, Weiler D, Abrams JS, Gleich GJ. Granulocyte/macrophage colony-stimulating factor and interleukin 3 release from human peripheral blood eosinophils and neutrophils. *J Exp Med* 1991;**174**:745–8.
42. Ohno I, Lea R, Finotto S, Marshall J, Denburg J, Dolovich J, et al. Granulocyte/macrophage colony-stimulating factor (GM-CSF) gene expression by eosinophils in nasal polyposis. *Am J Respir Cell Mol Biol* 1991;**5**:505–10.
43. Broide DH, Paine MM, Firestein GS. Eosinophils express interleukin 5 and granulocyte macrophage-colony-stimulating factor mRNA at sites of allergic inflammation in asthmatics. *J Clin Invest* 1992;**90**:1414–24.
44. Takanaski S, Nonaka R, Xing Z, O'byrne P, Dolovich J, Jordana M. Interleukin 10 inhibits lipopolysaccharide-induced survival and cytokine production by human peripheral blood eosinophils. *J Exp Med* 1994;**180**:711–15.
45. Girgis-Gabardo A, Kanai N, Denburg JA, Hargreave FE, Jordana M, Dolovich J. Immunocytochemical detection of granulocyte-macrophage colony-stimulating factor and eosinophil cationic protein in sputum cells. *J Allergy Clin Immunol* 1994;**93**:945–7.
46. Levi-Schaffer F, Lacy P, Severs NJ, Newman TM, North J, Gomperts B, et al. Association of granulocyte-macrophage colony-stimulating factor with the crystalloid granules of human eosinophils. *Blood* 1995;**85**:2579–86.
47. Lamkhioued B, Gounni AS, Aldebert D, Delaporte E, Prin L, Capron A, et al. Synthesis of type 1 (IFN γ) and type 2 (IL-4, IL-5, and IL-10) cytokines by human eosinophils. *Ann N Y Acad Sci* 1996;**796**:203–8.
48. Spencer LA, Szela CT, Perez SA, Kirchhoffer CL, Neves JS, Radke AL, et al. Human eosinophils constitutively express multiple Th1, Th2, and immunoregulatory cytokines that are secreted rapidly and differentially. *J Leukoc Biol* 2009;**85**:117–23.
49. Weller PF, Rand TH, Barrett T, Elovic A, Wong DT, Finberg RW. Accessory cell function of human eosinophils. HLA-DR-dependent, MHC-restricted antigen-presentation and IL-1 α expression. *J Immunol* 1993;**150**:2554–62.
50. Del Pozo V, De Andres B, Martin E, Maruri N, Zubeldia JM, Palomino P, et al. Murine eosinophils and IL-1:α IL-1 mRNA detection by in situ hybridization. Production and release of IL-1 from peritoneal eosinophils. *J Immunol* 1990;**144**: 3117–22.
51. Esnault S, Kelly EA, Nettenstrom LM, Cook EB, Seroogy CM, Jarjour NN. Human eosinophils release IL-1β and increase expression of IL-17A in activated $CD4^+$ T lymphocytes. *Clin Exp Allergy* 2012;**42**:1756–64.
52. Jung Y, Wen T, Mingler MK, Caldwell JM, Wang YH, Chaplin DD, et al. IL-1β in eosinophil-mediated small intestinal homeostasis and IgA production. *Mucosal Immunol* 2015;**8**:930–42.
53. Levi-Schaffer F, Barkans J, Newman TM, Ying S, Wakelin M, Hohenstein R, et al. Identification of interleukin-2 in human peripheral blood eosinophils. *Immunology* 1996;**87**:155–61.
54. Bosse M, Audette M, Ferland C, Pelletier G, Chu HW, Dakhama A, et al. Gene expression of interleukin-2 in purified human peripheral blood eosinophils. *Immunology* 1996;**87**:149–54.
55. Woerly G, Roger N, Loiseau S, Dombrowicz D, Capron A, Capron M. Expression of CD28 and CD86 by human eosinophils and role in the secretion of type 1 cytokines (interleukin 2 and interferon γ): inhibition by immunoglobulin a complexes. *J Exp Med* 1999;**190**:487–95.
56. Woerly G, Roger N, Loiseau S, Capron M. Expression of Th1 and Th2 immunoregulatory cytokines by human eosinophils. *Int Arch Allergy Immunol* 1999;**118**:95–7.
57. Fujisawa T, Fukuda S, Atsuta J, Ichimi R, Kamiya H, Sakurai M. Interferon-γ induces interleukin-3 release from peripheral blood eosinophils. *Int Arch Allergy Immunol* 1994;**104**(Suppl. 1):41–3.

58. Nonaka M, Nonaka R, Woolley K, Adelroth E, Miura K, Okhawara Y, et al. Distinct immunohistochemical localization of IL-4 in human inflamed airway tissues. IL-4 is localized to eosinophils in vivo and is released by peripheral blood eosinophils. *J Immunol* 1995;**155**:3234–44.
59. Moqbel R, Ying S, Barkans J, Newman TM, Kimmitt P, Wakelin M, et al. Identification of messenger RNA for IL-4 in human eosinophils with granule localization and release of the translated product. *J Immunol* 1995;**155**:4939–47.
60. Moller GM, De Jong TA, Van Der Kwast TH, Overbeek SE, Wierenga-Wolf AF, Thepen T, et al. Immunolocalization of interleukin-4 in eosinophils in the bronchial mucosa of atopic asthmatics. *Am J Respir Cell Mol Biol* 1996;**14**:439–43.
61. Nakajima H, Gleich GJ, Kita H. Constitutive production of IL-4 and IL-10 and stimulated production of IL-8 by normal peripheral blood eosinophils. *J Immunol* 1996;**156**:4859–66.
62. Barata LT, Ying S, Meng Q, Barkans J, Rajakulasingam K, Durham SR, et al. IL-4- and IL-5-positive T lymphocytes, eosinophils, and mast cells in allergen-induced late-phase cutaneous reactions in atopic subjects. *J Allergy Clin Immunol* 1998;**101**:222–30.
63. Bandeira-Melo C, Sugiyama K, Woods LJ, Weller PF. Cutting edge: eotaxin elicits rapid vesicular transport-mediated release of preformed IL-4 from human eosinophils. *J Immunol* 2001;**166**:4813–17.
64. Melo RC, Spencer LA, Perez SA, Ghiran I, Dvorak AM, Weller PF. Human eosinophils secrete preformed, granule-stored interleukin-4 through distinct vesicular compartments. *Traffic* 2005;**6**:1047–57.
65. Bandeira-Melo C, Weller PF. Mechanisms of eosinophil cytokine release. *Mem Inst Oswaldo Cruz* 2005;**100**(Suppl. 1):73–81.
66. Spencer LA, Melo RC, Perez SA, Bafford SP, Dvorak AM, Weller PF. Cytokine receptor-mediated trafficking of preformed IL-4 in eosinophils identifies an innate immune mechanism of cytokine secretion. *Proc Natl Acad Sci USA* 2006;**103**:3333–8.
67. Melo RC, Spencer LA, Dvorak AM, Weller PF. Mechanisms of eosinophil secretion: large vesiculotubular carriers mediate transport and release of granule-derived cytokines and other proteins. *J Leukoc Biol* 2008;**83**:229–36.
68. Desreumaux P, Janin A, Colombel JF, Prin L, Plumas J, Emilie D, et al. Interleukin 5 messenger RNA expression by eosinophils in the intestinal mucosa of patients with coeliac disease. *J Exp Med* 1992;**175**:293–6.
69. Desreumaux P, Janin A, Dubucquoi S, Copin MC, Torpier G, Capron A, et al. Synthesis of interleukin-5 by activated eosinophils in patients with eosinophilic heart diseases. *Blood* 1993;**82**:1553–60.
70. Dubucquoi S, Desreumaux P, Janin A, Klein O, Goldman M, Tavernier J, et al. Interleukin 5 synthesis by eosinophils: association with granules and immunoglobulin-dependent secretion. *J Exp Med* 1994;**179**:703–8.
71. Moller GM, De Jong TA, Overbeek SE, Van Der Kwast TH, Postma DS, Hoogsteden HC. Ultrastructural immunogold localization of interleukin 5 to the crystalloid core compartment of eosinophil secondary granules in patients with atopic asthma. *J Histochem Cytochem* 1996;**44**:67–9.
72. Ying S, Humbert M, Barkans J, Corrigan CJ, Pfister R, Menz G, et al. Expression of IL-4 and IL-5 mRNA and protein product by CD4 + and CD8 + T cells, eosinophils, and mast cells in bronchial biopsies obtained from atopic and nonatopic (intrinsic) asthmatics. *J Immunol* 1997;**158**:3539–44.
73. Roth N, Stadler S, Lemann M, Hosli S, Simon HU, Simon D. Distinct eosinophil cytokine expression patterns in skin diseases - the possible existence of functionally different eosinophil subpopulations. *Allergy* 2011;**66**:1477–86.
74. Hamid Q, Barkans J, Meng Q, Ying S, Abrams JS, Kay AB, et al. Human eosinophils synthesize and secrete interleukin-6, in vitro. *Blood* 1992;**80**:1496–501.

75. Melani C, Mattia GF, Silvani A, Care A, Rivoltini L, Parmiani G, et al. Interleukin-6 expression in human neutrophil and eosinophil peripheral blood granulocytes. *Blood* 1993;**81**:2744–9.
76. Lacy P, Levi-Schaffer F, Mahmudi-Azer S, Bablitz B, Hagen SC, Velazquez J, et al. Intracellular localization of interleukin-6 in eosinophils from atopic asthmatics and effects of interferon γ. *Blood* 1998;**91**:2508–16.
77. Shimbara A, Christodoulopoulos P, Soussi-Gounni A, Olivenstein R, Nakamura Y, Levitt RC, et al. IL-9 and its receptor in allergic and nonallergic lung disease: increased expression in asthma. *J Allergy Clin Immunol* 2000;**105**:108–15.
78. Gounni AS, Gregory B, Nutku E, Aris F, Latifa K, Minshall E, et al. Interleukin-9 enhances interleukin-5 receptor expression, differentiation, and survival of human eosinophils. *Blood* 2000;**96**:2163–71.
79. Fujisawa T, Katsumata H, Kato Y. House dust mite extract induces interleukin-9 expression in human eosinophils. *Allergol Int* 2008;**57**:141–6.
80. Huang L, Gebreselassie NG, Gagliardo LF, Ruyechan MC, Lee NA, Lee JJ, et al. Eosinophil-derived IL-10 supports chronic nematode infection. *J Immunol* 2014;**193**: 4178–87.
81. Minshall E, Chakir J, Laviolette M, Molet S, Zhu Z, Olivenstein R, et al. IL-11 expression is increased in severe asthma: association with epithelial cells and eosinophils. *J Allergy Clin Immunol* 2000;**105**:232–8.
82. Grewe M, Czech W, Morita A, Werfel T, Klammer M, Kapp A, et al. Human eosinophils produce biologically active IL-12: implications for control of T cell responses. *J Immunol* 1998;**161**:415–20.
83. Woerly G, Lacy P, Younes AB, Roger N, Loiseau S, Moqbel R, et al. Human eosinophils express and release IL-13 following CD28-dependent activation. *J Leukoc Biol* 2002;**72**:769–79.
84. Schmid-Grendelmeier P, Altznauer F, Fischer B, Bizer C, Straumann A, Menz G, et al. Eosinophils express functional IL-13 in eosinophilic inflammatory diseases. *J Immunol* 2002;**169**:1021–7.
85. Lim KG, Wan HC, Resnick M, Wong DT, Cruikshank WW, Kornfeld H, et al. Human eosinophils release the lymphocyte and eosinophil active cytokines, RANTES and lymphocyte chemoattractant factor. *Int Arch Allergy Immunol* 1995; **107**:342.
86. Lim KG, Wan HC, Bozza PT, Resnick MB, Wong DT, Cruikshank WW, et al. Human eosinophils elaborate the lymphocyte chemoattractants. IL-16 (lymphocyte chemoattractant factor) and RANTES. *J Immunol* 1996;**156**:2566–70.
87. Molet S, Hamid Q, Davoine F, Nutku E, Taha R, Page N, et al. IL-17 is increased in asthmatic airways and induces human bronchial fibroblasts to produce cytokines. *J Allergy Clin Immunol* 2001;**108**:430–8.
88. Wang YH, Angkasekwinai P, Lu N, Voo KS, Arima K, Hanabuchi S, et al. IL-25 augments type 2 immune responses by enhancing the expansion and functions of TSLP-DC-activated Th2 memory cells. *J Exp Med* 2007;**204**:1837–47.
89. Terrier B, Bieche I, Maisonobe T, Laurendeau I, Rosenzwajg M, Kahn JE, et al. Interleukin-25: a cytokine linking eosinophils and adaptive immunity in Churg-Strauss syndrome. *Blood* 2010;**116**:4523–31.
90. Tang W, Smith SG, Beaudin S, Dua B, Howie K, Gauvreau G, et al. IL-25 and IL-25 receptor expression on eosinophils from subjects with allergic asthma. *Int Arch Allergy Immunol* 2014;**163**:5–10.
91. Beil WJ, Weller PF, Tzizik DM, Galli SJ, Dvorak AM. Ultrastructural immunogold localization of tumor necrosis factor-α to the matrix compartment of eosinophil secondary granules in patients with idiopathic hypereosinophilic syndrome. *J Histochem Cytochem* 1993;**41**:1611–15.

92. Costa JJ, Matossian K, Resnick MB, Beil WJ, Wong DT, Gordon JR, et al. Human eosinophils can express the cytokines tumor necrosis factor-α and macrophage inflammatory protein-1α. *J Clin Invest* 1993;**91**:2673–84.
93. Tan X, Hsueh W, Gonzalez-Crussi F. Cellular localization of tumor necrosis factor (TNF)-α transcripts in normal bowel and in necrotizing enterocolitis. TNF gene expression by Paneth cells, intestinal eosinophils, and macrophages. *Am J Pathol* 1993;**142**:1858–65.
94. Legrand F, Driss V, Delbeke M, Loiseau S, Hermann E, Dombrowicz D, et al. Human eosinophils exert TNF-α and granzyme A-mediated tumoricidal activity toward colon carcinoma cells. *J Immunol* 2010;**185**:7443–51.
95. Izumi S, Hirai K, Miyamasu M, Takahashi Y, Misaki Y, Takaishi T, et al. Expression and regulation of monocyte chemoattractant protein-1 by human eosinophils. *Eur J Immunol* 1997;**27**:816–24.
96. Ying S, Meng Q, Taborda-Barata L, Corrigan CJ, Barkans J, Assoufi B, et al. Human eosinophils express messenger RNA encoding RANTES and store and release biologically active RANTES protein. *Eur J Immunol* 1996;**26**:70–6.
97. Lamkhioued B, Renzi PM, Abi-Younes S, Garcia-Zepada EA, Allakhverdi Z, Ghaffar O, et al. Increased expression of eotaxin in bronchoalveolar lavage and airways of asthmatics contributes to the chemotaxis of eosinophils to the site of inflammation. *J Immunol* 1997;**159**:4593–601.
98. Nakajima T, Yamada H, Iikura M, Miyamasu M, Izumi S, Shida H, et al. Intracellular localization and release of eotaxin from normal eosinophils. *FEBS Lett* 1998;**434**:226–30.
99. Gounni Abdelilah S, Wellemans V, Agouli M, Guenounou M, Hamid Q, Beck LA, et al. Increased expression of Th2-associated chemokines in bullous pemphigoid disease. Role of eosinophils in the production and release of these chemokines. *Clin Immunol* 2006;**120**:220–31.
100. Liu LY, Bates ME, Jarjour NN, Busse WW, Bertics PJ, Kelly EA. Generation of Th1 and Th2 chemokines by human eosinophils: evidence for a critical role of TNF-α. *J Immunol* 2007;**179**:4840–8.
101. Matsumoto K, Fukuda S, Hashimoto N, Saito H. Human eosinophils produce and release a novel chemokine, CCL23, in vitro. *Int Arch Allergy Immunol* 2011;**155** (Suppl. 1):34–9.
102. Persson-Dajotoy T, Andersson P, Bjartell A, Calafat J, Egesten A. Expression and production of the CXC chemokine growth-related oncogene-α by human eosinophils. *J Immunol* 2003;**170**:5309–16.
103. Persson T, Monsef N, Andersson P, Bjartell A, Malm J, Calafat J, et al. Expression of the neutrophil-activating CXC chemokine ENA-78/CXCL5 by human eosinophils. *Clin Exp Allergy* 2003;**33**:531–7.
104. Braun RK, Franchini M, Erard F, Rihs S, De Vries IJ, Blaser K, et al. Human peripheral blood eosinophils produce and release interleukin-8 on stimulation with calcium ionophore. *Eur J Immunol* 1993;**23**:956–60.
105. Miyamasu M, Hirai K, Takahashi Y, Iida M, Yamaguchi M, Koshino T, et al. Chemotactic agonists induce cytokine generation in eosinophils. *J Immunol* 1995;**154**:1339–49.
106. Simon HU, Yousefi S, Weber M, Simon D, Holzer C, Hartung K, et al. Human peripheral blood eosinophils express and release interleukin-8. *Int Arch Allergy Immunol* 1995;**107**:124–6.
107. Yousefi S, Hemmann S, Weber M, Holzer C, Hartung K, Blaser K, et al. IL-8 is expressed by human peripheral blood eosinophils. Evidence for increased secretion in asthma. *J Immunol* 1995;**154**:5481–90.
108. Cheung PF, Wong CK, Lam CW. Molecular mechanisms of cytokine and chemokine release from eosinophils activated by IL-17A, IL-17F, and

IL-23: implication for Th17 lymphocytes-mediated allergic inflammation. *J Immunol* 2008;**180**:5625–35.

109. Dajotoy T, Andersson P, Bjartell A, Lofdahl CG, Tapper H, Egesten A. Human eosinophils produce the T cell-attracting chemokines MIG and IP-10 upon stimulation with IFN-γ. *J Leukoc Biol* 2004;**76**:685–91.
110. Powell PP, Klagsbrun M, Abraham JA, Jones RC. Eosinophils expressing heparin-binding EGF-like growth factor mRNA localize around lung microvessels in pulmonary hypertension. *Am J Pathol* 1993;**143**:784–93.
111. Solomon A, Aloe L, Pe'er J, Frucht-Pery J, Bonini S, Levi-Schaffer F. Nerve growth factor is preformed in and activates human peripheral blood eosinophils. *J Allergy Clin Immunol* 1998;**102**:454–60.
112. Ohno I, Nitta Y, Yamauchi K, Hoshi H, Honma M, Woolley K, et al. Eosinophils as a potential source of platelet-derived growth factor B-chain (PDGF-B) in nasal polyposis and bronchial asthma. *Am J Respir Cell Mol Biol* 1995;**13**:639–47.
113. Hartman M, Piliponsky AM, Temkin V, Levi-Schaffer F. Human peripheral blood eosinophils express stem cell factor. *Blood* 2001;**97**:1086–91.
114. Wong DT, Weller PF, Galli SJ, Elovic A, Rand TH, Gallagher GT, et al. Human eosinophils express transforming growth factor α. *J Exp Med* 1990;**172**:673–81.
115. Todd R, Donoff BR, Chiang T, Chou MY, Elovic A, Gallagher GT, et al. The eosinophil as a cellular source of transforming growth factor α in healing cutaneous wounds. *Am J Pathol* 1991;**138**:1307–13.
116. Elovic A, Wong DT, Weller PF, Matossian K, Galli SJ. Expression of transforming growth factors-α and β1 messenger RNA and product by eosinophils in nasal polyps. *J Allergy Clin Immunol* 1994;**93**:864–9.
117. Egesten A, Calafat J, Knol EF, Janssen H, Walz TM. Subcellular localization of transforming growth factor-α in human eosinophil granulocytes. *Blood* 1996;**87**:3910–18.
118. Elovic AE, Ohyama H, Sauty A, Mcbride J, Tsuji T, Nagai M, et al. IL-4-dependent regulation of TGF-α and TGF-β1 expression in human eosinophils. *J Immunol* 1998;**160**:6121–7.
119. Wong DT, Elovic A, Matossian K, Nagura N, Mcbride J, Chou MY, et al. Eosinophils from patients with blood eosinophilia express transforming growth factor β1. *Blood* 1991;**78**:2702–7.
120. Ohno I, Lea RG, Flanders KC, Clark DA, Banwatt D, Dolovich J, et al. Eosinophils in chronically inflamed human upper airway tissues express transforming growth factor β1 gene (TGF β1). *J Clin Invest* 1992;**89**:1662–8.
121. Kadin M, Butmarc J, Elovic A, Wong D. Eosinophils are the major source of transforming growth factor-β1 in nodular sclerosing Hodgkin's disease. *Am J Pathol* 1993;**142**:11–16.
122. Wong DT, Donoff RB, Yang J, Song BZ, Matossian K, Nagura N, et al. Sequential expression of transforming growth factors α and β1 by eosinophils during cutaneous wound healing in the hamster. *Am J Pathol* 1993;**143**:130–42.
123. Ohno I, Nitta Y, Yamauchi K, Hoshi H, Honma M, Woolley K, et al. Transforming growth factor β1 (TGF β1) gene expression by eosinophils in asthmatic airway inflammation. *Am J Respir Cell Mol Biol* 1996;**15**:404–9.
124. Minshall EM, Leung DY, Martin RJ, Song YL, Cameron L, Ernst P, et al. Eosinophil-associated TGF-β1 mRNA expression and airways fibrosis in bronchial asthma. *Am J Respir Cell Mol Biol* 1997;**17**:326–33.
125. Shen ZJ, Esnault S, Rosenthal LA, Szakaly RJ, Sorkness RL, Westmark PR, et al. Pin1 regulates TGF-β1 production by activated human and murine eosinophils and contributes to allergic lung fibrosis. *J Clin Invest* 2008;**118**:479–90.
126. Horiuchi T, Weller PF. Expression of vascular endothelial growth factor by human eosinophils: upregulation by granulocyte macrophage colony-stimulating factor and interleukin-5. *Am J Respir Cell Mol Biol* 1997;**17**:70–7.

127. Hoshino M, Takahashi M, Aoike N. Expression of vascular endothelial growth factor, basic fibroblast growth factor, and angiogenin immunoreactivity in asthmatic airways and its relationship to angiogenesis. *J Allergy Clin Immunol* 2001;**107**: 295–301.
128. Simson L, Ellyard JI, Dent LA, Matthaei KI, Rothenberg ME, Foster PS, et al. Regulation of carcinogenesis by IL-5 and CCL11: a potential role for eosinophils in tumor immune surveillance. *J Immunol* 2007;**178**:4222–9.
129. Tedeschi A, Asero R, Marzano AV, Lorini M, Fanoni D, Berti E, et al. Plasma levels and skin-eosinophil-expression of vascular endothelial growth factor in patients with chronic urticaria. *Allergy* 2009;**64**:1616–22.
130. Llop-Guevara A, Chu DK, Walker TD, Goncharova S, Fattouh R, Silver JS, et al. A GM-CSF/IL-33 pathway facilitates allergic airway responses to sub-threshold house dust mite exposure. *PLoS One* 2014;**9**:e88714.
131. Ito R, Takahashi T, Katano I, Kawai K, Kamisako T, Ogura T, et al. Establishment of a human allergy model using human IL-3/GM-CSF-transgenic NOG mice. *J Immunol* 2013;**191**:2890–9.
132. Tai PC, Sun L, Spry CJ. Effects of IL-5, granulocyte/macrophage colony-stimulating factor (GM-CSF) and IL-3 on the survival of human blood eosinophils in vitro. *Clin Exp Immunol* 1991;**85**:312–16.
133. Anwar AR, Moqbel R, Walsh GM, Kay AB, Wardlaw AJ. Adhesion to fibronectin prolongs eosinophil survival. *J Exp Med* 1993;**177**:839–43.
134. Herz U, Lacy P, Renz H, Erb K. The influence of infections on the development and severity of allergic disorders. *Curr Opin Immunol* 2000;**12**:632–40.
135. Holt PG, Macaubas C, Stumbles PA, Sly PD. The role of allergy in the development of asthma. *Nature* 1999;**402**:B12–17.
136. Valerius T, Repp R, Kalden JR, Platzer E. Effects of IFN on human eosinophils in comparison with other cytokines. A novel class of eosinophil activators with delayed onset of action. *J Immunol* 1990;**145**:2950–8.
137. Bandeira-Melo C, Gillard G, Ghiran I, Weller PF. EliCell: a gel-phase dual antibody capture and detection assay to measure cytokine release from eosinophils. *J Immunol Methods* 2000;**244**:105–15.
138. Yamaguchi T, Kimura H, Kurabayashi M, Kozawa K, Kato M. Interferon-γ enhances human eosinophil effector functions induced by granulocyte-macrophage colony-stimulating factor or interleukin-5. *Immunol Lett* 2008;**118**:88–95.
139. Kanda A, Driss V, Hornez N, Abdallah M, Roumier T, Abboud G, et al. Eosinophil-derived IFN-γ induces airway hyperresponsiveness and lung inflammation in the absence of lymphocytes. *J Allergy Clin Immunol* 2009;**124**:573–82 , 582.e1–9.
140. Willart MA, Deswarte K, Pouliot P, Braun H, Beyaert R, Lambrecht BN, et al. Interleukin-1α controls allergic sensitization to inhaled house dust mite via the epithelial release of GM-CSF and IL-33. *J Exp Med* 2012;**209**:1505–17.
141. Whitcomb EA, Dinarello CA, Pincus SH. Differential effects of interleukin-1α and interleukin-1β on human peripheral blood eosinophils. *Blood* 1989;**73**:1904–8.
142. Bochner BS, Luscinskas FW, Gimbrone Jr. MA, Newman W, Sterbinsky SA, Derse-Anthony CP, et al. Adhesion of human basophils, eosinophils, and neutrophils to interleukin 1-activated human vascular endothelial cells: contributions of endothelial cell adhesion molecules. *J Exp Med* 1991;**173**:1553–7.
143. Walsh GM, Mermod JJ, Hartnell A, Kay AB, Wardlaw AJ. Human eosinophil, but not neutrophil, adherence to IL-1-stimulated human umbilical vascular endothelial cells is α 4β1 (very late antigen-4) dependent. *J Immunol* 1991;**146**:3419–23.
144. Shi HZ. Eosinophils function as antigen-presenting cells. *J Leukoc Biol* 2004;**76**:520–7.
145. Shi HZ, Humbles A, Gerard C, Jin Z, Weller PF. Lymph node trafficking and antigen presentation by endobronchial eosinophils. *J Clin Invest* 2000;**105**:945–53.

146. Wang HB, Ghiran I, Matthaei K, Weller PF. Airway eosinophils: allergic inflammation recruited professional antigen-presenting cells. *J Immunol* 2007;**179**:7585–92.
147. Akuthota P, Wang HB, Spencer LA, Weller PF. Immunoregulatory roles of eosinophils: a new look at a familiar cell. *Clin Exp Allergy* 2008;**38**:1254–63.
148. Lacy P, Stow JL. Cytokine release from innate immune cells: association with diverse membrane trafficking pathways. *Blood* 2011;**118**:9–18.
149. Eder C. Mechanisms of interleukin-1β release. *Immunobiology* 2009;**214**:543–53.
150. Krause K, Metz M, Makris M, Zuberbier T, Maurer M. The role of interleukin-1 in allergy-related disorders. *Curr Opin Allergy Clin Immunol* 2012;**12**:477–84.
151. Cosmi L, Liotta F, Maggi E, Romagnani S, Annunziato F. Th17 cells: new players in asthma pathogenesis. *Allergy* 2011;**66**:989–98.
152. Blanchard D, Gaillard C, Hermann P, Banchereau J. Role of CD40 antigen and interleukin-2 in T cell-dependent human B lymphocyte growth. *Eur J Immunol* 1994; **24**:330–5.
153. Hondowicz BD, An D, Schenkel JM, Kim KS, Steach HR, Krishnamurty AT, et al. Interleukin-2-dependent allergen-specific tissue-resident memory cells drive asthma. *Immunity* 2016;**44**:155–66.
154. Rand TH, Cruikshank WW, Center DM, Weller PF. CD4-mediated stimulation of human eosinophils: lymphocyte chemoattractant factor and other CD4-binding ligands elicit eosinophil migration. *J Exp Med* 1991;**173**:1521–8.
155. Kay AB, Ying S, Varney V, Gaga M, Durham SR, Moqbel R, et al. Messenger RNA expression of the cytokine gene cluster, interleukin 3 (IL-3), IL-4, IL-5, and granulocyte/macrophage colony-stimulating factor, in allergen-induced late-phase cutaneous reactions in atopic subjects. *J Exp Med* 1991;**173**:775–8.
156. Robinson DS, Hamid Q, Ying S, Tsicopoulos A, Barkans J, Bentley AM, et al. Predominant TH2-like bronchoalveolar T-lymphocyte population in atopic asthma. *N Engl J Med* 1992;**326**:298–304.
157. Durham SR, Ying S, Varney VA, Jacobson MR, Sudderick RM, Mackay IS, et al. Cytokine messenger RNA expression for IL-3, IL-4, IL-5, and granulocyte/macrophage-colony-stimulating factor in the nasal mucosa after local allergen provocation: relationship to tissue eosinophilia. *J Immunol* 1992;**148**:2390–4.
158. Humbert M, Ying S, Corrigan C, Menz G, Barkans J, Pfister R, et al. Bronchial mucosal expression of the genes encoding chemokines RANTES and MCP-3 in symptomatic atopic and nonatopic asthmatics: relationship to the eosinophil-active cytokines interleukin (IL)-5, granulocyte macrophage-colony-stimulating factor, and IL-3. *Am J Respir Cell Mol Biol* 1997;**16**:1–8.
159. Matsumoto T. Ongoing IgE synthesis by atopic B cells is enhanced by interleukin-3 and suppressed directly by interferon-γ in vitro. *Int Arch Allergy Appl Immunol* 1991; **95**:48–52.
160. Sugiyama H, Eda R, Hopp RJ, Bewtra AK, Townley RG. Importance of interleukin-3 on histamine release from human basophils. *Ann Allergy* 1993;**71**:391–5.
161. Clutterbuck EJ, Hirst EM, Sanderson CJ. Human interleukin-5 (IL-5) regulates the production of eosinophils in human bone marrow cultures: comparison and interaction with IL-1, IL-3, IL-6, and GMCSF. *Blood* 1989;**73**:1504–12.
162. Warringa RA, Koenderman L, Kok PT, Kreukniet J, Bruijnzeel PL. Modulation and induction of eosinophil chemotaxis by granulocyte-macrophage colony-stimulating factor and interleukin-3. *Blood* 1991;**77**:2694–700.
163. Hakansson L, Carlson M, Stalenheim G, Venge P. Migratory responses of eosinophil and neutrophil granulocytes from patients with asthma. *J Allergy Clin Immunol* 1990; **85**:743–50.
164. Rothenberg ME, Owen Jr. WF, Silberstein DS, Woods J, Soberman RJ, Austen KF, et al. Human eosinophils have prolonged survival, enhanced functional properties,

and become hypodense when exposed to human interleukin 3. *J Clin Invest* 1988;**81**:1986–92.

165. Dubois GR, Bruijnzeel-Koomen CA, Bruijnzeel PL. IL-4 induces chemotaxis of blood eosinophils from atopic dermatitis patients, but not from normal individuals. *J Invest Dermatol* 1994;**102**:843–6.
166. Chen L, Grabowski KA, Xin JP, Coleman J, Huang Z, Espiritu B, et al. IL-4 induces differentiation and expansion of Th2 cytokine-producing eosinophils. *J Immunol* 2004;**172**:2059–66.
167. Gessner A, Mohrs K, Mohrs M. Mast cells, basophils, and eosinophils acquire constitutive IL-4 and IL-13 transcripts during lineage differentiation that are sufficient for rapid cytokine production. *J Immunol* 2005;**174**:1063–72.
168. Conrad DH, Ben-Sasson SZ, Le Gros G, Finkelman FD, Paul WE. Infection with Nippostrongylus brasiliensis or injection of anti-IgD antibodies markedly enhances Fc-receptor-mediated interleukin 4 production by non-B, non-T cells. *J Exp Med* 1990;**171**:1497–508.
169. Voehringer D, Shinkai K, Locksley RM. Type 2 immunity reflects orchestrated recruitment of cells committed to IL-4 production. *Immunity* 2004;**20**:267–77.
170. Piehler D, Stenzel W, Grahnert A, Held J, Richter L, Kohler G, et al. Eosinophils contribute to IL-4 production and shape the T-helper cytokine profile and inflammatory response in pulmonary cryptococcosis. *Am J Pathol* 2011;**179**:733–44.
171. Padigel UM, Lee JJ, Nolan TJ, Schad GA, Abraham D. Eosinophils can function as antigen-presenting cells to induce primary and secondary immune responses to Strongyloides stercoralis. *Infect Immun* 2006;**74**:3232–8.
172. Jacobsen EA, Doyle AD, Colbert DC, Zellner KR, Protheroe CA, Lesuer WE, et al. Differential activation of airway eosinophils induces IL-13-mediated allergic Th2 pulmonary responses in mice. *Allergy* 2015;**70**:1148–59.
173. Shalit M, Sekhsaria S, Malech HL. Modulation of growth and differentiation of eosinophils from human peripheral blood CD34 + cells by IL5 and other growth factors. *Cell Immunol* 1995;**160**:50–7.
174. Yamaguchi Y, Suda T, Suda J, Eguchi M, Miura Y, Harada N, et al. Purified interleukin 5 supports the terminal differentiation and proliferation of murine eosinophilic precursors. *J Exp Med* 1988;**167**:43–56.
175. Kita H, Weiler DA, Abu-Ghazaleh R, Sanderson CJ, Gleich GJ. Release of granule proteins from eosinophils cultured with IL-5. *J Immunol* 1992;**149**:629–35.
176. Lee JJ, Jacobsen EA, Ochkur SI, McGarry MP, Condjella RM, Doyle AD, et al. Human versus mouse eosinophils: "that which we call an eosinophil, by any other name would stain as red". *J Allergy Clin Immunol* 2012;**130**:572–84.
177. Yamaguchi Y, Hayashi Y, Sugama Y, Miura Y, Kasahara T, Kitamura S, et al. Highly purified murine interleukin 5 (IL-5) stimulates eosinophil function and prolongs in vitro survival. IL-5 as an eosinophil chemotactic factor. *J Exp Med* 1988;**167**: 1737–42.
178. Nair P, Pizzichini MM, Kjarsgaard M, Inman MD, Efthimiadis A, Pizzichini E, et al. Mepolizumab for prednisone-dependent asthma with sputum eosinophilia. *N Engl J Med* 2009;**360**:985–93.
179. Ortega HG, Liu MC, Pavord ID, Brusselle GG, Fitzgerald JM, Chetta A, et al. Mepolizumab treatment in patients with severe eosinophilic asthma. *N Engl J Med* 2014;**371**:1198–207.
180. Castro M, Mathur S, Hargreave F, Boulet LP, Xie F, Young J, et al. Reslizumab for poorly controlled, eosinophilic asthma: a randomized, placebo-controlled study. *Am J Respir Crit Care Med* 2011;**184**:1125–32.
181. Castro M, Wenzel SE, Bleecker ER, Pizzichini E, Kuna P, Busse WW, et al. Benralizumab, an anti-interleukin 5 receptor α monoclonal antibody, versus placebo

for uncontrolled eosinophilic asthma: a phase 2b randomised dose-ranging study. *Lancet Respir Med* 2014;**2**:879–90.
182. Haldar P, Brightling CE, Hargadon B, Gupta S, Monteiro W, Sousa A, et al. Mepolizumab and exacerbations of refractory eosinophilic asthma. *N Engl J Med* 2009;**360**:973–84.
183. Neveu WA, Allard JL, Raymond DM, Bourassa LM, Burns SM, Bunn JY, et al. Elevation of IL-6 in the allergic asthmatic airway is independent of inflammation but associates with loss of central airway function. *Respir Res* 2010;**11**:28.
184. Yokoyama A, Kohno N, Fujino S, Hamada H, Inoue Y, Fujioka S, et al. Circulating interleukin-6 levels in patients with bronchial asthma. *Am J Respir Crit Care Med* 1995;**151**:1354–8.
185. Deetz DC, Jagielo PJ, Quinn TJ, Thorne PS, Bleuer SA, Schwartz DA. The kinetics of grain dust-induced inflammation of the lower respiratory tract. *Am J Respir Crit Care Med* 1997;**155**:254–9.
186. Dienz O, Rincon M. The effects of IL-6 on CD4 T cell responses. *Clin Immunol* 2009;**130**:27–33.
187. Vercelli D, Jabara HH, Arai K, Yokota T, Geha RS. Endogenous interleukin 6 plays an obligatory role in interleukin 4-dependent human IgE synthesis. *Eur J Immunol* 1989;**19**:1419–24.
188. Kung TT, Luo B, Crawley Y, Garlisi CG, Devito K, Minnicozzi M, et al. Effect of anti-mIL-9 antibody on the development of pulmonary inflammation and airway hyperresponsiveness in allergic mice. *Am J Respir Cell Mol Biol* 2001;**25**:600–5.
189. Erpenbeck VJ, Hohlfeld JM, Volkmann B, Hagenberg A, Geldmacher H, Braun A, et al. Segmental allergen challenge in patients with atopic asthma leads to increased IL-9 expression in bronchoalveolar lavage fluid lymphocytes. *J Allergy Clin Immunol* 2003;**111**:1319–27.
190. Louahed J, Zhou Y, Maloy WL, Rani PU, Weiss C, Tomer Y, et al. Interleukin 9 promotes influx and local maturation of eosinophils. *Blood* 2001;**97**:1035–42.
191. Gounni AS, Nutku E, Koussih L, Aris F, Louahed J, Levitt RC, et al. IL-9 expression by human eosinophils: regulation by IL-1β and TNF-α. *J Allergy Clin Immunol* 2000;**106**:460–6.
192. Kosaka S, Tamauchi H, Terashima M, Maruyama H, Habu S, Kitasato H. IL-10 controls Th2-type cytokine production and eosinophil infiltration in a mouse model of allergic airway inflammation. *Immunobiology* 2011;**216**:811–20.
193. Grunig G, Corry DB, Leach MW, Seymour BW, Kurup VP, Rennick DM. Interleukin-10 is a natural suppressor of cytokine production and inflammation in a murine model of allergic bronchopulmonary aspergillosis. *J Exp Med* 1997;**185**: 1089–99.
194. Kleinjan A, Dijkstra MD, Boks SS, Severijnen LA, Mulder PG, Fokkens WJ. Increase in IL-8, IL-10, IL-13, and RANTES mRNA levels (in situ hybridization) in the nasal mucosa after nasal allergen provocation. *J Allergy Clin Immunol* 1999;**103**:441–50.
195. Jeannin P, Lecoanet S, Delneste Y, Gauchat JF, Bonnefoy JY. IgE versus IgG4 production can be differentially regulated by IL-10. *J Immunol* 1998;**160**:3555–61.
196. Zhu Z, Lee CG, Zheng T, Chupp G, Wang J, Homer RJ, et al. Airway inflammation and remodeling in asthma. Lessons from interleukin 11 and interleukin 13 transgenic mice. *Am J Respir Crit Care Med* 2001;**164**:S67–70.
197. Bryan SA, O'connor BJ, Matti S, Leckie MJ, Kanabar V, Khan J, et al. Effects of recombinant human interleukin-12 on eosinophils, airway hyper-responsiveness, and the late asthmatic response. *Lancet* 2000;**356**:2149–53.
198. Nutku E, Gounni AS, Olivenstein R, Hamid Q. Evidence for expression of eosinophil-associated IL-12 messenger RNA and immunoreactivity in bronchial asthma. *J Allergy Clin Immunol* 2000;**106**:288–92.

199. Grewe M, Walther S, Gyufko K, Czech W, Schopf E, Krutmann J. Analysis of the cytokine pattern expressed in situ in inhalant allergen patch test reactions of atopic dermatitis patients. *J Invest Dermatol* 1995;**105**:407–10.
200. Wills-Karp M, Luyimbazi J, Xu X, Schofield B, Neben TY, Karp CL, et al. Interleukin-13: central mediator of allergic asthma. *Science* 1998;**282**:2258–61.
201. Punnonen J, Aversa G, Cocks BG, Mckenzie AN, Menon S, Zurawski G, et al. Interleukin 13 induces interleukin 4-independent IgG4 and IgE synthesis and CD23 expression by human B cells. *Proc Natl Acad Sci USA* 1993;**90**:3730–4.
202. Pope SM, Brandt EB, Mishra A, Hogan SP, Zimmermann N, Matthaei KI, et al. IL-13 induces eosinophil recruitment into the lung by an IL-5- and eotaxin-dependent mechanism. *J Allergy Clin Immunol* 2001;**108**:594–601.
203. De Boever EH, Ashman C, Cahn AP, Locantore NW, Overend P, Pouliquen IJ, et al. Efficacy and safety of an anti-IL-13 mAb in patients with severe asthma: a randomized trial. *J Allergy Clin Immunol* 2014;**133**:989–96.
204. Gauchat JF, Schlagenhauf E, Feng NP, Moser R, Yamage M, Jeannin P, et al. A novel 4-kb interleukin-13 receptor α mRNA expressed in human B, T, and endothelial cells encoding an alternate type-II interleukin-4/interleukin-13 receptor. *Eur J Immunol* 1997;**27**:971–8.
205. Walsh ER, Thakar J, Stokes K, Huang F, Albert R, August A. Computational and experimental analysis reveals a requirement for eosinophil-derived IL-13 for the development of allergic airway responses in C57BL/6 mice. *J Immunol* 2011;**186**:2936–49.
206. Laberge S, Ernst P, Ghaffar O, Cruikshank WW, Kornfeld H, Center DM, et al. Increased expression of interleukin-16 in bronchial mucosa of subjects with atopic asthma. *Am J Respir Cell Mol Biol* 1997;**17**:193–202.
207. Mashikian MV, Tarpy RE, Saukkonen JJ, Lim KG, Fine GD, Cruikshank WW, et al. Identification of IL-16 as the lymphocyte chemotactic activity in the bronchoalveolar lavage fluid of histamine-challenged asthmatic patients. *J Allergy Clin Immunol* 1998;**101**:786–92.
208. Bandeira-Melo C, Sugiyama K, Woods LJ, Phoofolo M, Center DM, Cruikshank WW, et al. IL-16 promotes leukotriene C(4) and IL-4 release from human eosinophils via CD4- and autocrine CCR3-chemokine-mediated signaling. *J Immunol* 2002;**168**:4756–63.
209. Wang YH, Wills-Karp M. The potential role of interleukin-17 in severe asthma. *Curr Allergy Asthma Rep* 2011;**11**:388–94.
210. Wakashin H, Hirose K, Maezawa Y, Kagami S, Suto A, Watanabe N, et al. IL-23 and Th17 cells enhance Th2-cell-mediated eosinophilic airway inflammation in mice. *Am J Respir Crit Care Med* 2008;**178**:1023–32.
211. Saenz SA, Taylor BC, Artis D. Welcome to the neighborhood: epithelial cell-derived cytokines license innate and adaptive immune responses at mucosal sites. *Immunol Rev* 2008;**226**:172–90.
212. Bisset LR, Schmid-Grendelmeier P. Chemokines and their receptors in the pathogenesis of allergic asthma: progress and perspective. *Curr Opin Pulm Med* 2005;**11**:35–42.
213. Rajakulasingam K, Hamid Q, O'brien F, Shotman E, Jose PJ, Williams TJ, et al. RANTES in human allergen-induced rhinitis: cellular source and relation to tissue eosinophilia. *Am J Respir Crit Care Med* 1997;**155**:696–703.
214. Allen JS, Eisma R, Lafreniere D, Leonard G, Kreutzer D. Characterization of the eosinophil chemokine RANTES in nasal polyps. *Ann Otol Rhinol Laryngol* 1998;**107**:416–20.
215. Rot A, Krieger M, Brunner T, Bischoff SC, Schall TJ, Dahinden CA. RANTES and macrophage inflammatory protein 1 α induce the migration and activation of normal human eosinophil granulocytes. *J Exp Med* 1992;**176**:1489–95.
216. Alam R, Stafford S, Forsythe P, Harrison R, Faubion D, Lett-Brown MA, et al. RANTES is a chemotactic and activating factor for human eosinophils. *J Immunol* 1993;**150**:3442–8.

217. Elsner J, Hochstetter R, Kimmig D, Kapp A. Human eotaxin represents a potent activator of the respiratory burst of human eosinophils. *Eur J Immunol* 1996;**26**:1919–25.
218. Melo RC, Perez SA, Spencer LA, Dvorak AM, Weller PF. Intragranular vesiculotubular compartments are involved in piecemeal degranulation by activated human eosinophils. *Traffic* 2005;**6**:866–79.
219. Stanley AC, Lacy P. Pathways for cytokine secretion. *Physiology* 2010;**25**:218–29.
220. Lacy P, Logan MR, Bablitz B, Moqbel R. Fusion protein vesicle-associated membrane protein 2 is implicated in IFNγ-induced piecemeal degranulation in human eosinophils from atopic individuals. *J Allergy Clin Immunol* 2001;**107**:671–8.
221. Sudhof TC, Rothman JE. Membrane fusion: grappling with SNARE and SM proteins. *Science* 2009;**323**:474–7.
222. Logan MR, Lacy P, Bablitz B, Moqbel R. Expression of eosinophil target SNAREs as potential cognate receptors for vesicle-associated membrane protein-2 in exocytosis. *J Allergy Clin Immunol* 2002;**109**:299–306.
223. Garcia-Zepeda EA, Rothenberg ME, Ownbey RT, Celestin J, Leder P, Luster AD. Human eotaxin is a specific chemoattractant for eosinophil cells and provides a new mechanism to explain tissue eosinophilia. *Nat Med* 1996;**2**:449–56.
224. Jose PJ, Griffiths-Johnson DA, Collins PD, Walsh DT, Moqbel R, Totty NF, et al. Eotaxin: a potent eosinophil chemoattractant cytokine detected in a guinea pig model of allergic airways inflammation. *J Exp Med* 1994;**179**:881–7.
225. Mould AW, Ramsay AJ, Matthaei KI, Young IG, Rothenberg ME, Foster PS. The effect of IL-5 and eotaxin expression in the lung on eosinophil trafficking and degranulation and the induction of bronchial hyperreactivity. *J Immunol* 2000;**164**:2142–50.
226. Ochkur SI, Jacobsen EA, Protheroe CA, Biechele TL, Pero RS, Mcgarry MP, et al. Coexpression of IL-5 and eotaxin-2 in mice creates an eosinophil-dependent model of respiratory inflammation with characteristics of severe asthma. *J Immunol* 2007;**178**:7879–89.
227. Rothenberg ME, Maclean JA, Pearlman E, Luster AD, Leder P. Targeted disruption of the chemokine eotaxin partially reduces antigen-induced tissue eosinophilia. *J Exp Med* 1997;**185**:785–90.
228. Kampen GT, Stafford S, Adachi T, Jinquan T, Quan S, Grant JA, et al. Eotaxin induces degranulation and chemotaxis of eosinophils through the activation of ERK2 and p38 mitogen-activated protein kinases. *Blood* 2000;**95**:1911–17.
229. Tenscher K, Metzner B, Schopf E, Norgauer J, Czech W. Recombinant human eotaxin induces oxygen radical production, Ca^{2+}-mobilization, actin reorganization, and CD11b upregulation in human eosinophils via a pertussis toxin-sensitive heterotrimeric guanine nucleotide-binding protein. *Blood* 1996;**88**:3195–9.
230. El-Shazly A, Masuyama K, Nakano K, Eura M, Samejima Y, Ishikawa T. Human eotaxin induces eosinophil-derived neurotoxin release from normal human eosinophils. *Int Arch Allergy Immunol* 1998;**117**(Suppl. 1):55–8.
231. Bochner BS, Hudson SA, Xiao HQ, Liu MC. Release of both CCR4-active and CXCR3-active chemokines during human allergic pulmonary late-phase reactions. *J Allergy Clin Immunol* 2003;**112**:930–4.
232. Borchers MT, Ansay T, Desalle R, Daugherty BL, Shen H, Metzger M, et al. In vitro assessment of chemokine receptor-ligand interactions mediating mouse eosinophil migration. *J Leukoc Biol* 2002;**71**:1033–41.
233. Jacobsen EA, Ochkur SI, Pero RS, Taranova AG, Protheroe CA, Colbert DC, et al. Allergic pulmonary inflammation in mice is dependent on eosinophil-induced recruitment of effector T cells. *J Exp Med* 2008;**205**:699–710.
234. Gavala ML, Bertics PJ, Gern JE. Rhinoviruses, allergic inflammation, and asthma. *Immunol Rev* 2011;**242**:69–90.

235. Sehmi R, Cromwell O, Wardlaw AJ, Moqbel R, Kay AB. Interleukin-8 is a chemoattractant for eosinophils purified from subjects with a blood eosinophilia but not from normal healthy subjects. *Clin Exp Allergy* 1993;**23**:1027–36.
236. Bochenska-Marciniak M, Kupczyk M, Gorski P, Kuna P. The effect of recombinant interleukin-8 on eosinophils' and neutrophils' migration in vivo and in vitro. *Allergy* 2003;**58**:795–801.
237. Jiang CK, Magnaldo T, Ohtsuki M, Freedberg IM, Bernerd F, Blumenberg M. Epidermal growth factor and transforming growth factor α specifically induce the activation- and hyperproliferation-associated keratins 6 and 16. *Proc Natl Acad Sci USA* 1993;**90**:6786–90.
238. Booth BW, Adler KB, Bonner JC, Tournier F, Martin LD. Interleukin-13 induces proliferation of human airway epithelial cells in vitro via a mechanism mediated by transforming growth factor-α. *Am J Respir Cell Mol Biol* 2001;**25**: 739–43.
239. Shi M, Zhu J, Wang R, Chen X, Mi L, Walz T, et al. Latent TGF-β structure and activation. *Nature* 2011;**474**:343–9.
240. Akdis M, Aab A, Altunbulakli C, Azkur K, Costa RA, Crameri R, et al. Interleukins (from IL-1 to IL-38), interferons, transforming growth factor β, and TNF-α: Receptors, functions, and roles in diseases. *J Allergy Clin Immunol* 2016;138:984–1010.
241. Bordon Y. Asthma and allergy: TGFβ—too much of a good thing? *Nat Rev Immunol* 2013;**13**:618–19.
242. Frischmeyer-Guerrerio PA, Guerrerio AL, Oswald G, Chichester K, Myers L, Halushka MK, et al. TGFβ receptor mutations impose a strong predisposition for human allergic disease. *Sci Transl Med* 2013;**5**:195ra94.
243. Halwani R, Al-Muhsen S, Al-Jahdali H, Hamid Q. Role of transforming growth factor-β in airway remodeling in asthma. *Am J Respir Cell Mol Biol* 2011;**44**:127–33.
244. Luttmann W, Franz P, Matthys H, Virchow Jr. JC. Effects of TGF-β on eosinophil chemotaxis. *Scand J Immunol* 1998;**47**:127–30.
245. Pazdrak K, Justement L, Alam R. Mechanism of inhibition of eosinophil activation by transforming growth factor-β. Inhibition of Lyn, MAP, Jak2 kinases and STAT1 nuclear factor. *J Immunol* 1995;**155**:4454–8.
246. Vignola AM, Chanez P, Chiappara G, Merendino A, Pace E, Rizzo A, et al. Transforming growth factor-β expression in mucosal biopsies in asthma and chronic bronchitis. *Am J Respir Crit Care Med* 1997;**156**:591–9.
247. Flood-Page P, Menzies-Gow A, Phipps S, Ying S, Wangoo A, Ludwig MS, et al. Anti-IL-5 treatment reduces deposition of ECM proteins in the bronchial subepithelial basement membrane of mild atopic asthmatics. *J Clin Invest* 2003; **112**:1029–36.
248. Levi-Schaffer F, Garbuzenko E, Rubin A, Reich R, Pickholz D, Gillery P, et al. Human eosinophils regulate human lung- and skin-derived fibroblast properties in vitro: a role for transforming growth factor β (TGF-β). *Proc Natl Acad Sci USA* 1999;**96**:9660–5.
249. Phipps S, Ying S, Wangoo A, Ong YE, Levi-Schaffer F, Kay AB. The relationship between allergen-induced tissue eosinophilia and markers of repair and remodeling in human atopic skin. *J Immunol* 2002;**169**:4604–12.
250. Berry M, Brightling C, Pavord I, Wardlaw A. TNF-α in asthma. *Curr Opin Pharmacol* 2007;**7**:279–82.
251. Brightling C, Berry M, Amrani Y. Targeting TNF-α: a novel therapeutic approach for asthma. *J Allergy Clin Immunol* 2008;**121**:5–10;quiz 11-12.
252. Slungaard A, Vercellotti GM, Walker G, Nelson RD, Jacob HS. Tumor necrosis factor α/cachectin stimulates eosinophil oxidant production and toxicity towards human endothelium. *J Exp Med* 1990;**171**:2025–41.

253. Tonnel AB, Gosset P, Molet S, Tillie-Leblond I, Jeannin P, Joseph M. Interactions between endothelial cells and effector cells in allergic inflammation. *Ann N Y Acad Sci* 1996;**796**:9–20.
254. Iwasaki M, Saito K, Takemura M, Sekikawa K, Fujii H, Yamada Y, et al. TNF-α contributes to the development of allergic rhinitis in mice. *J Allergy Clin Immunol* 2003;**112**:134–40.
255. Zeck-Kapp G, Czech W, Kapp A. TNF α-induced activation of eosinophil oxidative metabolism and morphology--comparison with IL-5. *Exp Dermatol* 1994;**3**:176–88.
256. Schwingshackl A, Duszyk M, Brown N, Moqbel R. Human eosinophils release matrix metalloproteinase-9 on stimulation with TNF-α. *J Allergy Clin Immunol* 1999;**104**:983–9.
257. Beil WJ, Weller PF, Peppercorn MA, Galli SJ, Dvorak AM. Ultrastructural immunogold localization of subcellular sites of TNFα in colonic Crohn's disease. *J Leukoc Biol* 1995;**58**:284–98.
258. Lee SD, Tontonoz P. Eosinophils in fat: pink is the new brown. *Cell* 2014;**157**:1249–50.
259. Qiu Y, Nguyen KD, Odegaard JI, Cui X, Tian X, Locksley RM, et al. Eosinophils and type 2 cytokine signaling in macrophages orchestrate development of functional beige fat. *Cell* 2014;**157**:1292–308.

CHAPTER

13

Cytokines in Hematopoietic Stem Cell Transplantation

Kate A. Markey[1] *and Geoffrey R. Hill*[2]

[1]QIMR Berghofer Medical Research Institute, Brisbane, QLD, Australia
[2]Royal Brisbane and Women's Hospital, Brisbane, QLD, Australia

OUTLINE

INTRODUCTION

Allogeneic bone marrow transplantation (BMT) is performed as curative therapy for hematological malignancy and bone marrow failure syndromes, and despite advances in the field it still carries a high

Cytokine Effector Functions in Tissues.
DOI: http://dx.doi.org/10.1016/B978-0-12-804214-4.00012-9

treatment-related mortality, largely attributable to graft-versus-host disease (GVHD). GVHD is a clinical syndrome effecting multiple organs and systems: In its acute form these are predominantly skin, liver, and the gastrointestinal tract and in chronic form is a debilitating fibrogenic pathology, characteristically involving skin (i.e., scleroderma) and lung (i.e., bronchiolitis obliterans). Both are significant contributors to morbidity and mortality after BMT, and, thus, effective strategies to mitigate the pathologic effects driven by both cellular immunity and cytokine effectors are needed. Traditionally, a temporal definition has been employed to delineate acute and chronic GVHD, with acute GVHD developing within 100 days of transplant and chronic symptoms developing later; however, it is now clear that they represent different entities at both clinical and pathophysiological levels.

Unfortunately, pathological GVHD is linked to protective immune-mediated antileukemic effects after transplant (termed graft-versus-leukemia or GVL), and the overarching goal within the field is to separate these two phenomena. Despite intensive research efforts, this goal remains largely unmet, with current strategies to prevent GVHD only able to reduce alloreactivity in general (and thus impacting both GVHD and GVL effects). GVL effects are largely mediated by the cognate interaction of donor T cells with leukemia, whereas inflammatory cytokines appear more important in GVHD than GVL.

Cytokines have represented a promising therapeutic target for the prevention and treatment of GVHD for many years, and strategies that focus on blocking cytokine-driven pathology are now a part of standard clinical practice. Knowledge regarding the role of cytokines in GVHD has been established using preclinical models (often utilizing "knock-out" mice, deficient in the cytokine or receptor of interest), studies of the genetic polymorphisms in cytokine genes which can be correlated with posttransplant outcome, and, of course, clinical trials of cytokine neutralizing antibodies in BMT patients.

This chapter aims to provide an overview of our current understanding of the role of cytokines in allogeneic transplantation, with a focus on GVHD, and the cytokines that can currently be targeted for clinical benefit in patients undergoing BMT (Fig. 13.1).

GVHD PATHOPHYSIOLOGY

BMT involves pretransplant chemotherapy and/or total body irradiation (TBI) in a process known as "conditioning" that aims to ablate recipient immunity and hematopoiesis (and so, debulk any residual malignancy). Conditioning results in gastrointestinal tract damage and

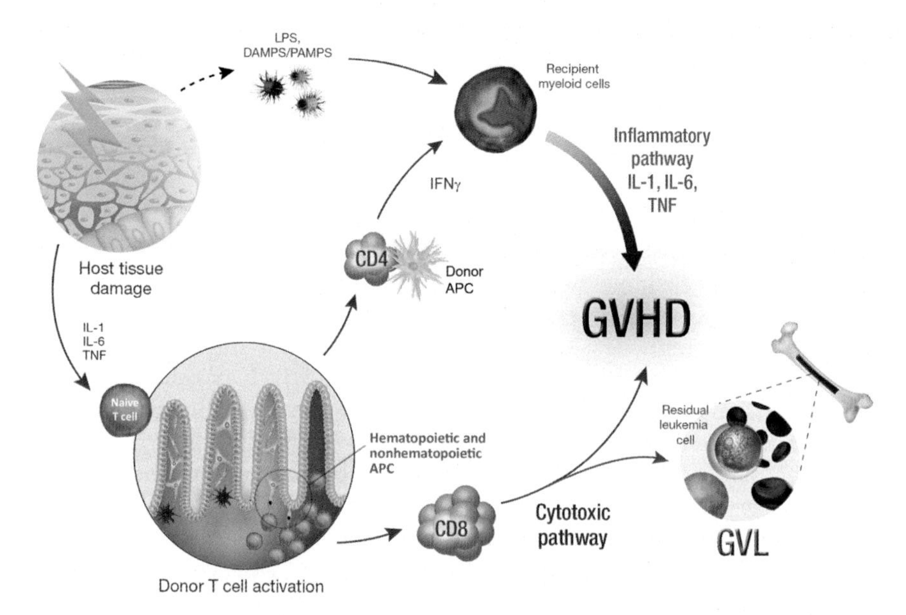

FIGURE 13.1 **Acute GVHD pathophysiology.** There are three phases in aGVHD pathophysiology. First, pretransplant conditioning results in host tissue damage. Second, antigen presentation by recipient hematopoietic and nonhematopoietic cells results in T cell activation and, finally, proliferation of T cells (propelled by donor APC) and acquisition of effector function. *IL-1*, Interleukin-1; *IL-6*, Interleukin-6; *TNF*, Tumor necrosis factor; *LPS*, Lipopolysaccharide; *DAMPs/PAMPs*, Damage/pathogen associated molecular patterns; *APC*, Antigen presenting cell; *GVHD*, Graft-versus-host disease; *GVL*, Graft-versus-leukemia effect. Illustration by Madeleine Kersting-Flynn, QIMR Berghofer Medical Research Institute.

translocation of damage- and pathogen-associated molecular patterns (DAMPs/PAMPs), which stimulate inflammatory cytokine production.[1,2] The cytokine response to conditioning (dominated by IL-6, IL-1, and TNF) represents the first step in the pathogenesis of GVHD.

The donor graft can be from a sibling or unrelated donor, and be derived from granulocyte-colony stimulating factor (G-CSF) mobilized peripheral blood, bone marrow, or umbilical cord blood, with the latter more common in pediatric transplantation. Human leukocyte antigen (HLA) matching between donor and host can either be complete (i.e., A, B, C, DR, DQ: 10/10), partially mismatched (e.g., 8 or 9/10) or haploidentical (from a parent or child, where only 5/10 HLA alleles are matched). In some centers, manipulated donor grafts are used (i.e., depleted of T cells or $CD34^+$ selected) or the patient is treated with agents prior to BMT to deplete T cells in vivo (e.g., ATG), although enthusiasm for these approaches differs geographically.

After infusion of the stem cell graft, any T cells present are exposed to recipient-type alloantigen (since even in the case where donor and host are 10/10 matched for HLA, all proteins in the body can be presented on class I and act as minor histocompatibility antigens). Both hematopoietic and non-hematopoietic, professional and non-professional, recipient and donor antigen presenting cells (APCs) invoke alloreactive T cell responses, with preclinical evidence supporting the concept that considerable redundancy exists for any particular APC in driving lethal GVHD.[3–5]

After engagement with APC, donor T cells become activated in an alloantigen specific fashion, which results in their proliferation and further cytokine production. The end result of this is a systemic cytokine storm, which is the hallmark of GVHD after BMT. Both donor and host cells, contribute to this proinflammatory cytokine milieu, which is generated as a response to tissue damage. Proinflammatory cytokines include tumor necrosis factor (TNF), interferon-gamma (IFNγ), and IL-1 and IL-6. This proinflammatory milieu is counter-regulated by the generation of "anti-inflammatory" cytokines (such as IL-10). Acute GVHD has been traditionally defined as predominantly Th1/Tc1 pathology (with accompanying classical cytokines from this group), but emerging evidence suggests that there is an important role for alternate cytokine pathways and T cell differentiation programs, particularly Th17/Tc17 differentiation.

Chronic GVHD is poorly understood as an entity compared with acute GVHD. Profibrogenic cytokines like TGFβ have a predominant role, and the characteristic tissue outcome is fibrosis, largely involving skin, oral mucosa, lacrimal and salivary glands, and liver. It is often gut sparing (compared with acute GVHD, where gastrointestinal pathology is often most severe and difficult to treat). Importantly, there is significant overlap in effector cytokines (e.g., IFNγ) and much still remains to be learned regarding cGVHD pathogenesis.

TISSUE-SPECIFIC CYTOKINE EFFECTOR FUNCTION

In addition to the characteristic systemic cytokine storm seen in GVHD, there are clearly also localized and tissue-specific effects of cytokines. The tissue tropism for a given cytokine is partly driven by receptor expression and likely also influenced by the local environment and migratory patterns of T cells after BMT. Insights into the tissue-specific effects of cytokine signaling have largely been gained from preclinical experiments in mice using genetically modified strains, where histological assessment of target tissues can be readily performed in a temporal fashion. In contrast, in the clinical setting, invasive tissue biopsies are not routinely performed in a serial fashion and are usually restricted to diagnostic procedures where results will guide clinical management (Table 13.1).

TABLE 13.1 Tissue-Specific Effects in GVHD

Target tissue	Key cytokines	Evidence and/or mechanism
Gastrointestinal tract	IFNγ	Causes crypt hypertrophy and villous atrophy[6,7] via direct signaling of the IFNGR expressed on recipient gut tissue[8]
	TNF	Directly induces apoptosis in host cells, promotes T cell proliferation
	IL-6	Direct induction of apoptosis and orchestration of T cell function[9]
Skin	IL-4	Present in skin in acute GVHD ("Th2" cytokine response)
	IL-13	
	IFNγ	Present in skin in both acute and chronic GVHD
	IL-17	Increased Th17 transcripts demonstrated in clinical cGVHD skin samples[10] Required for maximal skin GVHD in mice[11]
	TGFβ	Cause fibrosis in chronic GVHD[10]
Lung	IL-17	Th17 cells accumulate in lung and skin, driven by IL-6 from lung parenchyma, contributing to fatal IPS in mouse models[12,13]
	IL-4 IL-13	Drive lung pathology in mouse models, where cells are forced into Th2 differentiation[11]
	IFNγ	Protects from lung GVHD by signaling to parenchyma and preventing T cell infiltrate[8,11]

GRANULOCYTE-COLONY STIMULATING FACTOR

G-CSF has a physiological role in granulopoiesis and neutrophil egress from the marrow. In transplantation, it is used to mobilize donor hematopoietic stem cells into the peripheral blood, such that they can be collected via apheresis for transplantation, and it is commonly administered to patients posttransplant, with the goal of hastening count recovery. Peripheral blood stem cell (PBSC) grafts now represent approximately 90% of donor grafts transplanted in adult patients.[14]

Signaling occurs through the G-CSF receptor, which is expressed on hematopoietic cells, including pluripotent and myeloid progenitors, mature neutrophils, monocytes, selected lymphocyte subsets, and endothelial cells.[15,16] G-CSF driven granulocyte expansion leads to the secretion of proteases, which then disrupt adhesion molecules and chemokine receptors (e.g., VCAM-1, CXCL12) that control maintenance of stem/progenitor cells within the bone marrow. By this mechanism, high-dose G-CSF results in stem cell mobilization from the marrow into the peripheral blood.[17]

It is clear that immunomodulatory effects initiated during the brief period of donor exposure to G-CSF prior to harvesting influence GVHD outcome, with a decrease in acute GVHD but relative increase in cGVHD in the current era of predominantly G-CSF mobilized grafts.[18] Interestingly, PBSC grafts generate high levels of cGVHD[19] and G-CSF appears to provoke Th17 differentiation, which is likely, at least in part, the explanation for this.[12] More recently, G-CSF has been shown to potently induce autophagy, which is involved in stem cell mobilization and improved survival of regulatory T cells (T_{reg}).[20]

IMPORTANT CYTOKINES IN GVHD

Interferons

Type 1 Interferons

There are 16 members of the type-1 interferon family: 12 IFNα subtypes; a unique IFNβ molecule; and IFNε, IFNκ, and IFNω, which signal through the ubiquitously expressed IFNAR.[21,22] These cytokines can be produced by almost all cell types in response to viral infection; however, plasmacytoid dendritic cells (pDCs) are thought to be most prolific.[23] Type I IFN signaling (which is STAT1 dependent) is important in both the innate and adaptive immune responses. IFN results in antiproliferative effects as well as the upregulation of MHC I expression in response to viral infection, and in the posttransplant setting, type I interferons are important for the differentiation and subsequent expansion of effector cytolytic T cells (CTLs).[24–26] Preclinical studies have demonstrated that I-IFN signaling in recipients protects from CD4-dependent GVHD, whereas donor signaling augments protective GVL responses.[27] Carefully timed administration of IFNα in the clinical BMT setting may, therefore, be beneficial, though it is not currently part of standard practice.

Prior to the availability of tyrosine kinase inhibitors (e.g., imatinib) for chronic myeloid leukemia, IFNα was the mainstay of treatment because of its cytolytic effects and influence on T cell function.[28–30] This effect can also be harnessed in the posttransplant setting for patients with relapsed disease, although again, separation of GVHD and GVL can be challenging.[31]

Interferon-gamma

Interferon-γ (IFNγ), which is the only type-II interferon, is produced in large amounts by Th1 and Tc1 cells early after HSCT and is the classical "Th1" cytokine generated during GVHD. IFNγ plays a pathogenic role in gastrointestinal GVHD, but, interestingly, has a protective role in lung pathology (idiopathic pneumonia syndrome) after BMT, thus rendering IFNγ a difficult cytokine to target for therapeutic benefit. In vivo, its production is promoted by IL-12 and IL-18; downregulated by IL-4, IL-10, and TGF; and is not restricted to T cells, as IFNγ can also be made by B cells, NKT cells, and antigen presenting cells (APCs).

Signaling occurs through the IFNγ receptor (IFNγR) in a STAT1-dependent fashion[24] and downstream effects include: Promotion of T cell activation/Th1 differentiation, production of other pro-inflammatory cytokines (particularly TNF), inhibition of Th2 differentiation, and upregulation of MHC class II expression on APC. During GVHD development, IFNγ also enhances the sensitivity of macrophages to lipopolysaccharide (LPS), hence, increasing the production of other proinflammatory cytokines, particularly TNF.[1] IFNγ signaling in reconstituting myeloid compartment has been shown in preclinical models to impact posttransplant dendritic cell function, with improved exogenous antigen presentation when mice were unable to signal through the IFNγ-receptor.[32,33] Chronic IFNγ exposure likely contributes to impaired adaptive immune responses after BMT by acting on donor APC to decrease antigen presentation.[32]

TNF and Lymphotoxin

TNF and related TNF superfamily cytokine and lymphotoxin alpha (LTα3) are key inflammatory mediators of GVHD.[34–37] TNF is secreted by recipient myeloid cells after pretransplant conditioning[1] and by donor T cells upon activation.[37–39]

Mouse models of GVHD, using mice deficient in both the cytokine and its receptor have established the critical role of TNF in GVHD pathogenesis, predominantly in GVHD of the gastrointestinal tract and skin.[40,41] Engagement of the TNFR p55 receptor directly induces apoptosis and is thought to be the most important for mediating TNF-associated pathology.

Genetic polymorphisms in the genes coding both TNF and its receptors have been studied in BMT donors and recipients. These studies suggest that "high" donor TNF production and increased recipient responses to TNF due to polymorphisms of the TNFR contribute to GVHD risk.[42] High levels of serum TNF following BMT have been correlated with GVHD severity,[43] and, as such, blockade of TNF with Etanercept (TNFR:Fc construct) or infliximab (an anti-TNF monoclonal antibody) has become standard practice in GVHD therapy in many institutions.[44–47]

Interleukins

IL-1

IL-1 is a potent proinflammatory cytokine that has been implicated in GVHD and many other inflammatory diseases.[48,49] IL-1 signaling is dependent upon the intracellular adaptor protein MyD88[50] and results in the increased expression of several genes critical in GVHD, including TNF, IL-6, IL-12, G-CSF, TGF-β, the receptors for IL-2, c-kit, and the common beta chain required for IL-3, IL-5, and granulocyte macrophage-colony stimulating factor (GM-CSF) signaling.[51]

Early work using mouse models of allogeneic BMT confirmed that IL-1 is an important mediator of disease[48]; however, manipulation of this pathway could not be translated to the clinic, since phase III trials of IL-1 blockade in clinical BMT failed to demonstrate a benefit.[52] This is likely due to the redundancy in proinflammatory pathways such that isolated IL-1 blockade is insufficient.

IL-2

IL-2 is produced by T cells and acts in an autocrine fashion to promote T cell growth and proliferation. Strategies to prevent GVHD have, therefore, largely focused on blockade of IL-2 effects with various agents, using different strategies to prevent IL-2 driven T cell proliferation. The calcineurin inhibitors cyclosporine and tacrolimus form the basis of current GVHD prophylaxis in most centers. These drugs inhibit calcineurin phosphatase and, therefore, impair calcium dependent intracellular events including the synthesis of IL-2.[53]

In contrast, low-dose IL-2, when given as a therapy, has been shown to preferentially promote the expansion and survival of T_{reg}, both in animal studies and, recently, in clinical trials.[54–57] This preliminary trial reported successful control of cGVHD with daily subcutaneous injection of IL-2 cytokine, and this has since been confirmed in larger cohorts.[58,59]

Basiliximab (a human–mouse chimeric antibody against CD25, the IL-2 receptor) has been demonstrated to reduce acute allograft rejection in renal transplantation in randomized controlled trials.[60] RCT level evidence is not available in allogeneic transplantation, but case series support the use of basiliximab as a therapeutic option in steroid-refractory GVHD, particularly when other agents (e.g., cyclosporine) are poorly tolerated due to renal impairment.[61–63] Blockade of IL-2 signaling with this therapy is thought to have direct T cell antiproliferative effect. The impact on regulatory T cells is less clear, with conflicting evidence in the literature[64] highlighting that there is clearly a capacity for both positive and negative immunological effects with this agent.

IL-6

IL-6 promotes T cell proliferation, the differentiation of cytotoxic T lymphocyte populations, and, when present in combination with TGF-β, promotes Th-17 development.[65,66] Preclinical studies of IL-6 in GVHD and GVL confirm its key role as a pathogenic cytokine in GVHD. The absence of IL-6 in the donor T cell pool (using IL-6 deficient donor mice) or systemic blockade of IL-6 with an anti-IL-6R antibody results in decreased aGVHD with no loss of GVL effects in the models used.[9,67]

Recent data demonstrate that IL-6 is the major cytokine detectable in patient plasma early after BMT and that it appears to play a dominant role in conditioning-related pathology.[68] Blockade of IL-6 with tocilizumab (soluble IL-6R) has now progressed through a successful phase I/II clinical trial with low levels of acute GVHD in comparison to historical controls.[68] This represents a promising new strategy for GVHD prevention.

IL-10

IL-10 signals through the IL-10 receptor (IL-10R) and is a regulatory cytokine, as signaling suppresses macrophage and DC function and, subsequently, leads to attenuation of T cell responses. It is classically produced by T_{reg}, B cells, monocytes, and DCs.[69,70] IL-10 has the following functions relevant to GVHD:

- acts on naïve $CD4^+$ T cells to promote T_{reg} differentiation and licensing,[69]
- serves as a counter-regulator to TNF production,
- inhibits T cell proliferation,[71–73]
- contributes to the balance between the T_{reg}/Th17 pathway, by suppressing Th17 development,[74] and promoting $FoxP3^+CD25^+CD4^+$ T_{reg} development.

Mouse studies in GVHD have demonstrated that both donor and recipient-derived IL-10 is important for the control of acute GVHD.[75–77] Genetic polymorphisms in the IL-10 gene have been examined in both donor and recipient and are a clear prognostic factor in GVHD; however, functional studies that associate genotype and downstream IL-10 protein production are lacking, i.e., it is unclear if the polymorphisms identified drive a gain or loss of IL-10.[78,79]

IL-12

IL-12 is classically produced by APC, including macrophages, DC, and Langerhans' cells.[80,81] IL-12 signals through the IL-12R, and the signaling results in STAT4-dependent production of IFNγ and IL-18 synthesis. IL-12 stimulates proliferation, IFNγ secretion and cytotoxicity of NK cells and CTL.[82]

IL-17 and Related Cytokines

There is increasing evidence that IL-17 has an important role to play in acute and, particularly, later onset acute and chronic GVHD in skin and lung.[12,83] It is produced by Th17 CD4 T cells and Tc17 CD8 T cells. Indeed, T cells of this lineage produce a large number of other cytokines relevant to GVHD (e.g., IL-21, IL-22, TNF, GM-CSF, and IL-17F). As discussed above, IL-6 and TGF-β are required in combination for Th17 differentiation.[84]

Secretion of IL-23 by donor APC has been identified in preclinical models as an important contributor to GVHD, specifically within the gastrointestinal tract, where it is thought to contribute to LPS translocation and IFNγ production.[85,86] IL-23 also plays an important role in the IL-17 pathway, i.e., the presence of IL-23 is thought to promote a pathogenic Th17 phenotype, whereas when TGF-β and IL-6 are present in the absence of IL-23, immunomodulatory Th17 effects are seen.[87,88]

IL-21

IL-21 (another member of the common gamma chain cytokine family) is produced primarily by CD4 T cells of the T follicular helper (T_{FH}) lineage and has a broad range of functions, including the direct promotion of apoptosis and promotion of proliferation and gain of effector function in B cells, T cells, NK cells, and DC.[89] After BMT, T_{FH} cells drive marginal zone B cell expansion and aberrant B cell development. These B cells produce alloantibodies in a Bruton's tyrosine kinase (BTK) dependent fashion, which play an important and causative role in cGVHD.[90–94] IL-21 also induces IL-10 production and can, thus, have both regulatory and proinflammatory roles, depending on the other cytokines present. IL-21R deficient animals and specific blocking antibodies have been used to assess the role of IL-21 in GVHD, confirming it has a pathogenic role in the GI tract (with mice deficient in the receptor suffering less GI-specific GVHD).[95,96] Neutralization of IL-21 represents a promising clinical therapeutic strategy to prevent and treat GVHD that now requires clinical investigation.

Macrophage Colony Stimulating Factor-1

Macrophage colony stimulating factor (CSF-1) is a critical growth factor for macrophage development[97,98] and has been studied in both the context of donor and recipient macrophages, with opposing effects seen. Recipient CSF-1-dependent macrophages appear to regulate GVHD by controlling T cell activation and expansion.[99] In contrast, reconstituting and alternatively activated donor macrophages play a causative role in cGVHD,[100] and their depletion with an anti-CSF-1R blocking antibody results in a significant protection from fibrosis in skin and lung after BMT. Importantly, the infiltration of CSF-1R dependent macrophages and subsequent fibrosis seen in cGVHD is IL-17 dependent.[100] Thus, depletion of macrophages (on the basis of their CSF-1 dependency) late after BMT is a promising therapeutic strategy to prevent and/or treat cGVHD.

TABLE 13.2 Current Cytokine-Based Therapeutic Targets in GVHD

Cytokine	Pharmacological agent
TNF	***Etanercept (TNFR p75 fused to FC protein)***
	Infliximab (anti-TNF antibody)
	Etanercept neutralizes soluble TNF and LTα, binds to cells expressing membrane-bound TNF, inducing cell death
	Infliximab binds and neutralizes soluble TNF in isolation
IL-6	***Tocilizumab (humanized mAb to the IL-6 receptor alpha)***
	Neutralizes proinflammatory effects of IL-6
	Reduces GVHD compared to historical controls in Phase I/II study. RCT level evidence for use in rheumatoid arthritis[103]
IL-2	***Ultra low dose IL-2 cytokine therapy***
	Expansion of T_{reg} populations, due to their expression of the "high affinity" IL-2R and therefore susceptibility to the lowest physiological IL-2 concentrations[58]
	Basiliximab (anti-CD25, IL2R)
	Proven in phase III studies in renal transplantation to reduce acute rejection[60]
	Promising case series in HSCT, for steroid-refractory GVHD.[61–63] Some controversy surrounding effect on regulatory T cell populations
	Commonly used where other immunosuppression will not be tolerated due to renal impairment
IFNα	***Recombinant interferon alpha (including pegylated form)***
	Enhance sensitivity of tumor cells to cytolysis
	Enhances CTL function[104]

CYTOKINES AS DRUG TARGETS IN GRAFT-VERSUS-HOST DISEASE

Individual cytokines have been targeted in GVHD therapy as summarized in Table 13.2. Of note, these agents have had variable success, likely due to the considerable redundancy of any pathway in isolation when the proinflammatory cytokine milieu is so intense. Since the key cytokines responsible for GVHD signal via common intracellular signaling pathways, targeting these pathways (e.g., Jak-Stat signaling) may represent a more effective therapeutic strategy in the future. Ruxolitinib, a Jak2 inhibitor, currently used in treatment of myelofibrosis has shown significant promise in mouse studies and early phase clinical trials in steroid-refractory GVHD, and via effects on inhibiting IL-1, IL-6, TNF, and IFNγ may prove extremely effective in GVHD if future trials mimic early results.[101,102] Similarly, inhibition of both T cell and B cell dependent pathways using BTK inhibitors (e.g., Ibrutinib) appear encouraging in the treatment of cGVHD.[91]

CONCLUSIONS

Cytokine-driven pathology plays a key role in GVHD. Many cytokines contribute to disease, and mouse models of BMT have provided highly detailed insights into the tissue-specific effects of pathogenic cytokine signaling. IL-1, TNF, and IL-6 are the major initiators of acute pathology, and these are initially produced by recipient cells in response to initial pretransplant conditioning. Strategies targeting individual cytokines as therapy for GVHD have met with modest success (e.g., TNF and IL-1), and it is likely that future strategies to prevent and treat GVHD will involve the targeting of multiple cytokine signaling pathways, perhaps via the inhibition of downstream Jak-Stat or BTK signaling. In manipulating cytokine signaling, it is crucial to consider both the direct effects of cytokine signaling and the indirect impact on T cell subset development. Control of both the soluble and cellular effectors in GVHD is critical to gaining control over this syndrome and thus improving the safety and broadening the applicability of BMT as a therapeutic strategy.

References

1. Nestel FP, Price KS, Seemayer TA, Lapp WS. Macrophage priming and lipopolysaccharide-triggered release of tumor necrosis factor alpha during graft-versus-host disease. *J Exp Med* 1992;**175**(2):405–13.

2. Ferrara JL, Cooke KR, Pan L, Krenger W. The immunopathophysiology of acute graft-versus-host-disease. *Stem Cells* 1996;**14**(5):473–89.
3. Koyama M, Kuns RD, Olver SD, Raffelt NC, Wilson YA, Don AL, et al. Recipient nonhematopoietic antigen-presenting cells are sufficient to induce lethal acute graft-versus-host disease. *Nat Med* 2012;**18**(1):135–42.
4. Toubai T, Tawara I, Sun Y, Liu C, Nieves E, Evers R, et al. Induction of acute GVHD by sex-mismatched HY antigens in the absence of functional radiosensitive host hematopoietic-derived antigen presenting cells. *Blood* 2012;**119**(16):3844–53.
5. Li H, Demetris AJ, McNiff J, Matte-Martone C, Tan HS, Rothstein DM, et al. Profound depletion of host conventional dendritic cells, plasmacytoid dendritic cells, and B cells does not prevent graft-versus-host disease induction. *J Immunol* 2012;**188**(8):3804–11.
6. Mowat AM. Antibodies to IFN-gamma prevent immunologically mediated intestinal damage in murine graft-versus-host reaction. *Immunology* 1989;**68**(1):18–23.
7. Garside P, Reid S, Steel M, Mowat AM. Differential cytokine production associated with distinct phases of murine graft-versus-host reaction. *Immunology* 1994; **82**(2):211–14.
8. Burman AC, Banovic T, Kuns RD, Clouston AD, Stanley AC, Morris ES, et al. IFNgamma differentially controls the development of idiopathic pneumonia syndrome and GVHD of the gastrointestinal tract. *Blood* 2007;**110**(3):1064–72.
9. Tawara I, Koyama M, Liu C, Toubai T, Thomas D, Evers R, et al. Interleukin-6 modulates graft-versus-host responses after experimental allogeneic bone marrow transplantation. *Clin Cancer Res* 2011;**17**(1):77–88.
10. McCormick LL, Zhang Y, Tootell E, Gilliam AC. Anti-TGF-beta treatment prevents skin and lung fibrosis in murine sclerodermatous graft-versus-host disease: a model for human scleroderma. *J Immunol* 1999;**163**(10):5693–9.
11. Yi T, Chen Y, Wang L, Du G, Huang D, Zhao D, et al. Reciprocal differentiation and tissue-specific pathogenesis of Th1, Th2, and Th17 cells in graft-versus-host disease. *Blood* 2009;**114**(14):3101–12.
12. Hill GR, Olver SD, Kuns RD, Varelias A, Raffelt NC, Don AL, et al. Stem cell mobilization with G-CSF induces type 17 differentiation and promotes scleroderma. *Blood* 2010;**116**(5):819–28.
13. Varelias A, Gartlan KH, Kreijveld E, Olver SD, Lor M, Kuns RD, et al. Lung parenchyma-derived IL-6 promotes IL-17A-dependent acute lung injury after allogeneic stem cell transplantation. *Blood* 2015;**125**(15):2435–44.
14. Pasquini M, Zhu X. Current uses and outcomes of hematopoietic stem cell transplantation: CIBMTR summary slides. <http://www.cibmtr.org/>; 2015.
15. Demetri GD, Griffin JD. Granulocyte colony-stimulating factor and its receptor. *Blood* 1991;**78**(11):2791–808.
16. Bocchietto E, Guglielmetti A, Silvagno F, Taraboletti G, Pescarmona GP, Mantovani A, et al. Proliferative and migratory responses of murine microvascular endothelial cells to granulocyte-colony-stimulating factor. *J Cell Physiol* 1993;**155**(1):89–95.
17. Levesque JP, Hendy J, Takamatsu Y, Simmons PJ, Bendall LJ. Disruption of the CXCR4/CXCL12 chemotactic interaction during hematopoietic stem cell mobilization induced by GCSF or cyclophosphamide. *J Clin Invest* 2003;**111**(2):187–96.
18. Markey KA, MacDonald KP, Hill GR. The biology of graft-versus-host disease: experimental systems instructing clinical practice. *Blood* 2014;**124**(3):354–62.
19. Bensinger WI, Martin PJ, Storer B, Clift R, Forman SJ, Negrin R, et al. Transplantation of bone marrow as compared with peripheral-blood cells from HLA-identical relatives in patients with hematologic cancers. *N Engl J Med* 2001;**344**(3):175–81.
20. Le Texier L, Lineburg KE, Cao B, McDonald-Hyman C, Leveque-El Mouttie L, Nicholls J, et al. Autophagy-dependent regulatory T cells are critical for the control of graft-versus-host disease. *JCI Insight* 2016;**1**(15):e86850.

21. de Weerd NA, Samarajiwa SA, Hertzog PJ. Type I interferon receptors: biochemistry and biological functions. *J Biol Chem* 2007;**282**(28):20053–7.
22. Pestka S, Krause CD, Walter MR. Interferons, interferon-like cytokines, and their receptors. *Immunol Rev* 2004;**202**:8–32.
23. Gilliet M, Cao W, Liu YJ. Plasmacytoid dendritic cells: sensing nucleic acids in viral infection and autoimmune diseases. *Nat Rev Immunol* 2008;**8**(8):594–606.
24. Meraz MA, White JM, Sheehan KC, Bach EA, Rodig SJ, Dighe AS, et al. Targeted disruption of the Stat1 gene in mice reveals unexpected physiologic specificity in the JAK-STAT signaling pathway. *Cell* 1996;**84**(3):431–42.
25. Hwang SY, Hertzog PJ, Holland KA, Sumarsono SH, Tymms MJ, Hamilton JA, et al. A null mutation in the gene encoding a type I interferon receptor component eliminates antiproliferative and antiviral responses to interferons alpha and beta and alters macrophage responses. *Proc Natl Acad Sci USA* 1995;**92**(24):11284–8.
26. Webb IJ, Eickhoff CE, Elias AD, Ayash LJ, Wheeler CA, Schwartz GN, et al. Kinetics of peripheral blood mononuclear cell mobilization with chemotherapy and/or granulocyte-colony-stimulating factor: implications for yield of hematopoietic progenitor cell collections. *Transfusion* 1996;**36**(2):160–7.
27. Robb RJ, Hill GR. The interferon-dependent orchestration of innate and adaptive immunity after transplantation. *Blood* 2012;**119**(23):5351–8.
28. Klingemann HG, Grigg AP, Wilkie-Boyd K, Barnett MJ, Eaves AC, Reece DE, et al. Treatment with recombinant interferon (alpha-2b) early after bone marrow transplantation in patients at high risk for relapse [corrected]. *Blood* 1991;**78**(12):3306–11.
29. Steegmann JL, Casado LF, Tomas JF, Sanz-Rodriguez C, Granados E, de la Camara R, et al. Interferon alpha for chronic myeloid leukemia relapsing after allogeneic bone marrow transplantation. *Bone Marrow Transplant* 1999;**23**(5):483–8.
30. Giralt SA, Kantarjian HM, Talpaz M, Rios MB, Del Giglio A, Andersson BS, et al. Effect of prior interferon alfa therapy on the outcome of allogeneic bone marrow transplantation for chronic myelogenous leukemia. *J Clin Oncol* 1993;**11**(6):1055–61.
31. Collins Jr RH, Shpilberg O, Drobyski WR, Porter DL, Giralt S, Champlin R, et al. Donor leukocyte infusions in 140 patients with relapsed malignancy after allogeneic bone marrow transplantation. *J Clin Oncol* 1997;**15**(2):433–44.
32. Capitini CM, Herby S, Milliron M, Anver MR, Mackall CL, Fry TJ. Bone marrow deficient in IFN-γ signaling selectively reverses GVHD-associated immunosuppression and enhances a tumor-specific GVT effect. *Blood* 2009;**113**(20):5002–9.
33. Markey KA, Koyama M, Kuns RD, Lineburg KE, Wilson YA, Olver SD, et al. Immune insufficiency during GVHD is due to defective antigen presentation within dendritic cell subsets. *Blood* 2012;**119**(24):5918–30.
34. Hill GR, Teshima T, Gerbitz A, Pan L, Cooke KR, Brinson YS, et al. Differential roles of IL-1 and TNF-alpha on graft-versus-host disease and graft versus leukemia. *J Clin Invest* 1999;**104**(4):459–67.
35. Hill GR, Teshima T, Rebel VI, Krijanovski OI, Cooke KR, Brinson YS, et al. The p55 TNF-alpha receptor plays a critical role in T cell alloreactivity. *J Immunol* 2000; **164**(2):656–63.
36. Markey KA, Burman AC, Banovic T, Kuns RD, Raffelt NC, Rowe V, et al. Soluble lymphotoxin is an important effector molecule in GVHD and GVL. *Blood* 2010; **115**(1):122–32.
37. Schmaltz C, Alpdogan O, Muriglan SJ, Kappel BJ, Rotolo JA, Ricchetti ET, et al. Donor T cell-derived TNF is required for graft-versus-host disease and graft-versus-tumor activity after bone marrow transplantation. *Blood* 2003;**101**(6):2440–5.
38. Hill GR, Cooke KR, Brinson YS, Bungard D, Ferrara JL. Pretransplant chemotherapy reduces inflammatory cytokine production and acute graft-versus-host disease after allogeneic bone marrow transplantation. *Transplantation* 1999; **67**(11):1478–80.

39. Ewing P, Miklos S, Olkiewicz KM, Muller G, Andreesen R, Holler E, et al. Donor CD4 + T-cell production of tumor necrosis factor alpha significantly contributes to the early proinflammatory events of graft-versus-host disease. *Exp Hematol* 2007; **35**(1):155–63.
40. Speiser DE, Bachmann MF, Frick TW, McKall-Faienza K, Griffiths E, Pfeffer K, et al. TNF receptor p55 controls early acute graft-versus-host disease. *J Immunol* 1997; **158**(11):5185–90.
41. Hattori K, Hirano T, Miyajima H, Yamakawa N, Tateno M, Oshimi K, et al. Differential effects of anti-Fas ligand and anti-tumor necrosis factor alpha antibodies on acute graft-versus-host disease pathologies. *Blood* 1998;**91**(11):4051–5.
42. Wilson AG, Symons JA, McDowell TL, McDevitt HO, Duff GW. Effects of a polymorphism in the human tumor necrosis factor alpha promoter on transcriptional activation. *Proc Natl Acad Sci USA* 1997;**94**(7):3195–9.
43. Holler E, Kolb HJ, Moller A, Kempeni J, Liesenfeld S, Pechumer H, et al. Increased serum levels of tumor necrosis factor alpha precede major complications of bone marrow transplantation. *Blood* 1990;**75**(4):1011–16.
44. Levine JE, Paczesny S, Mineishi S, Braun T, Choi SW, Hutchinson RJ, et al. Etanercept plus methylprednisolone as initial therapy for acute graft-versus-host disease. *Blood* 2008;**111**(4):2470–5.
45. Kennedy GA, Butler J, Western R, Morton J, Durrant S, Hill GR. Combination antithymocyte globulin and soluble TNFalpha inhibitor (etanercept) +/− mycophenolate mofetil for treatment of steroid refractory acute graft-versus-host disease. *Bone Marrow Transplant* 2006;**37**(12):1143–7.
46. Herve P, Flesch M, Tiberghien P, Wijdenes J, Racadot E, Bordigoni P, et al. Phase I-II trial of a monoclonal anti-tumor necrosis factor alpha antibody for the treatment of refractory severe acute graft-versus-host disease. *Blood* 1992;**79**(12):3362–8.
47. Holler E, Kolb HJ, Wilmanns W. Treatment of GVHD—TNF-antibodies and related antagonists. *Bone Marrow Transplant* 1993;**12**(Suppl. 3):S29–31.
48. Abhyankar S, Gilliland DG, Ferrara JL. Interleukin-1 is a critical effector molecule during cytokine dysregulation in graft versus host disease to minor histocompatibility antigens. *Transplantation* 1993;**56**(6):1518–23.
49. Dinarello CA. Interleukin-1 in the pathogenesis and treatment of inflammatory diseases. *Blood* 2011;**117**(14):3720–32.
50. Weber A, Wasiliew P, Kracht M. Interleukin-1 (IL-1) pathway. *Sci Signal* 2010;**3** (105):cm1.
51. Dinarello CA. Biologic basis for interleukin-1 in disease. *Blood* 1996;**87**(6):2095–147.
52. Antin JH, Weisdorf D, Neuberg D, Nicklow R, Clouthier S, Lee SJ, et al. Interleukin-1 blockade does not prevent acute graft-versus-host disease: results of a randomized, double-blind, placebo-controlled trial of interleukin-1 receptor antagonist in allogeneic bone marrow transplantation. *Blood* 2002;**100**(10):3479–82.
53. Thomson AW, Bonham CA, Zeevi A. Mode of action of tacrolimus (FK506): molecular and cellular mechanisms. *Ther Drug Monit* 1995;**17**(6):584–91.
54. Sykes M, Romick ML, Hoyles KA, Sachs DH. In vivo administration of interleukin 2 plus T cell-depleted syngeneic marrow prevents graft-versus-host disease mortality and permits alloengraftment. *J Exp Med* 1990;**171**(3):645–58.
55. Sykes M, Romick ML, Sachs DH. Interleukin 2 prevents graft-versus-host disease while preserving the graft-versus-leukemia effect of allogeneic T cells. *Proc Natl Acad Sci USA* 1990;**87**(15):5633–7.
56. Sykes M, Harty MW, Szot GL, Pearson DA. Interleukin-2 inhibits graft-versus-host disease-promoting activity of CD4+ cells while preserving CD4- and CD8-mediated graft-versus-leukemia effects. *Blood* 1994;**83**(9):2560–9.

57. Wang MG, Szebeni J, Pearson DA, Szot GL, Sykes M. Inhibition of graft-versus-host disease by interleukin-2 treatment is associated with altered cytokine production by expanded graft-versus-host-reactive CD4 + helper cells. *Transplantation* 1995;**60**(5):481–90.
58. Koreth J, Matsuoka K, Kim HT, McDonough SM, Bindra B, Alyea III EP, et al. Interleukin-2 and regulatory T cells in graft-versus-host disease. *N Engl J Med* 2011;**365**(22):2055–66.
59. Koreth J, Kim HT, Jones KT, Lange PB, Reynolds CG, Chammas MJ, et al. Efficacy, durability, and response predictors of low-dose interleukin-2 therapy for chronic graft-versus-host disease. *Blood* 2016;**128**(1):130–7.
60. Nashan B, Moore R, Amlot P, Schmidt AG, Abeywickrama K, Soulillou JP. CHIB 201 International Study Group. Randomised trial of basiliximab versus placebo for control of acute cellular rejection in renal allograft recipients. *The Lancet* 1997; **350**(9086):1193–8.
61. Funke VA, de Medeiros CR, Setubal DC, Ruiz J, Bitencourt MA, Bonfim CM, et al. Therapy for severe refractory acute graft-versus-host disease with basiliximab, a selective interleukin-2 receptor antagonist. *Bone Marrow Transplant* 2006;**37**(10):961–5.
62. Massenkeil G, Rackwitz S, Genvresse I, Rosen O, Dorken B, Arnold R. Basiliximab is well tolerated and effective in the treatment of steroid-refractory acute graft-versus-host disease after allogeneic stem cell transplantation. *Bone Marrow Transplant* 2002; **30**(12):899–903.
63. Schmidt-Hieber M, Fietz T, Knauf W, Uharek L, Hopfenmuller W, Thiel E, et al. Efficacy of the interleukin-2 receptor antagonist basiliximab in steroid-refractory acute graft-versus-host disease. *Br J Haematol* 2005;**130**(4):568–74.
64. Chakupurakal G, Garcia-Marquez MA, Shimabukuro-Vornhagen A, Theurich S, Holtick U, Hallek M, et al. Immunological effects in patients with steroid-refractory graft-versus-host disease following treatment with basiliximab, a CD25 monoclonal antibody. *Eur J Haematol* 2016;**97**(2):121–7.
65. Kishimoto T, Akira S, Narazaki M, Taga T. Interleukin-6 family of cytokines and gp130. *Blood* 1995;**86**(4):1243–54.
66. Kishimoto T, Akira S, Taga T. Interleukin-6 and its receptor: a paradigm for cytokines. *Science* 1992;**258**(5082):593–7.
67. Chen X, Das R, Komorowski R, Beres A, Hessner MJ, Mihara M, et al. Blockade of interleukin-6 signaling augments regulatory T-cell reconstitution and attenuates the severity of graft-versus-host disease. *Blood* 2009;**114**(4):891–900.
68. Kennedy GA, Varelias A, Vuckovic S, Le Texier L, Gartlan KH, Zhang P, et al. Addition of interleukin-6 inhibition with tocilizumab to standard graft-versus-host disease prophylaxis after allogeneic stem-cell transplantation: a phase 1/2 trial. *The Lancet Oncology* 2014;**15**(13):1451–9.
69. Groux H, O'Garra A, Bigler M, Rouleau M, Antonenko S, de Vries JE, et al. A CD4+ T-cell subset inhibits antigen-specific T-cell responses and prevents colitis. *Nature* 1997;**389**(6652):737–42.
70. Rutella S, Danese S, Leone G. Tolerogenic dendritic cells: cytokine modulation comes of age. *Blood* 2006;**108**(5):1435–40.
71. Taga K, Tosato G. IL-10 inhibits human T cell proliferation and IL-2 production. *J Immunol* 1992;**148**(4):1143–8.
72. Moore KW, O'Garra A, de Waal Malefyt R, Vieira P, Mosmann TR. Interleukin-10. *Annu Rev Immunol* 1993;**11**:165–90.
73. Taga K, Mostowski H, Tosato G. Human interleukin-10 can directly inhibit T-cell growth. *Blood* 1993;**81**(11):2964–71.
74. Huber S, Gagliani N, Esplugues E, O'Connor W, Huber FJ, Chaudhry A, et al. Th17 cells express interleukin-10 receptor and are controlled by Foxp3 and Foxp3 + regulatory CD4+ T cells in an interleukin-10-dependent manner. *Immunity* 2011; **34**(4):554–65.

75. Cooke KR, Ferrara JL. A protective gene for graft-versus-host disease. *N Engl J Med* 2003;**349**(23):2183–4.
76. Rowe V, Banovic T, MacDonald KP, Kuns R, Don AL, Morris ES, et al. Host B cells produce IL-10 following TBI and attenuate acute GVHD after allogeneic bone marrow transplantation. *Blood* 2006;**108**(7):2485–92.
77. Tawara I, Sun Y, Liu C, Toubai T, Nieves E, Evers R, et al. Donor- but not host-derived interleukin-10 contributes to the regulation of experimental graft-versus-host disease. *J Leukoc Biol* 2012;**91**(4):667–75.
78. Lin MT, Storer B, Martin PJ, Tseng LH, Gooley T, Chen PJ, et al. Relation of an interleukin-10 promoter polymorphism to graft-versus-host disease and survival after hematopoietic-cell transplantation. *N Engl J Med* 2003;**349**(23):2201–10.
79. Lin MT, Storer B, Martin PJ, Tseng LH, Grogan B, Chen PJ, et al. Genetic variation in the IL-10 pathway modulates severity of acute graft-versus-host disease following hematopoietic cell transplantation: synergism between IL-10 genotype of patient and IL-10 receptor beta genotype of donor. *Blood* 2005;**106**(12):3995–4001.
80. Macatonia SE, Hosken NA, Litton M, Vieira P, Hsieh CS, Culpepper JA, et al. Dendritic cells produce IL-12 and direct the development of Th1 cells from naive CD4 + T cells. *J Immunol* 1995;**154**(10):5071–9.
81. Trinchieri G. Interleukin-12: a proinflammatory cytokine with immunoregulatory functions that bridge innate resistance and antigen-specific adaptive immunity. *Annu Rev Immunol* 1995;**13**:251–76.
82. Trinchieri G. Interleukin-12: a cytokine at the interface of inflammation and immunity. *Adv Immunol* 1998;**70**:83–243.
83. Carlson MJ, West ML, Coghill JM, Panoskaltsis-Mortari A, Blazar BR, Serody JS. In vitro-differentiated TH17 cells mediate lethal acute graft-versus-host disease with severe cutaneous and pulmonary pathologic manifestations. *Blood* 2009;**113**(6): 1365–74.
84. Bettelli E, Carrier Y, Gao W, Korn T, Strom TB, Oukka M, et al. Reciprocal developmental pathways for the generation of pathogenic effector TH17 and regulatory T cells. *Nature* 2006;**441**(7090):235–8.
85. Das R, Chen X, Komorowski R, Hessner MJ, Drobyski WR. Interleukin-23 secretion by donor antigen-presenting cells is critical for organ-specific pathology in graft-versus-host disease. *Blood* 2009;**113**(10):2352–62.
86. Das R, Komorowski R, Hessner MJ, Subramanian H, Huettner CS, Cua D, et al. Blockade of interleukin-23 signaling results in targeted protection of the colon and allows for separation of graft-versus-host and graft-versus-leukemia responses. *Blood* 2010;**115**(25):5249–58.
87. McGeachy MJ, Bak-Jensen KS, Chen Y, Tato CM, Blumenschein W, McClanahan T, et al. TGF-beta and IL-6 drive the production of IL-17 and IL-10 by T cells and restrain T(H)-17 cell-mediated pathology. *Nat Immunol* 2007;**8**(12):1390–7:[Epub 2007 Nov 11].
88. McGeachy MJ, Cua DJ. The link between IL-23 and Th17 cell-mediated immune pathologies. *Semin Immunol* 2007;**19**(6):372–6[Epub 2007 Dec 3].
89. Spolski R, Leonard WJ. Interleukin-21: basic biology and implications for cancer and autoimmunity. *Annu Rev Immunol* 2008;**26**:57–79.
90. Srinivasan M, Flynn R, Price A, Ranger A, Browning JL, Taylor PA, et al. Donor B-cell alloantibody deposition and germinal center formation are required for the development of murine chronic GVHD and bronchiolitis obliterans. *Blood* 2012;**119**(6): 1570–80.
91. Dubovsky JA, Flynn R, Du J, Harrington BK, Zhong Y, Kaffenberger B, et al. Ibrutinib treatment ameliorates murine chronic graft-versus-host disease. *J Clin Invest* 2014;**124** (11):4867–76.

92. Satterthwaite AB, Witte ON. The role of Bruton's tyrosine kinase in B-cell development and function: a genetic perspective. *Immunol Rev* 2000;**175**:120–7.
93. Flynn R, Du J, Veenstra RG, Reichenbach DK, Panoskaltsis-Mortari A, Taylor PA, et al. Increased T follicular helper cells and germinal center B cells are required for cGVHD and bronchiolitis obliterans. *Blood* 2014;**123**(25):3988–98.
94. Jin H, Ni X, Deng R, Song Q, Young J, Cassady K, et al. Antibodies from donor B cells perpetuate cutaneous chronic graft-versus-host disease in mice. *Blood* 2016;**127**(18):2249–60.
95. Hanash AM, Kappel LW, Yim NL, Nejat RA, Goldberg GL, Smith OM, et al. Abrogation of donor T-cell IL-21 signaling leads to tissue-specific modulation of immunity and separation of GVHD from GVL. *Blood* 2011;**118**(2):446–55.
96. Meguro A, Ozaki K, Hatanaka K, Oh I, Sudo K, Ohmori T, et al. Lack of IL-21 signal attenuates graft-versus-leukemia effect in the absence of CD8 T-cells. *Bone Marrow Transplant* 2011;**46**(12):1557–65.
97. Bonifer C, Hume DA. The transcriptional regulation of the Colony-Stimulating Factor 1 Receptor (csf1r) gene during hematopoiesis. *Front Biosci* 2008;**13**:549–60.
98. Stanley ER, Berg KL, Einstein DB, Lee PS, Pixley FJ, Wang Y, et al. Biology and action of colony-stimulating factor-1. *Mol Reprod Dev* 1997;**46**(1):4–10.
99. MacDonald KP, Palmer JS, Cronau S, Seppanen E, Olver S, Raffelt NC, et al. An antibody against the colony-stimulating factor 1 receptor depletes the resident subset of monocytes and tissue- and tumor-associated macrophages but does not inhibit inflammation. *Blood* 2010;**116**(19):3955–63.
100. Alexander KA, Flynn R, Lineburg KE, Kuns RD, Teal BE, Olver SD, et al. CSF-1-dependant donor-derived macrophages mediate chronic graft-versus-host disease. *J Clin Invest* 2014;**124**(10):4266–80.
101. Verstovsek S, Kantarjian H, Mesa RA, Pardanani AD, Cortes-Franco J, Thomas DA, et al. Safety and efficacy of INCB018424, a JAK1 and JAK2 inhibitor, in myelofibrosis. *N Engl J Med* 2010;**363**(12):1117–27.
102. Spoerl S, Mathew NR, Bscheider M, Schmitt-Graeff A, Chen S, Mueller T, et al. Activity of therapeutic JAK 1/2 blockade in graft-versus-host disease. *Blood* 2014;**123**(24):3832–42.
103. Singh JA, Beg S, Lopez-Olivo MA. Tocilizumab for rheumatoid arthritis. *Cochrane Database Syst Rev* 2010(7). Cd008331.
104. Molldrem JJ, Lee PP, Wang C, Felio K, Kantarjian HM, Champlin RE, et al. Evidence that specific T lymphocytes may participate in the elimination of chronic myelogenous leukemia. *Nat Med* 2000;**6**(9):1018–23.

PART IV

CYTOKINES AND THERAPEUTIC APPLICATIONS

CHAPTER

14

Cytokine Therapy in the Tumor Microenvironment: Old Players, New Tricks

Ruth Ganss

Harry Perkins Institute of Medical Research, Centre for Medical Research, The University of Western Australia, Perth, WA, Australia

OUTLINE

Cytokine Effector Functions in Tissues.
DOI: http://dx.doi.org/10.1016/B978-0-12-804214-4.00013-0

TUMOR MICROENVIRONMENT AND STROMAL TARGETING: A BALANCING ACT

Cancer comprises malignant cells enmeshed within a complex network of stromal cells, including immune cells, fibroblasts, and blood vessels. Tumor stroma, growth factors, and cytokines provide the framework for malignant cells and, importantly, shape the intratumoral immune environment. Thus, tumor–stromal interactions can restrict or promote cancer growth by regulating tumor cell survival, proliferation, and therapeutic responsiveness.[1] Since mutations in cancer cells drive cancer growth, the vast majority of anticancer drugs target cancer cell-intrinsic, growth-promoting pathways. More recently, strategies have emerged that delete stromal cells to starve tumor cells or alleviate immune suppression. However, data to date show that stromal cell elimination is a double-edged sword and can result in adaptive resistance and cancer relapse.[2–4] In contrast, reprogramming or reeducating tumor stroma has emerged as an alternative strategy to skew the balance from a tumor promoting to a tumoricidal environment without inducing therapeutic resistance.[5,6] Growth factors, chemokines, and cytokines by virtue of mediating reciprocal communication between malignant and stromal support cells are crucial polarizing factors. Although the beneficial role of cytokines in antitumor therapy remains confusing, many cytokines share the ability to activate innate and adaptive immune cells and have, thus, attracted renewed interest due to their ability to reprogram tumor stroma.

CYTOKINES IN ANTICANCER THERAPY: A HISTORIC PERSPECTIVE

Cytokines are messenger molecules, secreted by diverse cell types, which facilitate cell-to-cell communications and predominantly regulate tissue homeostasis, immunity, and inflammation. The immunomodulatory potential of cytokines has been explored in cancer therapy for several decades. In particular, interferons (IFN, type I and II), interleukins (IL), and growth factors such as tumor necrosis factor (TNF) have been extensively tested in animal tumor models, with some achieving clinical application. Early studies in the 1990s, focussed on cancer cells that were engineered to overexpress cytokines in murine transplantation models. These experiments showed surprising antitumor activities for a wide spectrum of cytokines in a mono-therapeutic setting and established the crucial role of cytotoxic T cells in growth inhibition.[7] Nevertheless, translation of cytokine therapy into the clinic has been slow, with unpredictable patient responses. For instance, systemic application of the promising cytokine IL12 had to be abandoned because of high toxicity and low overall efficacy.[8] In contrast, type I interferons (IFNα2a/b) have been clinically approved by the US Food and Drug Administration (FDA) since 1988/1995 for several hematological cancers and some solid tumors, including melanoma.[9] In 1992, high dose IL2 infusion was approved for metastatic renal cell carcinoma, followed by approval for metastatic melanoma in 1998[10]; systemic high dose IL2 treatment can induce complete remission in 5%–10% of cancer patients.[11] However, overall clinical benefits of systemic cytokine therapies are modest, not least because of short serum half-life and dose-limiting toxicity. Moreover, immune-activating functions of cytokines are often counterbalanced by immune suppression, for instance due to IL2-induced expansion of regulatory T cells (Treg).[10] Strategies to circumvent some of these problems include locally restricting cytokine applications, for instance isolated limb perfusion of high dose TNFα in sarcoma and melanoma patients,[12] stabilized (pegylated) cytokine formulations,[13] or mutant cytokines that activate fewer Tregs.[14] Since cytokine therapy often modulates both activating and inhibitory immune functions, their overall efficacy increases when combined with additional immune-boosting treatments. Indeed, IL2 therapy is significantly improved when co-administered with adoptively transferred autologous tumor infiltrating T cells (TILs). Moreover, response rates of up to 70% are achieved when IL2 and adoptive T cell therapy is preceded by a lymphodepletion protocol, which further stimulates effector T cell expansion.[11] Also, high dose TNFα limb infusion is only effective when combined with melphalan, an alkylating agent which induces immunogenic tumor cell death.[15] Currently, multiple clinical trials are under way to test the efficacy of systemic cytokine applications in combination with chemo-, radiation-, and immuno-therapies.[16] Emerging frontline therapies are combinations with immune checkpoint blocking antibodies targeting for instance T lymphocyte antigen 4 (CTLA4, FDA approved in 2011) and programmed cell death protein 1 (PD1, FDA approved in 2015). These checkpoint antibodies neutralize co-inhibitor receptors on target cells, enable sustained T cell activation, and have yielded noteworthy patient benefits, which are further improved in combination therapies.[17]

MECHANISM-GUIDED CYTOKINE THERAPY

Systemic cytokine applications are still being refined; in particular, to enhance other immunotherapies. Appropriate design of combination therapies, however, is the key to boost local immune effector functions. For instance, it has been known for some time that CTLA4 blockade shifts the

T cell effector/Treg ratio toward higher numbers of cytotoxic T cells. However, only recently has it been reported that blockade of IL2 signaling abolishes the beneficial effects of anti-CTLA4 treatment in murine models of sarcoma and colon cancer.[18] This implicates an intratumoral IL2 source as part of the effector arm. Indeed, anti-CTLA4 treatment repolarizes $CD4^+$ $FoxP3^+$ $IL10^+$ immunosuppressive cells in the tumor environment into IL2-secreting FoxP3-negative T cells. This finding perhaps explains why in a phase I/II study in melanoma patients no synergistic effects of CTLA4 blockade with systemic IL2 administration were observed.[19]

Intratumoral or tumor-associated macrophages (TAMs) are known for their plasticity, diversity, and capacity to determine therapeutic outcome. TAMs can secrete antitumorigenic cytokines, such as IFNγ and IL12 (M1 subtype) or pro-tumorigenic, antiinflammatory cytokines, such as IL10 (M2 subtype). Macrophages are also highly dependent on the growth factor colony-stimulating factor (CSF)-1, and indeed CSF-1 blockade repolarizes macrophages in preclinical cancer models.[20,21] Specifically, CSF-1 blockade blunts macrophage secretion of IL10. This in turn increases IL12 expression of intratumoral dendritic cells (DCs), T cell priming, and $CD8^+$ T cell-dependent tumor regression.[21] The effects of CSF-1 blockade can be mimicked by treatment with IL10R blocking antibodies, providing a strong rationale for cytokine blockade in combination with chemotherapy in breast cancer.[21] Recently, an intratumoral mechanism has been discovered, which underscores the significance of tumor-intrinsic type I interferons (IFNβ). Here, intratumoral antigen-presenting cells (APCs) sense tumor-derived DNA via the stimulator of interferon genes (STING) pathway, which triggers IFNβ secretion and primes effector T cell function. Interestingly, defective DNA sensing in STING-deficient mice abolishes therapeutic effects of checkpoint blockade antibodies, demonstrating the important synergistic effects of interferon and adoptive immunity.[22] Importantly, radiation treatment induces endogenous IFN secretion by APCs in a STING-dependent pathway and, thus, enhances adaptive anticancer immunity.[23] These studies demonstrate that local cytokine functions maximize the specific aspects of antitumor responses and, in the right combination treatment, may achieve higher anticancer efficacies with far less toxicity than systemic cytokine applications.

TARGETING THE MICROENVIRONMENT

Subtle and local changes in intratumoral cytokine profiles have profound effects and, indeed, form the basis for reeducating or reprogramming the tumor microenvironment.[5,6] However, targeting cytokines into cancers remains a technical challenge. Direct peri- or intra-tumoral

TABLE 14.1 Cytokine Targeting into the Tumor Microenvironment

Cancer targeting strategy	Homing mechanism or targeted moiety	Cytokine	Cancer type	Antitumor effects	Application	Reference
STEM CELLS						
Tie2+ hematopoietic precursors	Tumor tropism	IFNα	Glioma, breast carcinoma	Antiangiogenic; activation of innate/adaptive immunity	Preclinical	29,30
iPS[a]-derived myeloid cells	Tumor tropism	IFNα	Melanoma	Macrophage repolarization	Preclinical	1[b]
Mesenchymal stem cells	Tumor tropism	IFNα	Melanoma	T cell/NK cell activation	Preclinical	32
		IL12	Renal cell carcinoma	IFNγ production; NK cell activation	Preclinical	2[c]
Tumor cells	Tumor self-seeding	TNFα	Breast/lung carcinoma, melanoma	Vascular damage; cancer cell apoptosis	Preclinical	3[d]
T CELLS						
TRUCKs	gp100[e]	IL12 or inducible (i) IL12	Ovarian/colon cancer, melanoma, sarcoma, fibrosarcoma	Increased T cell persistence/cytotoxicity, IFNγ secretion, antigen presentation; myeloid cell reprogramming	Preclinical	37
	MUC16[f]					38
	CEA[g]					40
	VEGFR2					41

(*Continued*)

TABLE 14.1 (Continued)

Cancer targeting strategy	Homing mechanism or targeted moiety	Cytokine	Cancer type	Antitumor effects	Application	Reference
T cell–nanoparticle conjugates	gp100	IL15/ IL15Rα and IL21	Melanoma	Enhanced T cell persistence and memory induction	Preclinical	4[h]
TILs with cytokine transgene	Melanoma antigens	iIL12	Metastatic melanoma	60% Objective clinical response; toxicity due to high serum IL12	Phase I	42
ONCOLYTIC VIRUSES						
Herpes virus (T-Vec)	Intratumoral injection	GM-CSF	Advanced melanoma	Enhanced systemic antitumor immune response	FDA approved	46
IMMUNOCYTOKINES						
Antibody cytokine conjugates	EGFR	IFNβ	Breast cancer, melanoma	Enhanced T cell priming when combined with immunotherapy	Preclinical	49
	Tenascin C (F16) or Fibronectin (L19)	IL2	Breast cancer	Enhanced leukocyte infiltration when combined with chemotherapy	Preclinical	48
			Metastatic melanoma	28% objective clinical response when combined with chemotherapy; non toxic	Phase I	50
	Fibronectin (L19)	TNFα	Solid tumors	No objective response as single treatment modality	Phase I/II	51

PEPTIDE CYTOKINES						
Peptide cytokine conjugates	PDGFRβ (putative RGR ligand)	TNFα	Prostate cancer, melanoma, pancreatic cancer	Vessel activation; vessel normalization; enhanced T cell infiltration when combined with immuno- and/or chemotherapy	Preclinical	64
	CD13 (NGR ligand)					65
			Solid tumors/ mesothelioma	Low toxicity when combined with chemotherapy	Phase Ib and III	59 NCT01098266
			Advanced melanoma	Increase in intratumoral macrophages when combined with vaccine	Phase I	58

[a] *Inducible pluripotent stem cells.*

[b] *Miyashita A, Fukushima S, Nakahara S, Kubo Y, Tokuzumi A, Yamashita J, et al. Immunotherapy against metastatic melanoma with human iPS cell-derived myeloid cell lines producing type I interferons. Cancer Immunol Res 2016;4:248–58.*

[c] *Gao P, Ding Q, Wu Z, Jiang H, Fang Z. Therapeutic potential of human-mesenchymal stem cells producing IL-12 in a mouse xenograft model of renal cell carcinoma. Cancer Lett 2010;290:157–66.*

[d] *Dondossola E, Dobroff AS, Marchio S, Cardo-Vila M, Hosoya H, Libutti SK, et al. Self-targeting of TNF-releasing cancer cells in preclinical models of primary and metastatic tumors. Proc Natl Acad Sci USA 2016;113:2223–8.*

[e] *Glycoprotein 100 or melanocyte protein pmel.*

[f] *Member of the mucin family of glycoproteins.*

[g] *Carcinoembryonic antigen.*

[h] *Stephan MT, Moon JJ, Um SH, Bershteyn A, Irvine DJ. Therapeutic cell engineering with surface-conjugated synthetic nanoparticles. Nat Med 2010;16:1035–41.*

injections of recombinant cytokines, cytokine blocking antibodies, bacteria, viral expression vectors, or bio-scaffolds containing cytokines are convenient ways to analyze cytokine effects in murine transplantation tumors.[24–26] Some human cancers may be accessible by ultrasound-guided injections; however, most cancers are difficult to reach by injection routes. Nanoparticles are scavenged by peripheral and intratumoral monocytes/macrophages and may, therefore, be exploited as delivery platforms for cytokines.[27] In addition, a plethora of alternative strategies for intratumoral cytokine release has been developed, which exploit either natural tumor homing or active targeting, whereby ligands and antibodies are used to bind to cell surface targets. Some examples are summarized in Table 14.1 and discussed below.

GENE THERAPY AND CELLULAR VEHICLES FOR SPECIFIC CYTOKINE DELIVERY

Stem Cell Delivery Vehicles

Gene therapy is the transfer of genetic sequences, most commonly packaged into viral vectors, directly into tissue, or into cells ex vivo, followed by cell expansion and injection. Taking advantage of specific homing properties of transduced cells, they can be used as vehicles for therapeutic cytokine delivery into the tumor microenvironment. For instance, a unique subset of circulating hematopoietic progenitor cells, which expresses the angiopoietin receptor Tie2, is recruited into tumors to support the growth of new blood vessels in the process of angiogenesis.[28] Indeed, bone marrow-derived cells transduced with the IFNα gene under the control of the Tie2 enhancer/promoter home into tumors and locally deliver IFNα into murine brain and breast cancers. Consequently, a combination of antiangiogenic effects and stimulation of innate and adaptive immune responses impairs primary tumor growth and metastatic dissemination.[29,30]

Mesenchymal stem cells (MSC) are stromal progenitor cells, which can be isolated, for instance, from bone marrow, adipose tissue, and umbilical cord, and display multilineage potential. When systemically injected, MSCs specifically migrate to sites of inflammation which includes tumor tissue. Whilst MSC in the tumor environment support or inhibit tumor growth in a context dependent manner,[31] their therapeutic potential as tumor-homing cellular vehicles for cytokines has been established. For instance, IFNα-secreting MSCs when evaluated in B16 melanoma-bearing mice show antitumor effects that are dependent on $CD8^+$ T cell and NK cell activation.[32] While preclinical cancer models using gene-modified MSCs have been well studied, the safety of engineered MSCs for human anticancer therapy remains a concern and currently slows clinical translation.

Cytokine Delivery Via T Cells

There is clear evidence that at least some human cancers mount spontaneous T cell responses and the number of infiltrating T cells correlates with prognosis.[33] Therefore, using immune cells engineered to recognize tumor-specific antigens is an attractive targeting concept. In particular, T cells armed with cancer-specific T cell receptors (TCRs) or chimeric antigen receptors (CARs) may overcome the low frequency of naturally occurring TILs. CAR T cell therapy is based on human leukocyte antigen (HLA)-independent T cell activation via a chimeric receptor that has a single chain antibody external domain, a series of costimulatory molecules, and TCR signaling domains on the cytoplasmic tail.[34] So far, CAR technology in humans has been successfully employed in hematological malignancies with impressive results in early phase clinical trials.[35] Targeting of solid tumors, however, remains challenging, not least because of limited tumor penetration and effector function in a hostile tumor environment. To overcome these issues, fourth generation CARs or so called TRUCKs (T cells redirected for universal cytokine killing) have been employed that carry, for instance, an IL12 expression vector.[36–38] IL12 secretion in the vicinity of T cells greatly enhances cytotoxicity by reprogramming antigen-presenting cells in the tumor microenvironment. In particular, inducible IL12 TRUCK T cells, which secrete physiological amounts of IL12 in situ, upon activation of the TCR signaling complex overcome some of the problems associated with IL12 overproduction.[39–41] In an initial clinical trial, infusion of autologous TILs engineered to express inducible IL12 into metastatic melanoma patients has, however, been disappointing. Engineered T cells show a relatively low persistence in vivo, which may be related to antiproliferative IL12 activities. Moreover, toxicities due to high serum IL12 in some patients demonstrate the requirement to further refine the level of cytokine production in TILs and potentially TRUCK T cells.[42] This also demonstrates that high cytokine levels, even when produced locally, are not always beneficial.

ONCOLYTIC VIRUSES AS CYTOKINE CARRIERS

Oncolytic viruses have an inherent tropism for cancer cells but can also be engineered to specifically target antigens expressed on tumor cells such as integrins using the RGD (Arg-Gly-Asp) motif. Once in the tumor environment, they are designed to replicate and lyse malignant cells, which in turn promotes anticancer immunity. Gene engineered viruses can express cytokines to further enhance efficacy.[43] A variety of gene-modified viruses have been developed for clinical trials, which exploit intrinsic immunostimulatory properties of granulocyte-macrophage colony-stimulating factor (GM-CSF). GM-CSF augments antitumor immunity by stimulating antigen uptake by DCs and anticancer T cell responses and, thus, acts as an in situ, patient-specific, antitumor vaccine when combined with oncolytic viruses.[44] In mouse models, GM-CSF encoding viruses show significant antitumor effects and also confer protective immunity.[45] In humans, oncolytic herpes simplex virus type 1 (HSV-1) encoding human GM-CSF (talimogene laherparepvec, T-VEC, FDA approved in 2015) has been shown to evoke local and distant antitumor immune responses in melanoma patients.[46] Whilst these clinical data are encouraging, T-VEC and other viral particles are currently administered by intralesional injection, which ensures high-dose delivery at the tumor site, minimizes systemic reactions, and reduces complement and antibody inactivation of the viral particles, but also limits its application.

ANTIBODY–CYTOKINE FUSION THERAPY

Tumor and stromal cells provide a unique "address" by overexpressing cell surface antigens that can be targeted with monoclonal antibodies. So far, a handful of surface molecules and antibodies have been identified as suitable for the delivery of cytokines into solid tumors. Amongst those are antibodies that possess antitumor activity per se by targeting oncogenic signaling, such as the human epidermal growth factor receptor 2 (EGFR2, Her2), ganglioside GD2, or epithelial cell adhesion molecule (Ep-CAM). Alternatively, antibodies serve as carrier only, for instance, by targeting extra cellular matrix (ECM) splice variants of fibronectin (L19/F8) or tenascin C (F16), which are associated with highly angiogenic tumor blood vessels. Cytokines are fused to whole antibodies or antibody fragments via linkers using recombinant technology to generate so called immunocytokines.[47] Preclinical data in syngeneic mouse models are encouraging and demonstrate antitumor activities in correlation with targeting efficacy. Nevertheless, immunocytokines show best potential in combination with other treatment modalities such as chemotherapy and immunotherapy.[48,49] For instance, F16-IL2 increases efficacy of cytotoxic drugs, such as paclitaxel, in breast cancer models.[48] An EGFR-IFNβ fusion protein enhances the therapeutic effect of EGFR targeting by boosting antigen presentation in the tumor environment. However, as a direct consequence of high local IFNβ levels, PDL-1 is upregulated on tumor cells. Thus, in addition to EGFR-IFNβ, antiPDL-1 antibodies are required to overcome acquired resistance.[49] Some immunocytokines have entered clinical phase I/II trials, most notably in conjunction with IL2 and TNFα.[50,51] Thus far, results are encouraging, and it is predictable that the application of immunocytokines will increase with better knowledge of existing intratumoral immune profiles and expanded options for immune combination therapies.

HARNESSING PEPTIDES FOR TARGETED CYTOKINE DELIVERY

Short peptide sequences, commonly identified via in vivo phage display library screens, can also be employed as targeting moieties for tumor-selective delivery of therapeutics, including cytokines.[52] To date, the majority of peptides used in preclinical and clinical applications target cancer blood vessels.[53] Angiogenic blood vessels are morphologically and functionally different to the vasculature in normal tissues and, thus, provide tumor-selective targeting opportunities. This includes, amongst others, the integrin $\alpha_v\beta_3/\alpha_v\beta_5$ binding RGD peptide, NGR (Asn-Gly-Arg) peptide that binds the metalloprotease aminopeptidase N (CD13), and RGR (Arg-Gly-Arg), which is a putative PDGFRβ ligand.[54–56] In particular, antitumorigenic activities of TNFα-peptide fusion compounds have been extensively assessed in preclinical animal models and now in clinical trials.[57,58] In 2008, NGR-human TNFα was granted FDA orphan drug status for the treatment of mesothelioma and hepatocellular carcinoma.[59] Even though clinical advances have been moderate to date, low dose vascular delivery of cytokines is particularly attractive, since it may overcome problems common to targeting specific cancer antigens such as poor tissue penetration, intratumoral heterogeneity of antigen expression, and cancer-induced resistance mechanisms.[49]

VASCULAR TARGETING OF CYTOKINES IMPROVES BARRIER INTEGRITY AND T CELL INFLUX

Cytokine anticancer efficacy crucially depends not only on the existing immune milieu but also on the vascular barrier function.[6,33] Blood vessels in solid tumors are leaky, poorly perfused and limit lymphocyte adhesion and extravasation.[6] Consequently, delivery of drug- and cell-based therapies into solid tumors is often compromised. Strategies to "normalize" cancer blood vessels by improving vessel integrity aim to reduce vessel leakiness and, thus, improve perfusion and immune cell influx into tumors.[60,61] Importantly, selected cytokines when targeted in low, physiological quantities to cancer vessels can trigger normalization of the vascular bed. For instance, low levels of TNFα in the tumor environment reduce interstitial fluid pressure,[62] which in turn improves tumor perfusion and facilitates uptake of liposomes.[63] Indeed, targeting minute amounts of TNFα specifically to tumor vessels in a mouse model of pancreatic neuroendocrine tumors stabilizes the vascular network and improves vessel perfusion.[64] Moreover, RGR- or NGR-tagged TNFα improves lymphocyte–vessel wall interactions by upregulation of vascular adhesion molecules. Both, enhanced perfusion and endothelial activation, act as strong adjuvants to adoptive T cell therapy.[64,65] Thus, modulating the vascular barrier function locally with intratumoral cytokine therapy is potentially widely applicable and most effective in combination with other anticancer treatment modalities such as chemo- and immuno-therapies. It does not, however, work indiscriminately and is highly dependent on cytokine type, dose, and duration of cytokine exposure. For instance, therapeutically effective doses of IFNγ or IFNβ in the tumor environment induce angiogenic vessel death, which reduces tumor burden similar to antiangiogenic therapy but does not enhance immune cell influx.[64,66]

CYTOKINE AMPLIFICATION CASCADES

Cytokine gradients and paracrine signaling may only range across a few cell diameters. Moreover, cytokines are unstable and rapidly cleared from blood and tissues.[67] Nevertheless, a single cytokine can activate diverse cell types, stimulate further cytokine release and, thus, trigger pleiotropic and "amplified" antitumor effects (Fig. 14.1). Indeed, immune amplification in the tumor environment may explain some of the remarkable effects observed when targeting minute amounts of cytokines specifically to the tumor vasculature.[53] For instance, RGR-TNFα specifically accumulates around tumor vessels and attracts macrophages into the perivascular niche. These macrophages are reprogrammed from a M2 into M1-like phenotype, which alleviates T cell suppression. Moreover, macrophages secrete angiopoietin 2 (Ang2), a Tie2 tyrosine kinase receptor ligand that together with TNFα upregulates the expression of endothelial adhesion molecules and in turn facilitates leukocyte extravasation.[64] Similarly, RGR-mediated targeting of the TNF family member TNFSF14 or LIGHT to the tumor vasculature induces vessel normalization and enhances uptake of cytotoxic

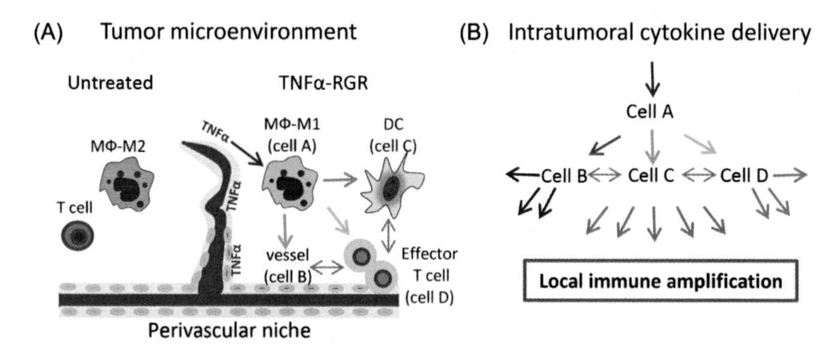

FIGURE 14.1 **Intratumoral cytokine signal amplification.** (A) Peptide targeting of TNFα (TNFα-RGR) to tumor vasculature induces macrophage repolarization, which in turn activates the vessel wall, increases antigen presentation and T effector cell influx/ function in a cascade of signaling events.[64] (B) Schematic presentation of signal amplification involving locally restricted delivery of a cytokine and subsequent signal diversification by neighboring cell types (cells A–D). Local immune amplification can lead to antitumor effects in combination with chemo- or immunotherapies.

drugs and T cell infiltration.[68] This is predominantly effected through induction of pericyte contractility and realignment with endothelial cells. Intriguingly, TNFSF14-stimulated macrophages secrete TGFβ exclusively in the perivascular niche, which in turn triggers pericyte maturation.[68] The dual role of TGFβ in the tumor environment highlights the exquisite spatial effects of cytokines and the need to harness locally amplified, short-range paracrine signaling. Targeting of physiological amounts of cytokines to specific cellular compartments within solid tumors will be required for optimal reeducation and therapeutic efficacy.

CONCLUDING REMARKS

Thus far, therapeutic success of cytokines in anticancer treatment has been variable. Cytokine therapy is hampered by the fact that the biological function of endogenous cytokines in the tumor environment is vastly distinct to the effects of systemically administered or ectopically expressed cytokines. Similarly, neutralizing antibodies or small molecule inhibitors indiscriminately deplete cytokines independent of their location, often triggering negative feedback mechanisms. As the field matures, it becomes clearer that cytokine effects crucially depend on cellular context, tumor stage, and dosing. High cytokine levels or prolonged exposure can nullify the beneficial effects of low dose application.[42,69] Importantly, cytokines are modulators of immune and stromal cells; they are not stand-alone anticancer therapeutics. As such, physiological amounts of cytokines and well-timed spatio-temporal application may pave the way for better drug efficacy. Cytokines are intimately involved in attracting immune cells into solid tumors and/or shifting the intratumoral immune profile and are, therefore, likely to be necessary tools to boost treatment efficacy. Combining intratumoral cytokines with immunotherapy, such as immune checkpoint blockade, may enhance durable response rates and induce sustained remission. Thus, the key to future successful cytokine applications are mechanistic insights in a 3D environment and careful evaluation of how to best apply cytokines in combination therapies.

Acknowledgments

The author acknowledges financial support from grants and fellowships from the Western Australian Cancer Council, the Australian National Health and Medical Research Council (NHMRC), Woodside Energy, and the Harry Perkins Institute of Medical Research.

References

1. McMillin DW, Negri JM, Mitsiades CS. The role of tumour–stromal interactions in modifying drug response: challenges and opportunities. *Nat Rev Drug Discov* 2013;**12**:217–28.
2. Hanahan D, Coussens LM. Accessories to the crime: functions of cells recruited to the tumor microenvironment. *Cancer Cell* 2012;**21**:309–22.
3. Ozdemir BC, Pentcheva-Hoang T, Carstens JL, Zheng X, Wu CC, Simpson TR, et al. Depletion of carcinoma-associated fibroblasts and fibrosis induces immunosuppression and accelerates pancreas cancer with reduced survival. *Cancer Cell* 2014;**25**:719–34.
4. Casazza A, Laoui D, Wenes M, Rizzolio S, Bassani N, Mambretti M, et al. Impeding macrophage entry into hypoxic tumor areas by Sema3A/Nrp1 signaling blockade inhibits angiogenesis and restores antitumor immunity. *Cancer Cell* 2013;**24**:695–709.
5. Quail DF, Joyce JA. Microenvironmental regulation of tumor progression and metastasis. *Nat Med* 2013;**19**:1423–37.
6. Johansson A, Hamzah J, Ganss R. More than a scaffold: stromal modulation of tumor immunity. *Biochim Biophys Acta* 2016;**1865**:3–13.
7. Hock H, Dorsch M, Kunzendorf U, Qin Z, Diamantstein T, Blankenstein T. Mechanisms of rejection induced by tumor cell-targeted gene transfer of interleukin 2, interleukin 4, interleukin 7, tumor necrosis factor, or interferon gamma. *Proc Natl Acad Sci USA* 1993;**90**:2774–8.
8. Leonard JP, Sherman ML, Fisher GL, Buchanan LJ, Larsen G, Atkins MB, et al. Effects of single-dose interleukin-12 exposure on interleukin-12-associated toxicity and interferon-gamma production. *Blood* 1997;**90**:2541–8.
9. Ortiz A, Fuchs SY. Anti-metastatic functions of type 1 interferons: foundation for the adjuvant therapy of cancer. *Cytokine* 2016. doi:10.1016/j.cyto.2016.01.010 [Epub ahead of print].
10. Liao W, Lin JX, Leonard WJ. Interleukin-2 at the crossroads of effector responses, tolerance, and immunotherapy. *Immunity* 2013;**38**:13–25.

11. Rosenberg SA. IL-2: the first effective immunotherapy for human cancer. *J Immunol* 2014;**192**:5451–8.
12. Mocellin S, Rossi CR, Pilati P, Nitti D. Tumor necrosis factor, cancer and anticancer therapy. *Cytokine Growth Factor Rev* 2005;**16**:35–53.
13. Eggermont AM, Suciu S, Santinami M, Testori A, Kruit WH, Marsden J, et al. Adjuvant therapy with pegylated interferon alfa-2b versus observation alone in resected stage III melanoma: final results of EORTC 18991, a randomised phase III trial. *Lancet* 2008;**372**:117–26.
14. Carmenate T, Pacios A, Enamorado M, Moreno E, Garcia-Martinez K, Fuente D, et al. Human IL-2 mutein with higher antitumor efficacy than wild type IL-2. *J Immunol* 2013;**190**:6230–8.
15. Lu X, Ding ZC, Cao Y, Liu C, Habtetsion T, Yu M, et al. Alkylating agent melphalan augments the efficacy of adoptive immunotherapy using tumor-specific CD4+ T cells. *J Immunol* 2015;**194**:2011–21.
16. Vacchelli E, Aranda F, Obrist F, Eggermont A, Galon J, Cremer I, et al. Trial watch: immunostimulatory cytokines in cancer therapy. *OncoImmunology* 2014;**3**:e29030.
17. Mahoney KM, Rennert PD, Freeman GJ. Combination cancer immunotherapy and new immunomodulatory targets. *Nat Rev Drug Discov* 2015;**14**:561–84.
18. Hannani D, Vetizou M, Enot D, Rusakiewicz S, Chaput N, Klatzmann D, et al. Anticancer immunotherapy by CTLA-4 blockade: obligatory contribution of IL-2 receptors and negative prognostic impact of soluble CD25. *Cell Res* 2015;**25**:208–24.
19. Maker AV, Phan GQ, Attia P, Yang JC, Sherry RM, Topalian SL, et al. Tumor regression and autoimmunity in patients treated with cytotoxic T lymphocyte-associated antigen 4 blockade and interleukin 2: a phase I/II study. *Ann Surg Oncol* 2005;**12**:1005–16.
20. Pyonteck SM, Akkari L, Schuhmacher AJ, Bowman RL, Sevenich L, Quail DF, et al. CSF-1R inhibition alters macrophage polarization and blocks glioma progression. *Nat Med* 2013;**19**:1264–72.
21. Ruffell B, Chang-Strachan D, Chan V, Rosenbusch A, Ho CM, Pryer N, et al. Macrophage IL-10 blocks CD8+ T cell-dependent responses to chemotherapy by suppressing IL-12 expression in intratumoral dendritic cells. *Cancer Cell* 2014;**26**:623–37.
22. Woo SR, Fuertes MB, Corrales L, Spranger S, Furdyna MJ, Leung MY, et al. STING-dependent cytosolic DNA sensing mediates innate immune recognition of immunogenic tumors. *Immunity* 2014;**41**:830–42.
23. Deng L, Liang H, Xu M, Yang X, Burnette B, Arina A, et al. STING-Dependent Cytosolic DNA sensing promotes radiation-induced type I interferon-dependent antitumor immunity in immunogenic tumors. *Immunity* 2014;**41**:843–52.
24. Van der Jeught K, Joe PT, Bialkowski L, Heirman C, Daszkiewicz L, Liechtenstein T, et al. Intratumoral administration of mRNA encoding a fusokine consisting of IFN-beta and the ectodomain of the TGF-beta receptor II potentiates antitumor immunity. *Oncotarget* 2014;**5**:10100–13.
25. Kim J, Li WA, Choi Y, Lewin SA, Verbeke CS, Dranoff G, et al. Injectable, spontaneously assembling, inorganic scaffolds modulate immune cells in vivo and increase vaccine efficacy. *Nat Biotechnol* 2015;**33**:64–72.
26. Kiselyov A, Bunimovich-Mendrazitsky S, Startsev V. Treatment of non-muscle invasive bladder cancer with bacillus Calmette-Guerin (BCG): biological markers and simulation studies. *BBA Clin* 2015;**4**:27–34.
27. Christian DA, Hunter CA. Particle-mediated delivery of cytokines for immunotherapy. *Immunotherapy* 2012;**4**:425–41.
28. De Palma M, Venneri MA, Galli R, Sergi LS, Politi LS, Sampaolesi M, et al. Tie2 identifies a hematopoietic lineage of proangiogenic monocytes required for tumor vessel

formation and a mesenchymal population of pericyte progenitors. *Cancer Cell* 2005;**8**:211–26.
29. De Palma M, Mazzieri R, Politi LS, Pucci F, Zonari E, Sitia G, et al. Tumor-targeted interferon-alpha delivery by Tie2-expressing monocytes inhibits tumor growth and metastasis. *Cancer Cell* 2008;**14**:299–311.
30. Escobar G, Moi D, Ranghetti A, Ozkal-Baydin P, Squadrito ML, Kajaste-Rudnitski A, et al. Genetic engineering of hematopoiesis for targeted IFN-alpha delivery inhibits breast cancer progression. *Sci Transl Med* 2014;**6**:217ra3.
31. Moniri MR, Dai LJ, Warnock GL. The challenge of pancreatic cancer therapy and novel treatment strategy using engineered mesenchymal stem cells. *Cancer Gene Ther* 2014;**21**:12–23.
32. Xu C, Lin L, Cao G, Chen Q, Shou P, Huang Y, et al. Interferon-alpha-secreting mesenchymal stem cells exert potent antitumor effect in vivo. *Oncogene* 2014;**33**(42):5047–52.
33. Gajewski TF, Woo SR, Zha Y, Spaapen R, Zheng Y, Corrales L, et al. Cancer immunotherapy strategies based on overcoming barriers within the tumor microenvironment. *Curr Opin Immunol* 2015;**25**:268–76.
34. Gill S, Maus MV, Porter DL. Chimeric antigen receptor T cell therapy: 25 years in the making. *Blood Rev* 2015. doi:10.1016/j.blre.2015.10.003 [Epub ahead of print].
35. Cheadle EJ, Gornall H, Baldan V, Hanson V, Hawkins RE, Gilham DE. CAR T cells: driving the road from the laboratory to the clinic. *Immunol Rev* 2014;**257**:91–106.
36. Chmielewski M, Hombach AA, Abken H. Of CARs and TRUCKs: chimeric antigen receptor (CAR) T cells engineered with an inducible cytokine to modulate the tumor stroma. *Immunol Rev* 2014;**257**:83–90.
37. Kerkar SP, Goldszmid RS, Muranski P, Chinnasamy D, Yu Z, Reger RN, et al. IL-12 triggers a programmatic change in dysfunctional myeloid-derived cells within mouse tumors. *J Clin Invest* 2011;**121**:4746–57.
38. Koneru M, Purdon TJ, Spriggs D, Koneru S, Brentjens RJ. IL-12 secreting tumor-targeted chimeric antigen receptor T cells eradicate ovarian tumors. *OncoImmunology* 2015;**4**:e994446.
39. Zhang L, Kerkar SP, Yu Z, Zheng Z, Yang S, Restifo NP, et al. Improving adoptive T cell therapy by targeting and controlling IL-12 expression to the tumor environment. *Mol Ther* 2011;**19**:751–9.
40. Chmielewski M, Kopecky C, Hombach AA, Abken H. IL-12 release by engineered T cells expressing chimeric antigen receptors can effectively muster an antigen-independent macrophage response on tumor cells that have shut down tumor antigen expression. *Cancer Res* 2011;**71**:5697–706.
41. Chinnasamy D, Yu Z, Kerkar SP, Zhang L, Morgan RA, Restifo NP, et al. Local delivery of interleukin-12 using T cells targeting VEGF receptor-2 eradicates multiple vascularized tumors in mice. *Clin Cancer Res* 2012;**18**:1672–83.
42. Zhang L, Morgan RA, Beane JD, Zheng Z, Dudley ME, Kassim SH, et al. Tumor-infiltrating lymphocytes genetically engineered with an inducible gene encoding interleukin-12 for the immunotherapy of metastatic melanoma. *Clin Cancer Res* 2015;**21**:2278–88.
43. Kaufman HL, Kohlhapp FJ, Zloza A. Oncolytic viruses: a new class of immunotherapy drugs. *Nat Rev Drug Discov* 2015;**14**:642–62.
44. Kaufman HL, Ruby CE, Hughes T, Slingluff Jr. CL. Current status of granulocyte-macrophage colony-stimulating factor in the immunotherapy of melanoma. *J Immunother Cancer* 2014;**2**:11.
45. Liu BL, Robinson M, Han ZQ, Branston RH, English C, Reay P, et al. ICP34.5 deleted herpes simplex virus with enhanced oncolytic, immune stimulating, and anti-tumour properties. *Gene Ther* 2003;**10**:292–303.

46. Andtbacka RH, Kaufman HL, Collichio F, Amatruda T, Senzer N, Chesney J, et al. Talimogene laherparepvec improves durable response rate in patients with advanced melanoma. *J Clin Oncol* 2015;**33**:2780–8.
47. Pasche N, Neri D. Immunocytokines: a novel class of potent armed antibodies. *Drug Discov Today* 2012;**17**:583–90.
48. Marlind J, Kaspar M, Trachsel E, Sommavilla R, Hindle S, Bacci C, et al. Antibody-mediated delivery of interleukin-2 to the stroma of breast cancer strongly enhances the potency of chemotherapy. *Clin Cancer Res* 2008;**14**:6515–24.
49. Yang X, Zhang X, Fu ML, Weichselbaum RR, Gajewski TF, Guo Y, et al. Targeting the tumor microenvironment with interferon-beta bridges innate and adaptive immune responses. *Cancer Cell* 2014;**25**:37–48.
50. Eigentler TK, Weide B, de Braud F, Spitaleri G, Romanini A, Pflugfelder A, et al. A dose-escalation and signal-generating study of the immunocytokine L19-IL2 in combination with dacarbazine for the therapy of patients with metastatic melanoma. *Clin Cancer Res* 2011;**17**:7732–42.
51. Spitaleri G, Berardi R, Pierantoni C, De Pas T, Noberasco C, Libbra C, et al. Phase I/II study of the tumour-targeting human monoclonal antibody–cytokine fusion protein L19-TNF in patients with advanced solid tumours. *J Cancer Res Clin Oncol* 2013;**139**: 447–55.
52. Ruoslahti E, Bhatia SN, Sailor MJ. Targeting of drugs and nanoparticles to tumors. *J Cell Biol* 2010;**188**:759–68.
53. Johansson A, Hamzah J, Ganss R. License for destruction: tumor-specific cytokine targeting. *Trends Mol Med* 2014;**20**:16–24.
54. Pasqualini R, Koivunen E, Ruoslahti E. α Integrins as receptors for tumor targeting by circulating ligands. *Nature Biotechnol* 1997;**15**:542–6.
55. Pasqualini R, Koivunen E, Kain R, Lahdenranta J, Sakamoto M, Stryhn A, et al. Aminopeptidase N is a receptor for tumor-homing peptides and a target for inhibiting angiogenesis. *Cancer Res* 2000;**60**:722–7.
56. Joyce JA, Laakkonen P, Bernasconi M, Bergers G, Ruoslahti E, Hanahan D. Stage-specific vascular markers revealed by phage display in a mouse model of pancreatic islet tumorigenesis. *Cancer Cell* 2003;**4**:393–403.
57. Corti A, Pastorino F, Curnis F, Arap W, Ponzoni M, Pasqualini R. Targeted drug delivery and penetration into solid tumors. *Med Res Rev* 2012;**32**:1078–91.
58. Parmiani G, Pilla L, Corti A, Doglioni C, Cimminiello C, Bellone M, et al. A pilot Phase I study combining peptide-based vaccination and NGR-hTNF vessel targeting therapy in metastatic melanoma. *OncoImmunology* 2014;**3**:e963406.
59. Gregorc V, Zucali PA, Santoro A, Ceresoli GL, Citterio G, De Pas TM, et al. Phase II study of asparagine-glycine-arginine-human tumor necrosis factor alpha, a selective vascular targeting agent, in previously treated patients with malignant pleural mesothelioma. *J Clin Oncol* 2010;**28**:2604–11.
60. Hamzah J, Jugold M, Kiessling F, Rigby P, Manzur M, Marti HH, et al. Vascular normalization in Rgs5-deficient tumours promotes immune destruction. *Nature* 2008;**453**:410–14.
61. Jain RK. Antiangiogenesis strategies revisited: from starving tumors to alleviating hypoxia. *Cancer Cell* 2014;**26**:605–22.
62. Kristensen CA, Nozue M, Boucher Y, Jain RK. Reduction of interstitial fluid pressure after TNF-alpha treatment of three human melanoma xenografts. *Br J Cancer* 1996;**74**: 533–6.
63. Seynhaeve AL, Hoving S, Schipper D, Vermeulen CE, de Wiel-Ambagtsheer G, van Tiel ST, et al. Tumor necrosis factor alpha mediates homogeneous distribution of liposomes in murine melanoma that contributes to a better tumor response. *Cancer Res* 2007;**67**:9455–62.

64. Johansson A, Hamzah J, Payne CJ, Ganss R. Tumor-targeted TNFalpha stabilizes tumor vessels and enhances active immunotherapy. *Proc Natl Acad Sci USA* 2012;**109**: 7841–6.
65. Calcinotto A, Grioni M, Jachetti E, Curnis F, Mondino A, Parmiani G, et al. Targeting TNF-alpha to neoangiogenic vessels enhances lymphocyte infiltration in tumors and increases the therapeutic potential of immunotherapy. *J Immunol* 2012; **188**:2687–94.
66. Spaapen RM, Leung MY, Fuertes MB, Kline JP, Zhang L, Zheng Y, et al. Therapeutic activity of high-dose intratumoral IFN-beta requires direct effect on the tumor vasculature. *J Immunol* 2014;**193**:4254–60.
67. Thurley K, Gerecht D, Friedmann E, Hofer T. Three-dimensional gradients of cytokine signaling between T Cells. *PLoS Comput Biol* 2015;**11:**e1004206.
68. Johansson-Percival A, Li ZJ, Lakhiani DD, He B, Wang X, Hamzah J, et al. Intratumoral LIGHT restores pericyte contractile properties and vessel integrity. *Cell Rep* 2015;**13**:2687–98.
69. Yang ZZ, Grote DM, Ziesmer SC, Niki T, Hirashima M, Novak AJ, et al. IL-12 upregulates TIM-3 expression and induces T cell exhaustion in patients with follicular B cell non-Hodgkin lymphoma. *J Clin Invest* 2012;**122**:1271–82.

CHAPTER

15

Cytokines From Mesenchymal Stem Cells Induce Immunosuppressive Cells

Dobroslav Kyurkchiev

Medical University of Sofia, Sofia, Bulgaria

OUTLINE

MESENCHYMAL STEM CELLS AND THEIR CYTOKINE SECRETION

The classical definition of Pittenger describes the mesenchymal stem cells (MSCs) with their capacity to adhere to plastic surface, fibroblast-like morphology, self-renewal capacity, and potential do differentiate in different mesenchymal lineages.[1] MSCs are localized in cell niches

Cytokine Effector Functions in Tissues.

DOI: http://dx.doi.org/10.1016/B978-0-12-804214-4.00014-2

predominantly in the perivascular spaces thus outlining the vessels of the microcirculation at the outer side.[2,3] This localization of MSCs makes possible their communications with numerous different cells in practically all tissues of the body.[4] MSCs can accomplish a process of homing under the influence of chemokine gradients,[5,6] and the chemokine receptors expressed on their surface include CCR1, CCR2, CCR3, CCR4, CCR7, CCR8, CCR9, CCR10, CXCR4, CXCR5, and CXCR6.[7]

The major MSCs feature known as "self-renewal" is defined as the capability to reproduce cells identical to the maternal cell due to blocked differentiation. Thus a rather homogeneous cell population is formed.[8] This way self-renewal and differentiation seem to be two opposing processes and the block of differentiation is a condition sine qua non for intensive self-renewal (expansion) which is characteristics for the MSCs.

The minimal criteria for the expression of phenotypic markers characterizing any cell population as MSCs is the positive expression of CD105, CD73, CD90 by more than 95% of the cells and negative staining for CD45, CD34, CD14 or CD11b, CD79a or CD19, HLA-DR as recommended by the International Society of Cell Therapy (ISCT).[9]

The ability of the MSCs to dedifferentiate as well as their capacity to set the formation of various progenitor cells in the ranges of the same germinal layer is defined the MSCs plasticity, while the term transdifferentiation is used to designate an extreme form of plasticity when MSC originating from one germinal layer differentiates and gives cell lineages from another germinal layer. Due to their plasticity, the MSCs are multipotent in their potential to differentiate in different mesenchymal tissues: Bone, cartilage, adipose tissue, tendons, and muscles. Even more the MSCs can differentiate in nonmesenchymal tissues such as epithelia and nerve tissue which shows that MSCs can undergo transdifferentiation.[10,11] To prove that some cells are MSCs, it is necessary to demonstrate their potential to differentiate in vitro as osteogenic, chondrogenic, and adipogenic cells documented by specific staining methods.[9]

The MSCs and the immunocompetent cells have multiple contacts and not one-sided but a numerous interactions have been described. MSCs influence the immune cells predominantly suppressing their activities and many factors have been reported to be engaged in this process. Most generally, the suppressive effects of MSCs are exerted by direct cell-to-cell contacts or secreted factors and in most cases both types of activity act simultaneously. MSCs secrete cytokines with predominantly immunosuppressive effect as varieties of secreted factors have been reported in the literature published on this topic. The number of MSCs secreted cytokines is quite high but most commonly reported cytokines engaged in the immunosuppression are IL-10, TGFβ, IL-6, prostaglandin E2 (PGE2), indoleamine-2,3-dioxygenase (IDO) as well as some chemokines (CCL2/MCP-1, CXCL12/Groα) and other biologically active factors such as HLA-G5.[12] Generally, it is accepted that MSCs isolated from different sources (bone marrow, adipose tissue, etc.) secrete similar cytokines and most of the authors do not report any different profiles of secreted cytokines or find some quantitative differences, only.[13–16]

It is noteworthy that MSCs secrete cytokines either "spontaneously" or after "activation" by other cytokines and the most important of the inducers are IFNγ, TNFα, and IL-1β.[17–19] Activation of MSCs via their Toll-like receptors (TLRs) is another possible way to modulate the secretory activity of MSCs. The data on this mechanism are rather divergent but still attempts are published to divide the MSCs into MSC1 and MSC2 in dependence of the stimulation of their TLRs and the subsequent cytokine secretion.[20]

The basic idea for the "activated" MSCs is that these cells function in "inflammatory milleu" because of the local presence of IFNγ, IL-1β, TNFα, and activation of certain TLRs. All these factors form and stimulate the immunosuppressive potential of the MSCs present. Still, it is not clear whether the "inflammatory milleu" is absolutely necessary for the MSCs to exert their immunosuppressive activity because there are quite many papers reporting that the MSCs can have immunosuppressive effect without any activation.

The present chapter is focused on the effects of some biologically active factors secreted by the MSCs on the activities of T lymphocytes and monocyte derived dendritic cells (mDCs). It is known that the MSCs have immunomodulatory effects on all cells of the immune system, but T cells and mDCs are selected for detailed discussion because these cells under the influence of the MSCs secreted cytokines can exert immunosuppressive effects themselves. In other words, the discussion will be focused on the potential of the MSCs by their cytokine secretion to induce suppressor cells.

IMMUNOSUPPRESSIVE EFFECTS OF MSCs ON T LYMPHOCYTES

MSCs Suppress the Activation and Cytokine Secretion of T Cells

Initially the immunosuppressive mechanisms of MSCs have been described in details in relation to their effects on the activities of T lymphocytes. Much data piled up from the beginning of this century show without any doubt that MSCs suppress the activation and the proliferation of the T cells.[21,22] Parallel to that T helper lymphocytes change their phenotype and cytokine secretion under the influence of MSCs. The capacity of MSCs to suppress the T cell proliferation has been proven both in experimental animal models and in many in vitro experiments. For an example, it has been reported in an experimental mice model that injection of allogeneic MSCs leads to the survival of skin graft for more than 7 days and a suppressed T lymphocyte proliferation and lower levels of IL-2 and IFNγ were found in the graft. Studies on the effects of the MSCs on the T lymphocytes both in vitro and in vivo clearly show that the process of inhibition of the T cells by the MSCs is exerted in antigen independent way. The molecular mechanisms of suppression are not dependent on the major histocompatability complex and can be realized of auto-, allo-, and even xenogeneic MSCs and also, there is no dependence on the factor which has induced the T cell proliferation.[21,23] The effects of MSCs affect the cytokine secretion of the T cells as the most pronounced effect has been demonstrated with the Th1 subpopulation of the T helper cells. All in vitro results published unanimously demonstrate that the MSCs cause a substantial suppression of the secretion of the major Th1 determining cytokines such as TNFα[24–28] and IFNγ.[26–33] An inhibited secretion of IFNγ by Th1 cells under the influence of MSCs has been described in experimental animal models as for example, in experimental autoimmune colitis the application of MSCs causes a suppression of the disease activity accompanied by lower secretion of IFNγ by the Th1 lymphocytes.[34] Similar effects have been observed and recorded in streptozotocin induced diabetes in rats and in NOD mice.[35] Histological studies of biopsies from skin grafts rejected after allogeneic transplantation in the presence of MSCs (longer survival time) find lower numbers of Th1 cells and lower expression of IFNγ and IL-2.[32]

Studies on the effect of MSCs on the secretion of Th2 cytokines are not so numerous and the results are somewhat divergent. Some authors report an increased numbers of Th2 cells and corresponding increased secretion of IL-4 in the presence of MSCs.[36–38] It was reported that in NOD mice application of allogeneic MSCs caused a switch to Th2 immune response.[39] In other experiments recovery after limb paralysis in experimental autoimmune encephalomyelitis (EAE) model was described after treatment with MSCs accompanied by increased induction of Th2 cells secreting IL-4 and IL-5.[40] On the other hand, in mice model of OVA-induced equivalent of human bronchial asthma, treatment with MSCs leads to differentiation of T helper cells in Th1 direction and lower secretion of IL-4 and IL-13.[41,42] These results are substantiated by the results found after MSCs treatment of patients with chronic form of graft-versus-host disease (GVHD) as lower numbers of Th2 secreting lymphocytes have been recorded.[43]

A suppression of the proinflammatory Th17 lymphocytes and inhibited secretion of their products IL-17A, IL-17F, IL-21 have been also found after treatment with MSCs.[25,32,44–46]

The suppression caused by the MSCs influence also cytotoxic T cell populations.[22] Multiple data demonstrate such effect of MSCs (Krampera et al., 2005),[27,47] and this finding is valid for both naïve and memory Tc lymphocytes.[48] Even more, some authors claim that the suppression of TNFα, IFNγ, and IL-2 secretion caused by the MSCs is better expressed with Tc cells in comparison to the T helper populations.[46] The effect of the MSCs is demonstrated by the inhibition of the cytotoxic T lymphocytes formation in the mixed lymphocyte culture (MLC). A specific feature of this model is that the MSCs are effective just in case they are added to the culture during the stimulation phase while their addition during the effector phase has no effect.[49]

Secretory Factors Mediating Immune Suppression on T Cells

The majority of the authors consider the secretory mechanism as the basic one in the process of the MSCs immune suppression on T cells activity[25,27,28,36] as the major evidence for this belief are the results from experiments carried out in transwell system when suppression of T cell activity is documented without any cell-to-cell contact.[47,48] Testing the suppressive effects of MSCs conditioned medium for their effects on T cells support these notion as well.[37,48,50] Convincing support of the idea that secreted products of MSCs are the major agents in suppressing the T cell activity is rendered by the experiments using MSCs treated with EDCI (an agent blocking the cytokine secretion while preserving the cell surface molecules). Thus treated MSCs do not have suppressive effect on T lymphocytes in a mouse model.[50]

More than 30 MSCs secreted products are discussed and blamed for the suppressive effect on the T lymphocytes and among them are IDO, PGE2, TGFβ, CCL2/MCP-1, hepatocyte growth factor (HGF), IL-10 galectins 1 and 3, HLA-G5, etc.[23] However, most numerous and convincing results have been published from studies on four factors secreted by MSCs which exert suppressive effect on T cells and they are HLA-G5, PGE2, IDO, and CCL/MCP-1.

Human Leukocyte Antigen G5 (HLA-G5)

Human leukocyte antigen G5 (HLA-G5) is the secreted form of the nonclassic class one HLA-G antigen. In reference to its effect on the T cells, it is currently accepted that the effect is not a direct one but depends on other factors such as IDO,[51] direct cell-to-cell contact and presence of IL-10.[52] The HLA-G5 secreted by the MSCs inhibits the

T cell proliferation after binding to its specific receptor immunoglobulin-like transcript 2 (ILT-2) on the T cell surface. It was established that neutralizing HLA-G5 causes at least partially restoration of the inhibited T cell proliferation capacity.[52]

Prostaglandin E2 (PGE2)

PGE2 is considered to be one of the basic factors involved in the MSCs suppression on the T lymphocytes activity.[38] It role is determined by the inhibition of mitogenesis and the IL-2 secretion.[53] When MSCs are cocultured with T lymphocytes an increased level of PGE2 is found in the culture supernatant and the subsequent inhibition of PGE2 by addition of indomethacin and E prostanoid 2 (EP2) antagonist leads to partial restoration of the proliferative capacities of the suppressed T lymphocytes. The same effect is observed in in vivo experiments on induced GVHD model in mice.[27,28,36] Contrary to that the addition of EP2 agonist increases the immunosuppressive effect of MSCs on the T cells.[28] Some authors report that the effect of the indomethacin is manifested on activated T lymphocytes in MLC and cocultured with MSCs but such an effect is not observed with phytohaemagglutinin (PHA) activated T cells cocultured with MSCs.[48] There are other authors that directly reject the effect of indomethacin for the restoration of the pre-suppressed T cells.[51]

Indoleamine 2,3-dioxygenase (IDO)

Indoleamine 2,3-dioxygenase (IDO) is seemingly the most convincing candidate for a crucial role in MSCs suppression of T cell proliferation. It is cytoplasmic enzyme which basic function is in the kinurenine pathway of tryptophan metabolism and under its action kinurenine is synthesized instead of tryptophan. MSCs do not synthesize IDO when are in steady state but after "activation" under the influence of IFNγ mostly, the cells start to produce IDO and it affects the surrounding cells. As far as the T lymphocytes are concerned one of the effects of IDO is related to the suppression of their proliferation by blocking the translation of many transcribed already factors.[51,54] According to some data blocking of IDO activity completely abolish the inhibiting effect of MSCs on the T cell proliferation.[51] When various doses of tryptophan were added in these experiments it has been found blocking of the MSCs immunosuppressive action although kinurenine was present in the cultures. The suppressing effect of the MSCs is restored by adding kunirenine in higher doses. However this effect is observed with MSCs from some but not from all donors which made the authors speculate that the sensibility to kunerenin is donor specific effect.[51]

The Chemokine (C-C Motif) Ligand 2 (CCL2/MCP-1)

The chemokine (C-C motif) ligand 2 (CCL2/*MCP-1*) generally has a role in apoptosis induced in T cells[55] and also it is engaged in MSCs induced in vivo suppression of Th17 cells by inhibition of STAT-3 phosphorylation. In mouse model of EAE CCL2 produced by MSCs causes limitation of the clinical symptoms while the injection of MSCs isolated from CCL2 knock-out mice in EAE mice does not lead to therapeutic effect.[56]

Processes in T Lymphocytes Induced by MSCs

The suppressed proliferation and altered cytokine secretion of T lymphocytes induced by MSCs secreted factors can be the expression of three basic immunological phenomena: Apoptosis, anergy, and suppression.

Apoptosis

In general, apoptosis and the resulting deletion of selected T cell clones is rejected as a process in T cells caused by the MSCs,[24,57] as evidence for that statement are based on the facts that T lymphocytes have the capacity to restore their proliferative potential after discarding the MSCs or discarding some of the MSCs factors leading to T cell suppression. Still, there are a few papers published supporting the possibility MSCs to induce apoptosis in T cells by various mechanisms including the role of CCL2 in this process as discussed above.[51,55]

Anergy

Induction of T cell anergy seems to be better proven mechanism for the effect of MSCs by secreted factors[57] although the role of this process is not accepted by some authors.[24] Mechanism of T cell anergy is thought to be exerted by initial direct intercellular contact followed by secretion of IDO from MSCs and this increases the synthesis of $p27^{Kip1}$ in T cells which inhibits the expression of cyclin D and as a result the T cells are arrested in Go/G1 phase of the cell cycle.[31]

Generation of Suppressive Cells (Suppression)

The third molecular mechanism involved in T lymphocyte suppression intensively discussed is generation of suppressive cells.[24,48] Generation of Tregs (CD4+ FoxP3+) induced by MSCs secreted factors is the major player in this process as documented in a number of published studies.[52,58–64] A general finding leading to a consensus is that MSCs increase both the numbers and the activity of Tregs.[65,66] It has been established that Tregs are induced in cocultures of T helpers and allogeneic MSCs and after separation of the both cell population the Tregs induced demonstrate their immunosuppressive activities.[67] Also, Tregs secreting IL-10 are formed when T lymphocytes isolated from patients with rheumatoid arthritis are cocultured with MSCs and after the isolation of the Tregs induced it is shown that these cells suppress the proliferation and secretion of IFNγ by collagen stimulated T lymphocytes.[68] Induction of Tregs by MSCs has been proven in animal models and under the influence of MSCs an infiltration of Tregs is detected in the skin graft[32] and also an infiltration of Tregs is seen in the skin and muscles of transplanted limbs after treatments with MSCs.[50] Similarly, in other models in vivo allo-transplantation of heart,[69] liver,[68] or kidney[70] in combination with MSCs application leads to induction of Tregs. It is accepted that MSCs represent a "homeostatic niches" for Tregs as they attract these cells and regulate their activity.[71] Major factors commented in the literature about de novo generation of Tregs induced by MSCs are TGFβ, HLA-G5, PGE2, IL-10, IDO.[62]

TGFβ

Stimulation of T cellular receptor (TCR) in the presence of TGFβ leads to differentiation of "naïve" CD4+ T lymphocytes as Tregs expressing FoxP3[72] as well as TGFβ and IL-2 are engaged in stimulation of the proliferation of already locally present Tregs.[73] It has been shown that blocking the TGFβ in experimental animals hampers the generation of Tregs.[74] TGFβ acts through a molecular mechanism which includes interactions between the transcription factors STAT-3 and the nuclear factor of activated T cells (NFAT)[75] which binds to CNS1 enhancer in FoxP3 locus and induces its transcription.[76] TGFβ also limits the activation of DNA methyltransferase I in the FoxP3 locus and this is an enzyme known to inhibit the expression of FoxP3 after TCR stimulation.[77] An important point for the role of TGFβ in generation and sustenance of the Tregs is its participation together with IL-6 in the regulation and control of correlations between Tregs and Th17 cell populations.[78] In studies of naïve T cells isolated from umbilical cord blood and cultured under Th17 inducing conditions, it has been established that the addition of MSCs suppress the secretion of IL-17 and induces FoxP3 expression via TGFβ, and the same effect has been documented in experiments with differentiated Th17 cells. The major findings from these studies have been substantiated by the results from in vivo experiments which show that MSCs via TGFβ regulate the balance Th17/Tregs populations. This leads to suppression of the EAE progress and the process is accompanied by suppression of IL-17 production in the central nervous system.[79] Inhibition of the Th17 generation and the switch to Tregs has been demonstrated also in animal models of diabetes type I,[80] collagen induced arthritis[81] and experimentally induced myasthenia gravis.[82]

HLA-G5

The soluble form 5 of HLA-G secreted by the MSCs is considered to be one of the factors leading to generation of Tregs. It is supposed that a pre-requisition for the HLA-G5 secretion is a direct cell-to-cell contact between the MSCs and T cells which should differentiate into Tregs.[52,59,60,63] HLA-G5 is a key player in a complicated cascade of cellular and molecular interactions. Intracellular synthesis and secretion of HLA-G5 are potentiated by the effect of IL-10. Initially the hemoxigenase-1 produced by the MSCs leads to generation of regulatory Tr1 cells which through IL-10 secretion potentiate the HLA-G5 secretion by the MSCs and the latter is consequently involved in Tregs generation by binding the ILT-2 receptors of the lymphocytes.[52,83]

PGE2

The role of the PGE2 is related with the modulation of the FoxP3 protein which a marker of the Tregs.[53] It is supposed that PGE2 has a basic role in the transformation of Th17 into Tregs by an increased cycloxigenase-2 activity and secretion of adenosine.[21,23,52]

IL-10

As discussed later the ability of MSCs to secrete IL-10 is very disputable but their capacity to induce generation of IL-10 secreting cells is a key element in the MSCs immunosuppressive effect.[64] Generation of Tregs induced by the MSCs is associated with "cell mediators" secreting IL-10 such as Tr1, macrophages, tolerogenic dendritic cells.[23,52,83] Very similar to this process has been known earlier as "infectious tolerance."[84]

IDO

IDO and I-309 have been described as factors capable to induce generation and activation of Tregs and their secretion by MSCs is possible solely after stimulation of MSCs by proinflammatory cytokines (inflammatory milleu).[19,64]

IL-6

IL-6 secreted by the MSCs has been suspected as a factor inducing generation of Tregs[15] but still there is not clear cut evidence for such a role. It has been shown undoubtedly that IL-6 leads to generation of a specific CD8+FoxP3+ cell population.[85] There are some discussions whether their suppressive activity is equivalent to that of the classic

Tregs. According to some studies the CD8+FoxP3+ cells just partially suppress the proliferation of T cells and secretion of IFNγ, while other authors accept that the CD8+FoxP3+ cells are effective suppressors generated under the effect of IL-6 which controls their formation and functions.[85] Parallel to this IL-6 has a major role in synthesis of PGE2 and thus influences indirectly the generation of Tregs.[21,23,52]

In regard to cytokine secretion by Tregs generated under the instructions of MSCs produced factors the results published in the literature definitely show that Tregs are able to secrete IL-10[15,61,63] while the data about TGFβ are rather contradictory.[12,61,63]

EFFECTS OF MSCs ON THE MONOCYTE-DERIVED DENDRITIC CELLS

MSCs Induce Generation of Tolerogenic Dendritic Cells

Partially differentiated and immature mDCs have a major role in induction of immune tolerance. These cells are characterized by high capacity for phagocytosis and low expression of costimulating molecules which makes it possible to engulf and present antigens without the process of cosimulation, and this causes induction of anergy in the T cells communicating with the mDCs.[86–89]

Suppression of the differentiation/maturation of dendritic cells under the influence of the MSCs has been described in mice[90] and rat models[91] and in human cells.[92–96] Markers of this process are the preserved expression of monocyte marker CD14[93–96] and suppression of the expression CD83 marker which is typical for mature DCs.[92,94,96] Also, there is a suppressed expression of basic costimulatory molecules such as B7 complex,[90–94,96] MHC-II,[90–92,94,96] CD1a,[94,96] and CD40.[90,92] Parallel to these changes, functional changes have been described such as suppressed secretion of IL-12, an increased secretion of IL-10[91–93] and decreased ability of mDCs to activate T lymphocytes.[90,92,95] The MSCs are able to inhibit the secretion of CCL3 and CCL4 as well which are key chemokines for the dendritic cells[93] and thus suppress their potential to attract T lymphocytes, NK cells, eosinophils, and macrophages.[97,98] As a result of these changes induced by the influence of the MSCs are formed dendritic cells with tolerogenic phenotype which are actively engaged with the process of immune suppression due to the lack of costimulatory molecules and key chemokines. The secretion of IL-10 is a typical feature for these cells as well.

The assumption that MSCs modulate all these processes via secreted factors is substantiated by a number of published facts. Many research groups have studied the MSCs effects on mDCs in experiments using transwell systems[91,94,96] and have established that both cocultures and transwell systems induce similar quantitative parameters showing suppressed differentiation/maturation of mDCs.[95]

Factors Modulating the Generation of Tolerogenic mDCS

IL-10

The question whether MSCs secrete IL-10 has not a one direction answer because there are very contradictory opinions have been published. An overview of the published papers reveals that almost half of the authors detect IL-10 in the culture supernatants of MSCs,[18,53,60,99,100] while others do not find Il-10 in such supernatants[15,16,19,101–105] although using similar laboratory methods. In spite of this an undisputable fact is that the MSCs induce secretion of IL-10 by other cell populations as the most important of them are Tregs and tolerogenic dendritic cells. The paracrine (secretion by MSCs) or autocrine (induced by the MSCs but secreted by the mDCs themselves) effects of the IL-10 on the immature DC cause formation of cells with tolerogenic phenotype and the mDCs do not secrete proinflammatory cytokines such as IL-12, TNFα, and IL-1β,[106] do not up-regulate the expression of costimulatory molecules such as CD80, CD86, CD40[107] and do not stimulate T cell proliferation.[86,108]

The basic molecular mechanisms of IL-10 induces suppression that are discussed most actively include the activation of phosphoinositol-3 kinase (PI3K) and STAT-3 which block the activation of the NFκB which is the key element regulating the DCs development, maturation, cytokine secretion and expression of costimulating molecules. A supplementary mechanism is the expression of the membrane ILT-3 and ILT-4 by the mDCs which interact with the T cells and induce differentiation as Tregs as these cells secrete IL-10 by themselves (Wallet et al., 2005). The described molecular mechanism is a classic example of the earlier termed "infectious tolerance" as mentioned above.[84] Thus, in a cascade one immunoregulatory cell (MSC) induces generation of another immunoregulatory cell (DC) which on its hand induces the formation of a third immunoregulatory cell (Treg). It was mentioned earlier that the MSCs can directly induce the generation of Tregs as well. In that manner the MSCs independently to its secretion of IL-10 initiate interaction between mDCs and Tregs mutually sustaining their immunosuppressive activities via paracrine and autocrine secretion of IL-10. According to the model proposed by[91] and based on the idea that the MSCs secrete IL-10, the interactions between MSCs and mDCs strongly resemble the interactions between Tregs and mDCs in regard to the positive feedback effect of the IL-10. However, it should be mentioned that there are alternative concepts supporting the idea that other cytokines secreted by the MSCs lead to secretion of IL-10 by the dendritic cells.

IL-6

Contrary to the IL-10 the secretion of IL-6 by MSCs raises no doubts and discussions,[15,100,101,104,105] and it is a major representative of the cytokines inducing IL-10 secretion by DCs which directs their differentiation as tolerogenic in an autocrine manner.[60,109] Together with its indirect effect IL-6 has a direct effect on the generation of tolerogenic dendritic cells because under its influence the signals for mDC maturation are blocked and the expression of CD80, CD86, CD40, and MHC II is suppressed which strongly reduces the mDCs capacity to activate the T lymphocytes.[110] This activity of the IL-6 is dose-dependent.[90] It has been established that the purposeful activation of the genes associated with the IL-6 production leads to generation of tolerogenic DCs, while in knockout IL-6 mice, a significantly higher number of mature immunogenic DCs are found.[111] Results from other experiments with application of monoclonal antibodies against IL-6 show that the dendritic cells have lost expression of CD14, while contrary to that, after the application of MSCs, the secreted IL-6 leads to preserved expression of CD14 as a sign of immaturity and respectively tolerogenicity of the DCs.[90,94,95] It is known that the IL-6 binds to gp130 on surface of the dendritic cells or their precursors and as a consequence gp130 interacts with the JAK-STAT-3.[112] Using a mouse model with knockout IL-6 and gp130 mice it has been shown that following the pathway mentioned above the IL-6 is directly engaged in suppression of the differentiation of DCs expressed in inhibition of the expression of MHC-II, CD80/CD86, CD40, and IL-12 secretion.[113] Data have been published that IL-6 acts together with other biologically active factors such as PGE2 which have an additive effect and potentiate its effect.[90] Similarly, the HLA-G5 secreted by the MSCs binds to the ILT-4 receptor expressed on the surface of the dendritic cells and the binding results in activation of two tyrosine phosphatases—SHP-1 and SHP-2 which induces production of IL-6. The newly synthesized IL-6 on its hand acts in an autocrine manner on the dendritic cells.[111]

TGFβ

TGFβ is another cytokine secreted by the MSCs which causes inhibition of the differentiation/maturation of the DCs. Its action on the DCs leads to blocking of the expression of CD80 and CD83 which attenuates negatively the DCs capacity to stimulate T lymphocytes.[114] TGFβ causes a down regulation of the expression of TLR-4 which results in a lower susceptibility of the DCs to factors inducing maturation and particularly LPS.[115] Furthermore TGFβ diminishes the capacity of the DCs to secrete

IL-12,[116] and inhibits the expression of CCR7 on the DCs surface, and this has a negative impact on the DCs traffic in the lymph nodes.[117] The precise mechanism of the action of TGFβ in the direction of tolerogenic dendritic cells has not been established yet and there are some convincing data that a possible target of the TGFβ is the activation of the transcription factor known as RunX3.[118]

PGE2

PGE2 stimulates the formation of DCs with tolerogenic phenotype most commonly in chronic inflammation and in combination with TNFα, as the resulting DCs secrete IL-10.[119,120] PGE2 is also associated with the suppressed expression of CCL3 and CCL4 by the dendritic cells.[97]

Along with the cytokines discussed above data have been publish about other MSCs secreted factors which cause an inhibited differentiation of the DCs and an increased DCs secretion of IL-10 such as IDO[91] and VEGF.[121–123] The chemokine CCL2/MCP-1 which stimulates the secretion of IL-10 by the tolerogenic DCs is still another player studies in attempts to clarify the mechanism of the immunodulating effect of MSCs on the dendritic cells.[124] Some recent data demonstrate that under the influence of MSCs secreted factors the progenitors of the dendritic cells alter the direction of their differentiation from DCs to myeloid derived suppressor cells secreting IL-10. These authors suppose that the reasons for this effect are the Gro (growth regulated oncogene) chemokines which are members of the IL-8 family and some of them are secreted by the MSCs.[15]

In conclusion by the secretion of immunomodulatory cytokines the MSCs influence multiple cells of the immune system predominantly inducing their suppression. Among the targets of influence the T lymphocytes and the dendritic cells are of particular significance because they are transformed into cells with immunosuppressive properties such as Tregs and tolerogenic dendritic cells. These immunosuppressive cells sustain their properties by mutual influence predominantly by secreting IL-10 and thus the existence is constantly perpetuated of immunosuppressive cell populations which have their impact on other cells of the immune system. Cytokines secreted by the MSCs "create" immunosuppressors, and this is an indirect mechanism of the inhibitory action of the MSCs on the immune system.

References

1. Pittenger M, Mackay A, Beck S, Jaiswal R, Douglas R, Mosca J, et al. Multilineage potential of adult human mesenchymal stem cells. *Science* 1999;**284**(5411):143–7.

2. Blashki D, Short B, Bertoncello I, Simmons PJ, Brouard N. Identification of stromal MSC candidates from multiple adult mouse tissues. In: *International society for stem cell research 4th annual meeting*; 2006. p. 206.
3. Shi S, Gronthos S. Perivascular niche of postnatal mesenchymal stem cells in human bone marrow and dental pulp. *J Bone Miner Res* 2003;**18**:696–704.
4. Tuli R, Tuli S, Nandi S, Huang X, Manner P, Hozack W, et al. Transforming growth factor-β-mediated chondrogenesis of human mesenchymal progenitor cells involves N-cadherin and mitogen-activated protein kinase and Wnt signaling cross-talk. *J Biol Chem* 2003;**278**:41227–36.
5. Salem H, Thiemermann C. Mesenchymal stem cells: current understanding and clinical status. *Stem Cells* 2010;**28**:585–96.
6. Honczarenko M, Le Y, Swierkowski M, Ghiran I, Glodek A, Silberstein L. Human bone marrow stromal cells express a distinct set of biologically functional chemokine receptors. *Stem Cells* 2006;**24**:1030–41.
7. Chamberlain G, Fox J, Ashton B, Meddleton J. Concise review: mesenchymal stem cells: their phenotype, differentiation capacity, immunological features, and potential of homing. *Stem Cells* 2007;**25**:2739–49.
8. Bianco P, Robey P, Simmons P. Mesenchymal stem cells: revisiting history, concepts, and assays. *Cell Stem Cells* 2008;**2**:313–19.
9. Dominici M, Le Blanc K, Mueller I, Slaper-Cortenbach I, Martin F, Krause D, et al. Minimal criteria for defining multipotent mesenchymal stromal cells. The International Society for Cellular Therapy position statement. *Cytotherapy* 2006;**8**:315–17.
10. Zipori D. Mesenchymal stem cells: harnessing cell plasticity to tissue and organ repair. *Blood Cells Mol Dis* 2004;**33**:211–15.
11. Roufosse C, Direkze N, Otto W, Wright N. Circulating mesenchymal stem cells. *Int J Biochem Cell Biol* 2004;**36**:585–97.
12. Kyurkchiev D, Ivanova-Todorova E, Bochev I, Mourdjeva M, Kyurkchiev S. Differences between adipose tissue derived mesenchymal stem cells and bone marrow-derived mesenchymal stem cells as regulators of the immune response. In: Hayat MA, editor. *Stem cells and cancer stem cells*, vol. 10. Netherlands: Springer; 2013. p. 71–84.
13. Chan CK, Wu KH, Lee YS, Hwang SM, Lee MS, Liao SK, et al. The comparison of interleukin 6-associated immunosuppressive effects of human ESCs, fetal-type MSCs, and adult-type MSCs. *Transplantation* 2012;**94**:132–8.
14. Elman JS, Li M, Wang F, Gimble JM, Parekkadan B. A comparison of adipose and bone marrow-derived mesenchymal stromal cell secreted factors in the treatment of systemic inflammation. *J Inflamm (Lond)* 2014;**11**:1.
15. Ivanova-Todorova E, Bochev I, Dimitrov R, Belemezova K, Mourdjeva M, Kyurkchiev S, et al. Conditioned medium from adipose tissue derived mesenchymal stem cells induces CD4+ FOXP3+ cells and increases IL-10 secretion. *J Biomed Biotechnol* 2012; **2012**:295167.
16. Park CW, Kim KS, Bae S, Son HK, Myung PK, Hong HJ, et al. Cytokine secretion profiling of human mesenchymal stem cells by antibody array. *Int J Stem Cells* 2009; **2**:59–68.
17. Dazzi F, Krampera M. Mesenchymal stem cells and autoimmune diseases. *Best Pract Res Clin Haematol* 2011;**24**:49–57.
18. Bernardo ME, Fibbe WE. Mesenchymal stromal cells: sensors and switchers of inflammation. *Cell Stem Cell* 2013;**13**:392–402.
19. Newman RE, Yoo D, LeRoux MA, Danilkovitch-Miagkova A. Treatment of inflammatory diseases with mesenchymal stem cells. *Inflamm Allergy Drug Targets* 2009; **8**:110–23.

20. Waterman R, Tomchuck S, Henkle S, Betancourt A. A new mesenchymal stem cell (MSC) paradigm: polarization into a pro-inflmmatory MSC1 or an immunosuppressive MSC2 phenotype. *PLoS One* 2010;**5**(4) [Article IDe10088].
21. Duffy M, Ritter T, Ceredig R, Griffin M. Mesenchymal stem cell effects on T-cell effector pathways. *J Stem Cell Res Ther* 2011;**2**:34–43.
22. Rasmusson I, Le Blanc K, Sundberg B, Ringden O. Mesenchymal stem cells stimulate antibody secretion in human Bcells. *Scand J Immunol* 2007;**65**:336–43.
23. Haddad R, Saldanha-Araujo F. Mechanisms of T-cell immunosuppression by mesenchymal stromal cells: what do we know so far? *BioMed Res Int* 2014 [Article ID 216806].
24. Tobin L, Healy M, English K, Mahon B. Human mesenchymal stem cells suppress donor CD4+ T cell proliferation and reduce pathology in a humanized mouse model of acute graft-versus-host disease. *Clin Exp Immunol* 2012;**172**:333–48.
25. Yanez R, Lamana M, Garcia-Castro J, Colmenero I, Ramire M, Bueren J. Adipose tissue-derived mesenchymal stem cells have in vivo immunosuppressive properties applicable for the control of the graft-versus-host disease. *Stem Cells* 2006; **24**:2582–91.
26. Zappia E, Casazza S, Pedemonte E, Benvenuto F, Bonanni I, Gerdoni E, et al. Mesenchymal stem cells ameliorate experimental autoimmune encephalomyelitis inducing T-cell anergy. *Blood* 2005;**106**:1755–61.
27. DelaRosa O, Lombardo E, Beraza A, Mancheno-Corvo P, Ramirez C, Menta R, et al. Requirement of IFN-γ-mediated indoleamine 2,3-dioxygenase expression in the modulation of lymphocyte proliferation by human adipose-derived stem cells. *Tissue Eng* 2009;**15**(10):2795–806.
28. Auletta J, Eid S, Wuttisarnwattana P, Silva I, Metheny L, Keller M, et al. Human mesenchymal stromal cells attenuate graft-versus-host disease and maintain graft-versus-leukemia activity following experimental allogenic bone marrow transplantation. *Stem Cells* 2014;**33**(2):601–14.
29. Sioud M, Mobergslien A, Boudabous A, Floisand Y. Mesenchymal stem cell-mediated T cell suppression occur through secreted galectins. *Int J Oncol* 2011;**38**:385–90.
30. Bochev I, Elmadjian G, Kyurkchiev D, Tzvetanov L, Altanokova I, Tivchev P, et al. Mesenchymal stem cells from bone marrow or adipose tissue differently modulate mitogen-stimulated B cells immunoglobulin production in vitro. *Cell Biol Int* 2008;**32**:384–93.
31. Glennie S, Soeiro I, Dyson P, Lam E, Dazzi F. Bone marrow mesenchymal stem cells induce division arrest anergy of activated T cells. *Blood* 2005;**105**:2821–7.
32. Larocca R, Moraes-Vieira P, Bassi E, Semedo P, de Almeida D, da Silva M, et al. Adipose tissue-derived mesenchymal stem cells increase skin allograft survival and inhibit Th17 immune response. *PLoS One* 2013;**8**(10):e76396.
33. vanRhijn M, Khairoun M, Korevaar S, Lievers E, Leuning D, Baan C, et al. Human bone marrow and adipose tissue derived mesenchymal stromal cells are immunosuppressive *in vitro* and in humanized allograft rejection model. *J Stem Cell Res Ther* 2013;**6**(1):20780.
34. Gonzalez M, Gonzalez-Rey E, Rico L, Buscher D, Delgado M. Adipose derived mesenchymal stem cells alleviate experimental colitis by inhibiting inflammatory and autoimmune responses. *Gastroenterology* 2009;**136**:978–89.
35. Boumaza I, Srinivasan S, Witt W, Feghali-Bostwick C, Dai Y, Garcia-Ocana A, et al. Autologous bone marrow-derived rat mesenchymal stem cells promote PDX-1 and insulin expression in the islets, alter T cell cytokine pattern and preserve regulatory T cells in the periphery and induce sustained normoglycemia. *J Autoimmun* 2009;**32**:33–42.
36. Aggarwal S, Pittenger M. Human mesenchymal stem cells modulate allogeneic immune cell responses. *Blood* 2005;**105**:1815–22.

37. Groh ME, Maitra B, Szekely E, Koç ON. Human mesenchymal stem cells require monocyte-mediated activation to suppress alloreactive T cells. *Exp Hematol* 2005;**33**:928–34.
38. Peng W, Gao T, Yang Z, Zhang S, Ren M, Wang Z, et al. Adipose derived stem cells induced dendritic cells undergo tolerance and inhibit Th1 polarization. *Cell Immunol* 2012;**278**:152–7.
39. Fiorina P, Jurewicz M, Augello A, Vergani A, Dada S, La Rosa S, et al. Immunomodulatory function of bone marrow-derived mesenchymal stem cells in experimental autoimmune type 1 diabetes. *J Immunol* 2009;**183**:993–1004.
40. Bai L, Lennon D, Eaton V, Maier K, Caplan AI, Miller S, et al. Human bone marrow-derived mesenchymal stem cells induce Th2-polarized immune response and promote endogenous repair in animal models of multiple sclerosis. *Glia* 2009;**57**:1192–203.
41. Goodwin M, Sueblinvong V, Eisenhauer P, Ziats N, LeClair L, Poynter M, et al. Bone marrow-derived mesenchymal stromal cells inhibit Th2-mediated allergic airways inflammation in mice. *Stem Cells* 2011;**29**:1137–48.
42. Kavanagh H, Mahon B. Allogeneic mesenchymal stem cells prevent allergic airway inflammation by inducing murine regulatory T cells. *Allergy* 2011;**66**:523–31.
43. Zhou H, Guo M, Bian C, Sun Z, Yang Z, Zeng Y, et al. Efficacy of bone marrow-derived mesenchymal stem cells in the treatment of sclerodermatous chronic graft-versus-host disease: clinical report. *Biol Blood Marrow Transplant* 2010;**16**:403–12.
44. Darlington P, Boivin M, Renoux C, Francois M, Galipeau J, Freedman M, et al. Reciprocal Th1 and Th17 regulation by mesenchymal stem cells: implication for multiple sclerosis. *Ann Neurol* 2010;**68**:540–5.
45. Carrion F, Nova E, Luz P, Apablaza F, Figueroa F. Opposing effect of mesenchymal stem cells on Th1 and Th17 cell polarization according to the state of CD4(+) T cell activation. *Immunol Lett* 2011;**135**:10–16.
46. Laranjeira P, Pedrosa M, Pedeiro S, Gomes J, Martinho A, Antunes B, et al. Effect of human bone marrow mesemchymal stromal cells on cytokine production by peripheral blood naïve, memory and effector T cells. *Stem Cell Res Ther* 2015;6 [in press].
47. Di Nicola M, Carlo-Stella C, Magni M, Milanesi M, Longoni PD, Matteucci P, et al. Human bone marrow stromal cells suppress T-lymphocyte proliferation induced by cellular or nonspecific mitogenic stimuli. *Blood* 2002;**99**:3838–43.
48. Rasmusson I, Ringden O, Sundberg B, Le Blanc K. Mesenchymal stem cells inhibit lymphocyte proliferation by mitogens and alloantigens by different mechanisms. *Exp Cell Res* 2005;**305**:33–41.
49. Rasmusson I, Ringden O, Sundberg B, Le Blanc K. Mesenchymal stem cells inhibit the formation of cytotoxic T lymphocytes, but not activated cytotoxic T lymphocytes or natural killer cells. *Transplantation* 2003;**76**:1208–13.
50. Jeong S, Ji Y, Yoon E. Immunosuppressive activity of adipose tissue derived mesenchymal stem cells in a rat model of hind limb allotransplantant. *Transplant Proc* 2014;**46**:1606–14.
51. Menta R, Mancheno-Corvo P, Del Rio B, Ramirez C, Delarosa O, Dalemans W, et al. Tryptophan concentration is the main mediator of the capacity of adipose mesenchymal stromal cells to inhibit T-lymphocyte proliferation *in vitro*. *Cytotherapy* 2014;**16** (12):1679–91.
52. Selmani Z, Naji A, Zidi I, Favier I, Gaiffe E, Obert L, et al. Human leucocyte antigen-G5 secretion by human mesenchymal stem cells is required to suppress T lymphocyte and natural killer function and to induce CD4+CD25highFoxP3+ regulatory T cells. *Stem Cells* 2008;**26**:212–22.
53. Engela A, Baan C, Dor F, Weimar W, Hoogduijn M. On the interaction between mesenchymal stem cells and regulatory T cells for immunomodulation in transplantation. *Front Immunol* 2012;3 [article 126].

54. Krampera M, Cosmi L, Angeli R, Pasini A, Liotta F, Andreini A, et al. Role for interferon-gamma in the immunomodulatory activity of human bone marrow mesenchymal stem cells. *Stem Cells* 2006;**24**:386–98.
55. Akiyama K, Chen C, Wang D, Xu X, Qu C, Yamaza T, et al. Mesenchymal-stem-cell-induced immunoregulation involves FAS-ligand-/FAS-mediated T cell apoptosis. *Cell Stem Cell* 2012;**10**:544–55.
56. Rafei M, Hsieh J, Fortier S, Li M, Yuan S, Birman E, et al. Mesenchymal stromal cell-derived CCL2 suppresses plasma cell immunoglobulin production via STAT3 inactivation and PAX5 induction. *Blood* 2008;**112**:4991–8.
57. Krampera M, Glennie S, Dyson J, Scott D, Laylor R, Simpson E, et al. Bone marrow mesenchymal stem cells inhibit the response of naive and memory antigen-specific T cells to their cognate peptide. *Blood* 2003;**101**:3722–9.
58. Ye Z, Wang Y, Xie H, Zheng S. Immunossupressive effects of rat mesenchymal stem cells: involvement of CD4+ CD25+ regulatory T cells. *Hepatobiliary Pancreat Dis Int* 2008;**7**:608–14.
59. Bassi EJ, Aita CA, Câmara NO. Immune regulatory properties of multipotent mesenchymal stromal cells: where do we stand? *World J Stem Cells* 2011;**3**:1–8.
60. Ben-Ami E, Berrih-Aknin S, Miller A. Mesenchymal stem cells as an immunomodulatory therapeutic strategy for autoimmune diseases. *Autoimmun Rev* 2011;**10**:410–15.
61. Prevosto C, Zancolli M, Canevali P, Zocchi M, Poggi A. Generation of CD4+ or CD8+ regulatory T cells upon mesenchymal stem cell-lymphocyte interaction. *Hematol J* 2007;**92**(07):881–8.
62. Engela A, Hoogduijn M, Boer K, Litjens N, Betjes M, Weimar W, et al. Human adipose-tissue derived mesenchymal stem cells induce functional de-novo regulatory T cells with methylated FoxP3 gene DNA. *Clin Exp Immunol* 2013;**173**:343–54.
63. Luz-Crawford P, Kurte M, Bravo-Alegria J, Contreras R, Nova-Lamperti E, Tejedor G, et al. Mesenchymal stem cells generate a CD4+ CD25+ FoxP3+ regulatory T cell population during the differentiation process of Th1 and Th17. *Stem Cell Res Ther* 2013;**4**:1–12.
64. Eggenhofer E, Lik F, Dahlke M, Hoogduijn M. The life and fate of mesenchymal stem cells. *Front Immunol* 2014;5 [article 148].
65. English K, French A, Wood K. Mesenchymal stromal cells: facilitators of successful transplantation? *Cell Stem Cell* 2010;**7**:431–42.
66. Griffin M, Ritter T, Mahon B. Immunological aspects of allogeneic mesenchymal stem cell therapies. *Hum Gen Ther* 2010;**21**:1641–55.
67. English K, Ryan J, Tobin L, Murphy M, Barry F, Mahon B. Cell contact, prostaglandin E(2) and transforming growth factor beta 1 play nonredundant roles in human mesenchymal stem cell induction of CD4+ CD25(high) forkhead box P3+ regulatory T cells. *Clin Exp Immunol* 2009;**156**:149–60.
68. Wang Y, Zhang A, Ye Z, Xie H, Zheng S. Bone marrow-derived mesenchymal stem cells inhibit acute rejection of rat liver allografts in association with regulatory T-cell expansion. *Transplant Proc* 2009;**41**:4352–6.
69. Casiraghi F, Azzollini N, Cassis P, Imberti B, Morigi M, Cugini D, et al. Pretransplant infusion of mesenchymal stem cells prolongs the survival of a semiallogeneic heart transplant through the generation of regulatory T cells. *J Immunol* 2008;**181**:3933–46.
70. Ge W, Jiang J, Baroja M, Arp J, Zassoko R, Liu W, et al. Infusion of mesenchymal stem cells and rapamycin synergize to attenuate alloimmune responses and promote cardiac allograft tolerance. *Am J Transplant* 2009;**9**:1760–72.
71. Di Iannia M, Del Papaa B, De Ioannia M, Morettia L, Bonifacioa E, Cecchinia D, et al. Mesenchymal cells recruit and regulate T regulatory cells. *Exp Hematol* 2008;**36**(3):309–18.

72. Chen W, Jin W, Hardegen N, Lei K, Li L, Marinos N, et al. Conversion of peripheral CD4+ CD25− naive T cells to CD4+ CD25+ regulatory T cells by TGF-beta induction of transcription factor Foxp3. *J Exp Med* 2003;**198**:1875–86.
73. Ghiringhelli F, Puig P, Roux S, Parcellier A, Schmitt E, Solary E, et al. Tumor cells convert immature myeloid dendritic cells into TGF-β secreting cells inducing CD4+ CD25+ regulatory T cell proliferation. *J Exp Med* 2005;**202**:919–29.
74. Mucida D, Kutchukhidze N, Erazo A, Russo M, Lafaille J, de Lafaille M. Oral tolerance in the absence of naturally occurring Tregs. *J Clin Invest* 2005;**115**:1923–33.
75. Josefowicz S, Rudensky A. Control of regulatory T cell lineage commitment and maintenance. *Immunity* 2009;**30**:616–25.
76. Tone Y, Furuuchi K, Kojima Y, Tykocinski M, Greene M, Tone M. Smad3 and NFAT cooperate to induce Foxp3 expression through its enhancer. *Nat Immunol* 2008;**9**:194–202.
77. Josefowicz S, Wilson C, Rudensky A. Cutting edge: TCR stimulation is sufficient for induction of Foxp3 expression in the absence of DNA methyltransferase1. *J Immunol* 2009;**182**:6648–52.
78. Xu L, Kitani A, Fuss I, Strober W. Cutting edge: regulatory T cells induce CD4+ CD25-Foxp3- T cells or are self-induced to become Th17 cells in the absence of exogenous TGF-beta. *J Immunol* 2007;**178**:6725–9.
79. Wang J, Wang G, Sun B, Li H, Mu L, Wang Q, et al. Interleukin-27 suppresses experimental autoimmune encephalomyelitis during bone marrow stromal cell treatment. *J Autoimmun* 2008;**30**:222–9.
80. Zhao W, Wang Y, Wang D, Sun B, Wang G, Wang J, et al. TGF-beta expression by allogeneic bone marrow stromal cells ameliorates diabetes in NOD mice through modulating the distribution of CD4+ T cell subsets. *Cell Immunol* 2008;**253**:23–30.
81. Bouffi C, Bony C, Courties G, Jorgensen C, Noel D. IL-6-dependent PGE2 secretion by mesenchymal stem cells inhibits local inflammation in experimental arthritis. *PLoS One* 2010;**5**:e14247.
82. Kong Q, Sun B, Wang G, Zhai D, Mu L, Wang D, et al. BM stromal cells ameliorate experimental autoimmune myasthenia gravis by altering the balance of Th cells through the secretion of IDO. *Eur J Immunol* 2009;**39**:800–9.
83. Mougiakakos D, Jitschin R, Johansson C, Okita R, Kiessling R, Le Blanc K. The impact of inflammatory licensing on heme oxygenase-1-mediated induction of regulatory T cells by human mesenchymal stem cells. *Blood* 2011;**117**(18):4826–35.
84. Gershon R, Kondo K. Infectious immunological tolerance. *Immunology* 1971;**21**:903–14.
85. Nakagawa T, Tsuruoka M, Ogura H, Okuyama Y, Arima Y, Hirano T, et al. IL-6 positively regulates FoxP3+ CD8+ T cells *in vivo*. *Int Immunol* 2009;**22**(2):129–39.
86. Jonuleit H, Schmitt E, Schuler G, Knop J, Enk A. Induction of interleukin 10-producing, nonproliferating CD4(+) T cells with regulatory properties by repetitive stimulation with allogeneic immature human dendritic cells. *J Exp Med* 2000;**192**:1213–22.
87. Lutz M, Kukutsch N, Menges M, Rossner S, Schuler G. Culture of bone marrow cells in GM-CSF plus high doses of lipopolysaccharide generates exclusively immature dendritic cells which induce alloantigen-specific CD4 T cell anergy *in vitro*. *Eur J Immunol* 2000;**30**:1048–52.
88. Manicassamy S, Pulendran B. Dendritic cell control of tolerogenic responses. *Immunol Rev* 2011;241:206–27.
89. Reis e Sousa C. Dendritic cells in a mature age. *Nat Rev Immunol* 2006;**6**:476–83.
90. Djouad F, Charbonneier L-M, Bouffi C, Louis-Plence P, Bony C, Apparailly F, et al. Mesenchymal stem cells inhibit the differentiation of dendritic cells through an interleukin-6-dependent mechanism. *Stem Cells* 2007;**25**:2025–31.

91. Liu WH, Liu JJ, Wu J, Zhang LL, Liu F, Yin L, et al. Novel mechanism of inhibition of dendritic cells maturation by mesenchymal stem cells via interleukin-10 and the JAK1/STAT3 signaling pathway. *PLoS One* 2013;**8**:e55487.
92. Chen H, Chen H, Wang L, Wang F, Fang L, Lai H, et al. Mesenchymal stem cells tune the development of monocyte- derived dendritic cells toward a myeloid-derived suppressive phenotype through growth-regulated oncogene chemokines. *J Immunol* 2013;**190**(10):5065–77.
93. Ivanova-Todorova E, Bochev I, Mourdjeva M, Dimitrov R, Bukarev D, Kyurkchiev S, et al. Adipose tissue derived mesenchymal stem cells are more potent suppressors of dendritic cells differentiation compared with bone-marrow derived mesenchymal stem cells. *Immunol Lett* 2009;**126**:37–42.
94. Jiang X-X, Zhang Y, Liu B, Zhang S-X, Wu Y, Yu X-D, et al. Human mesenchymal stem cells inhibit differentiation and function of monocyte derived dendritic cells. *Blood* 2005;**105**:4120–6.
95. Nauta A, Kruisselbrink A, Lurvink E, Willemze R, Fibbe W. Mesenchymal stem cells inhibit generation and function of both CD34+ -derived and monocyte- derived dendritic cells. *J Immunol* 2006;**177**(4):2080–7.
96. Saeidi M, Masoud A, Shakiba Y, Hadjati J, Bonab M, Nicknam M, et al. Immunomodulatory effects of human mesenchynal stem cells on differentiation, maturation and andocytosis of monocyte derived dendritic cells. *Iran J Allergy Asthma Immunol* 2013;**12**(1):37–49.
97. Jing H, Vassiliou E, Ganea D. Prostaglandin E2 inhibits production of the inflammatory chemokines CCL-3 and CCL-4 in dendritic cells. *J Leuk Biol* 2003;**74**:868–79.
98. Piqueras B, Connolly J, Freitas H, Palucka A, Banchereau J. Upon viral exposure, myeloid and plasmacytoid dendritic cells produce 3 waves of distinct chemokines to recruit immune effectors. *Blood* 2006;**107**(7):2613–18.
99. Ma S, Xie N, Li W, Yuan B, Shi Y, Wang Y. Immunobiology of mesenchymal stem cells. *Cell Death Differ* 2014;**21**:216–25.
100. Blaber SP, Webster RA, Hill CJ, Breen EJ, Kuah D, Vesey G, et al. Analysis of in vitro secretion profiles from adipose-derived cell populations. *J Transpl Med* 2012;**10**:172.
101. Salgado AJ, Reis RL, Sousa NJ, Gimble JM. Adipose tissue derived stem cells secretome: Soluble factors and their roles in regenerative medicine. *Curr Stem Cell Res Ther* 2010;**5**:103–10.
102. Solovyeva VV, Salafutdinov II, Martynova EV, Khaiboullina SF, Rizvanov AA. Human adipose derived stem cells do not alter cytokine secretion in response to the genetic modification with pEGFP-N2 plasmid DNA. *World Appl Sci J* 2013;**26**:968–72.
103. Djouad F, Bouffi C, Ghannam S, Noël D, Jorgensen C. Mesenchymal stem cells: innovative therapeutic tools for rheumatic diseases. *Nat Rev Rheumatol* 2009;**5**:392–9.
104. Kilroy GE, Foster SJ, Wu X, Ruiz J, Sherwood S, Heifetz A, et al. Cytokine profile of human adipose-derived stem cells: expression of angiogenic, hematopoietic, and pro-inflammatory factors. *J Cell Physiol* 2007;**212**:702–9.
105. Perrini S, Ficarella R, Picardi E, Cignarelli A, Barbaro M, Nigro P, et al. Differences in gene expression and cytokine release profiles highlight the heterogeneity of distinct subsets of adipose tissue derived stem cells in the subcutaneous and visceral adipose tissue in humans. *PLoS One* 2013;**8**:e57892.
106. De Smedt T, Van Mechelen M, DeBecker G, Urbain J, Leo O, Moser M. Effect of interleukin-10 on dendritic cell maturation and function. *Eur J Immunol* 1997;**27**:1229–35.
107. McBride J, Jung T, de Vries J, Aversa G. IL-10 alters DC function via modulation of cell surface molecules resulting in impaired T-cell responses. *Cell Immunol* 2002;**215**:162–72.

108. Moore K, DeWaal Malefyt R, Coffman R, O'Garra A. Interleukin-10 and the interleukin-10 receptor. *Annu Rev Immunol* 2001;**19**:683–765.
109. Steensberg A, Fischer C, Keller C, Moller K, Pedersen B. IL-6 enhances IL-1ra, IL-10 and cortisol in humans. *Am J Physiol Endocrinol Metab* 2003;**285**:433–7.
110. Takahashi M, Kobayashi Y. Cytokine production in association with phagocytosis of apoptotic cells by immature dendritic cells. *Cell Immunol* 2003;**226**:105–15.
111. Liang S, Ristich V, Arase H, Dausset J, Carosella E, Horuzsko A. Modulation of dendritic cell differentiation by HLA-G and ILT4 requires the IL-6-STAT3 signaling pathway. *Proc Natl Acad Sci USA* 2008;**105**(24):8357–62.
112. Kishimoto T. IL-6: from its discovery to clinical applications. *Int Immunol* 2010;**22**:347–52.
113. Park S, Nakagawa T, Kitamura H, Atsumi T, Kamon H, Sawa S, et al. IL-6 regulates in vivo dendritic cell differentiation through STAT3 activation. *J Immunol* 2004;**173**:3844–54.
114. Strobl H, Knapp W. TGF-beta1 regulation of dendritic cells. *Microbes Infect* 1999;**1**:1283–90.
115. Mou H, Lin M, Cen H, Yu J, Meng X. TGF-beta1 treated murine dendritic cells are maturation resistant and down-regulate Toll-like receptor 4 expression. *J Zhejiang Univ Sci* 2004;**5**:1239–44.
116. Tada Y, Asahina A, Fujita H, Mitsui H, Torii H, Watanabe T, et al. Differential effects of LPS and TGF-beta on the production of IL-6 and IL-12 by Langerhans cells, splenic dendritic cells, and macrophages. *Cytokine* 2004;**25**:155–61.
117. Ogata M, Zhang Y, Wang Y, Itakura M, Zhang Y, Harada A, et al. Chemotactic response toward chemokines and its regulation by transforming growth factor-beta1 of murine bone marrow hematopoietic progenitor cell-derived different subset of dendritic cells. *Blood* 1999;**93**:3225–32.
118. Wallet M, Sen P, Tisch R. Immunoregulation of dendritic cells. *Clin Med Res* 2005;**3** (3):166–75.
119. Kalinski P, Hilkens C, Snijders A, Snijdewint F, Kapsenberg M. IL-12-deficient dendritic cells, generated in the presence of prostaglandin E2, promote type 2 cytokine production in maturing human naïve T helper cells. *J Immunol* 1997;**159**:28–35.
120. Popov A, Schultze J. IDO-expressing regulatory dendritic cells in cancer and chronic infection. *J Mol Med (Berl)* 2008;**86**:145–60.
121. Spaggiari GM, Abdelrazik H, Becchetti F, Moretta L. MSCs inhibit monocyte-derived DC maturation and function by selectively interfering with the generation of immature DCs: central role of MSC-derived prostaglandin E2. *Blood* 2009;**113**:6576–83.
122. Dikov MM, Ohm JE, Ray N, Tchekneva EE, Burlison J, Moghanaki D, et al. Differential roles of vascular endothelial growth factor receptors 1 and 2 in dendritic cell differentiation. *J Immunol* 2005;**174**:215–22.
123. Lin YL, Liang YC, Chiang BL. Placental growth factor down-regulates type 1 T helper immune response by modulating the function of dendritic cells. *J Leukoc Biol* 2007;**82**:1473–80.
124. Michielsen A, Hogan A, Marry J, Tosetto M, Cox F, Hyland J, et al. Tumour tissue microenvironment can inhibit dendritic cell maturation in colorectal cancer. *PLoS One* 2011;**6**:e27944.

Index

Note: Page numbers followed by "*f*" and "*t*" refer to figures and tables, respectively.

B

F

G

J

K

L

M

O

U

V

W

Printed in the United States
By Bookmasters